Recent Advances in VLSI Design

Recent Advances in VLSI Design

Edited by **Martin Limestone**

*C*LANRYE
INTERNATIONAL

New Jersey

Published by Clanrye International,
55 Van Reypen Street,
Jersey City, NJ 07306, USA
www.clanryeinternational.com

Recent Advances in VLSI Design
Edited by Martin Limestone

© 2015 Clanrye International

International Standard Book Number: 978-1-63240-442-8 (Hardback)

Printed in the United States of America.

Contents

Preface

This book is a result of research of several months to collate the most relevant data in the field.

Very-large-scale integration (VLSI) is the procedure of making an integrated circuit by joining several transistors into a single chip. This book presents an overview on current developments in design nanometer VLSI chips. The vital information described in this book discusses frequently encountered complications and challenges covering significant topics such as novel post-silicon devices, GPU-based parallel computing, design tools, antenna design and rising 3D integration. The book covers the following major aspects of VLSI Design: 3D integrated circuits design for 1000-core processors, Algorithms for CAD tools VLSI design, VLSI design for multi-sensor smart systems on a chip, multilevel mimetic algorithm for large SAT-encoded complications and Parallel symbolic analysis of large analog circuits on GPU platforms.

When I was approached with the idea of this book and the proposal to edit it, I was overwhelmed. It gave me an opportunity to reach out to all those who share a common interest with me in this field. I had 3 main parameters for editing this text:

1. Accuracy – The data and information provided in this book should be up-to-date and valuable to the readers.

2. Structure – The data must be presented in a structured format for easy understanding and better grasping of the readers.

3. Universal Approach – This book not only targets students but also experts and innovators in the field, thus my aim was to present topics which are of use to all.

Thus, it took me a couple of months to finish the editing of this book.

I would like to make a special mention of my publisher who considered me worthy of this opportunity and also supported me throughout the editing process. I would also like to thank the editing team at the back-end who extended their help whenever required.

Editor

Part 1

VLSI Design

VLSI Design for Multi-Sensor Smart Systems on a Chip

Louiza Sellami[1] and Robert W. Newcomb[2]
[1]*Electrical and Computer Engineering Department,*
US Naval Academy, Annapolis, MD
[2]*Electrical and Computer Engineering Department,*
University of Maryland, College Park, MD
USA

1. Introduction

Sensors are becoming of considerable importance in several areas, particularly in healthcare. Therefore, the development of inexpensive and miniaturized sensors that are highly selective and sensitive, and for which control and analysis is present all on one chip is very desirable. These types of sensors can be implemented with micro-electro-mechanical systems (MEMS), and because they are fabricated on a semiconductor substrate, additional signal processing circuitry can easily be integrated into the chip, thereby readily providing additional functions, such as multiplexing and analog-to-digital conversion. Here we present a general framework for the design of a multi-sensor system on a chip, which includes intelligent signal processing, as well as a built-in self test and parameter adjustment units. Specifically, we outline the system architecture, and develop a transistorized bridge biosensor for monitoring changes in the dielectric constant of a fluid, which could be used for in-home monitoring of kidney function of patients with renal failure.

In a number of areas it would be useful to have available smart sensors which can determine the properties of a fluid and from those make a reasoned decision. Among such areas of interest might be ecology, food processing, and health care. For example, in ecology it is important to preserve the quality of water for which a number of parameters are of importance, including physical properties such as color, odor, PH, as well as up to 40 inorganic chemical properties and numerous organic ones (DeZuane, 1990). Therefore, in order to determine the quality of water it would be extremely useful if there were a single system on a chip which could be used in the field to measure the large number of parameters of importance and make a judgment as to the safety of the water. For such, a large number of sensors is needed and a means of coordinating the readouts of the sensors into a user friendly output from which human decisions could be made. As another example, the food processing industry needs sensors to tell if various standards of safety are met. In this case it is important to measure the various properties of the food, for example the viscosity and thermal conductivity of cream or olive oil (Singht & Helman, 1984).

In biomedical engineering, biosensors are becoming of considerable importance. General theories of different types of biosensors can be found in (Van der Shoot & Berveld, 1988; Eggins, 1996; Scheller & Schubert,1992) while similar devices dependent upon temperature sensing are introduced in (Herwarden et al, 1994). Methods for the selective determination of compounds in fluids, such as blood, urine, and saliva, are indeed very important in clinical analysis. Present methods often require a long reaction time and involve complicated and delicate procedures. One valuable application in the health care area is that of the use of multiple sensors for maintaining the health of astronauts where presently an array of eleven sensors is used to maintain the quality of recycled air (Turner et al, 1987), although separate control is effected by the use of an external computer. Therefore, it is desirable to develop inexpensive and miniaturized sensors that are highly selective and sensitive, and for which control and analysis is available all on the same chip. These sensors can be implemented with micro-electro-mechanical systems (MEMS). Since they are fabricated on a semiconductor substrate, additional signal processing units can easily be integrated into the chip thereby readily providing functions such as multiplexing and analog-to-digital conversion. In numerous other areas one could find similar uses for a smart multi-sensor array from which easy measurements can be made with a small portable device. These are the types of systems on a chip (SOC) that this chapter addresses.

2. System on a chip architecture

The architecture of these systems is given in Fig. 2.1 where there are multiple inputs, sensors, and outputs. In between are smart signal processing elements including built-in self-test (BIST). In this system there may be many classes of input signals (for example, material [as a fluid] and user [as indicator of what to measure]). On each of the inputs there may be many sensors (for example, one material may go to several sensors each of which

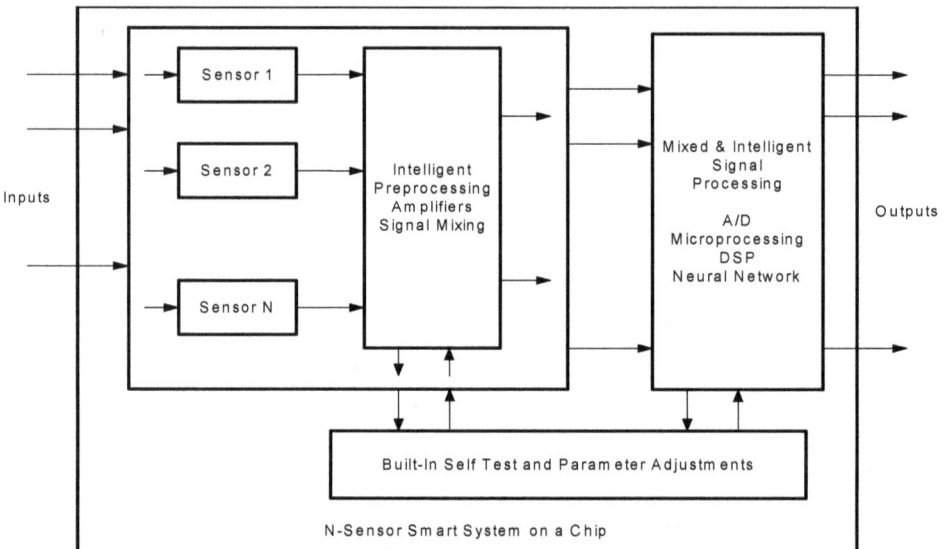

Fig. 2.1. Architecture for N-Sensor Smart System on a Chip

senses a different property [as dielectric constant in one and resistivity in another]). The sensor signals are treated as an N-vector and combined as necessary to obtain the desired outputs, of which there may be many (such as an alarm for danger and indicators for different properties). For example, a patient with kidney disease may desire a system on a chip which gives an indication of when to report to the hospital. For this an indication of deviation of dielectric constant from normal and spectral properties of peritonal fluid may be sensed and combined to give the presence of creatinine (a protein produced by the muscles and released in the blood) in the fluid, with the signal output being the percent of creatinine in the fluid and an alarm when at a dangerous level.

3. Dielectric constant and resistivity sensor

The fluid sensing transistor in this sensor can be considered as a VLSI adaptation of the CHEMFET (Turner et al, 1987) which we embed in a bridge to allow for adjustment to a null (Sellami & Newcomb, 1999). The sensor is designed for ease of fabrication in standard VLSI processing with an added glass etch step. A bridge is used such that a balance can be set up for a normal dielectric constant, with the unbalance in the presence of a body fluid being used to monitor the degree of change from the normal. The design chosen leads to a relatively sensitive system, for which on-chip or off-chip balance detection can occur. In the following we present the basic sensor bridge circuit, its layout with a cross section to show how the chip is cut to allow measurements on the fluid, and simulation results from the Spice extraction of the layout that indicate the practicality of the concept.

Figure 3.1 shows a schematic of the sensor circuit. This is a capacitive-type bridge formed from four CMOS transistors, the two upper ones being diode connected PMOS and the two lower NMOS, one diode connected and the other with a gate voltage control. The output is taken between the junction of the PMOS and NMOS transistors, and as such is the voltage across the midpoint with the circuit being supplied by the bias supply. As the two upper and the lower right transistors are diode connected, they operate in the saturation region

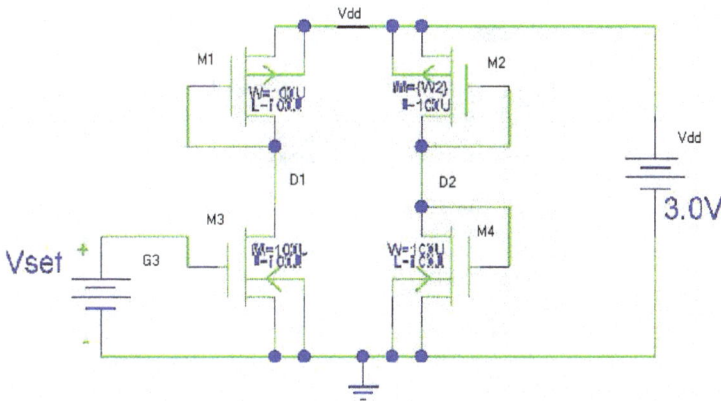

Fig. 3.1. Circuit Schematic of a Fluid Biosensor.

while the gate (the set node) of the lower left transistor, M3, is fed by a variable DC supply allowing that transistor to be adjusted to bring the bridge into balance. The upper right transistor, M2, has cuts in its gate to allow fluid to enter between the silicon substrate and the polysilicon gate. In so doing the fluid acts as the gate dielectric for that transistor. Because the dielectric constants of most fluids are a fraction of that of silicon dioxide, the fraction for water being about 1/4, M2 is actually constructed out of several transistors, four in the case of water, with all of their terminals (source, gate, drain) in parallel to effectively multiply the Spice gain constant parameter KP which is proportional to the dielectric constant.

The sensor relies upon etching out much of the silicon dioxide gate dielectric. This can be accomplished by opening holes in protective layers by using the overglass cut available in MEMS fabrications. Since, in the MOSIS processing that is readily available, these cuts should be over an n-well, the transistor in which the fluid is placed is chosen as a PMOS one. And, since we desire to maintain a gate, only portions are cut open so that a silicon dioxide etch can be used to clear out portions of the gate oxide, leaving the remaining portions for mechanical support. To assist the mechanical support we also add two layers of metal, metal-1 and metal-2, over the polysilicon gate.

A preliminary layout of the basic sensor is shown in Fig. 3.2 for M2 constructed from four subtransistors, this layout having been obtained using the MAGIC layout program. As the latter can be used with different lambda values to allow for different technology sizes, this layout can be used for different technologies and thus should be suitable for fabrications presently supported by MOSIS. Associated with Fig. 3.2 is Fig. 3.3 where a cross section is shown cut through the upper two transistors in the location seen on the upper half of the figure. The section shows that the material over the holes in the gate is completely cut away so that an etching of the silicon dioxide can proceed to cut horizontally under the remaining portions of the gate. The two layers of metal can also be seen as adding mechanical support to maintain the cantilevered portions of the gate remaining after the silicon dioxide etch.

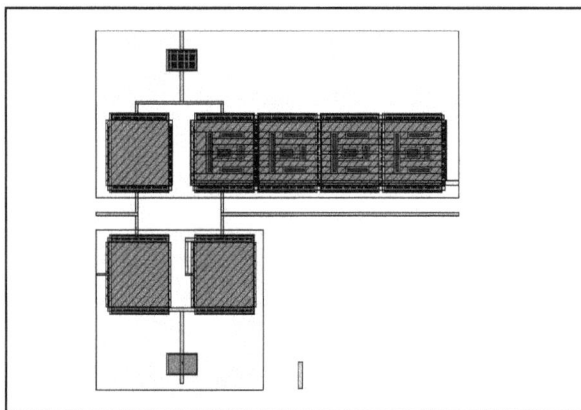

Fig. 3.2. Biosensor VLSI Layout

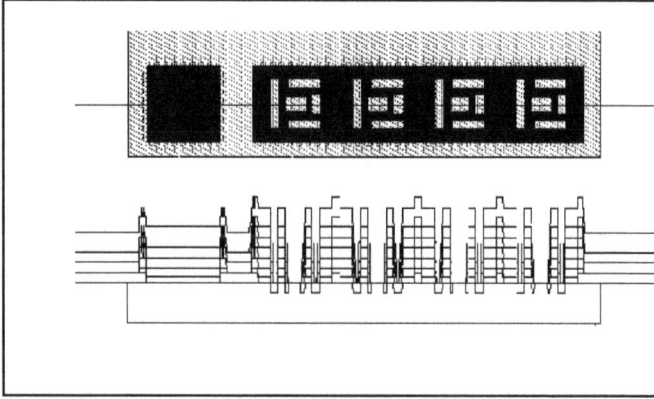

Fig. 3.3. Cross Section of Upper Transistors

To study the operation of the sensor we turn to the describing equations. Under the valid assumption that no current is externally drawn from the sensor, the drain currents of M1 and M3 are equal and opposite, ID3=-ID1, and similarly for M2 and M4, ID4=-ID2. Assuming that all transistors are operating above threshold, since M1, M3, and M4 are in saturation they follow a square law relationship while the law for M3 we designate through a function f(Vset,VD1) which is controlled by Vset. Thus,

$$-ID1 = \beta 1 \cdot (Vdd-VD1- |Vthp|)^2(1+\lambda p \cdot [Vdd-VD2]) \qquad (3.1a)$$

$$= \beta 3 \cdot [f(Vset,VD1) \cdot (1+\lambda n \cdot VD1) = ID3 \qquad (3.1b)$$

$$-ID2 = \varepsilon \cdot \beta 2 \cdot (Vdd-VD2- |Vthp|)^2(1+\lambda p \cdot [Vdd-VD2]) \qquad (3.2a)$$

$$= \beta 4 \cdot (VD2-Vthn)^2(1+\lambda n \cdot VD2) = ID4 \qquad (3.2b)$$

where, for the ith transistor,

$$\beta i = KPi \cdot Wi/2Li, \ i=1,2,3,4 \qquad (3.3)$$

and

$$f(x,y) = \{(x-Vthn)^2 \text{ if } x-Vthn<y, \ 2(x-Vthn)y-y^2 \text{ if } x-Vthn\geq y\} \qquad (3.4)$$

Here Vth, KP, and λ are Spice parameters for silicon transistors, all constants in this case, with the n or p denoting the NMOS or PMOS case, and epsilon is the ratio of the dielectric constant of the fluid to that of silicon dioxide,

$$\varepsilon = \varepsilon\_fluid / \varepsilon\_Sio2. \qquad (3.5)$$

In order to keep the threshold voltages constant we have tied the source nodes to the bulk material in the layout. In our layout we also choose the widths and lengths of M1, M3, and M4 to be all equal to 100μ and L2/W2 to approximate ε. Under the reasonable assumption that the λ's are negligibly small, an analytic solution for the necessary Vset to obtain a balance can be obtained. When M3 is in saturation the solution is

$$VD1 = Vdd - |Vthp| - (\beta3/\beta1)^{1/2} \cdot (Vset - Vthn) \tag{3.6}$$

while irrespective of the state of M3

$$VD2 = \{Vthn + (\varepsilon.\beta2/\beta4)^{1/2} \cdot (Vdd - |Vthp|) / [1 + \varepsilon.\beta2/\beta4]^{1/2}\} \tag{3.7}$$

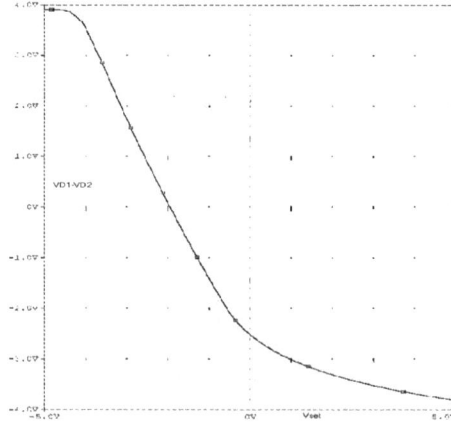

Fig. 3.4. Extracted circuit output voltage versus Vset

Balance is obtained by setting VD1=VD2. Still assuming that M3 is in saturation the value of Vset needed to obtain balance is obtained from equations (3.6) and (3.7) as

$$Vset = Vthn + \{(\beta1/\beta3)\}^{1/2} \cdot (Vdd - |Vthp| - Vthn) / [1 + (\varepsilon.\beta2/\beta4)]^{1/2}\} \tag{3.8}$$

At this point we can check the condition for M3 to be in saturation, this being that VDS $\geq$ VGS-Vthn; since VDS=VD1 and VGS = Vset, the use of Equation (3.6) gives

$$Vthn < Vset\{sat\} \leq Vthn + (Vdd - |Vthp|) / [1 + (\beta3/\beta1)^{1/2}] \tag{3.9}$$

Substituting the value of Vset at balance, Equation (3.8), shows that the condition for M3 to be in saturation at balance is $\varepsilon.\beta2 \geq \beta3$; this normally would be satisfied but can be guaranteed by making M2 large enough.

Several things are added to the sensor itself per Fig. 2.1. Among these is a differential pair for direct current mode readout followed by a current mode pulse coded neural network to do smart preprocessing to insure the integrity of the signals. Finally a built in test circuit is included to detect any breakdown in the sensor operation.

From the layout of Fig. 3.1 a Spice extraction was obtained. On incorporating the BiCMOSIS transistor models (Sellami & Newcomb, 1999; Moskowtitz et al, 1999) the extracted circuit file was run in PSpice with the result for the output difference voltage versus Vset shown in Fig. 3.4. As can be seen, adjustment can be made over the wide range of -5V<Vset<5V

Thus, it is seen that a sensor that is sensitive to the dielectric constant of a fluid over an 11 to 1 range of dielectric constant most likely can be incorporated into a multi-sensor chip. Using standard analog VLSI-MEMS processing one can use the bridge for anomalies in a fluid by obtaining Vset for the normal situation and then comparing with Vset found for the anomalous situation. This could be particularly useful for determining progress of various diseases. For example, one way to determine kidney function and dialysis adequacy is through the clearance test of creatinine. The latter tests for the amount of blood that is cleared of creatinine per time period, which is usually expressed in ml per minute. For a healthy adult the creatinine clearance is 120 ml/min.

A renal adult patient will need dialysis because symptoms of kidney failure appear at a clearance of less than 10 ml/min. Creatinine clearance is measured by urine collection, usually 12 or 24 hours. Therefore a possible use for the proposed sensor could be as a creatinine biosensor device for individual patient to monitor the creatinine level at home. An alternate to the proposed biosensor is based on biologically sensitive coatings, often enzymes, which could be used on M2 transistor in a technology that is used for urea biosensors which are presently marketed for end stage renal disease patients (Eggins, 1996). The advantage of the sensor presented here is that it should be able to be used repetitively whereas enzyme based coatings have a relatively short life. The same philosophy of a balanced bridge constructed in standard VLSI processing can be carried over to the measurement of resistivity of a fluid. In this case the bridge will be constructed of three VLSI resistors with the fourth arm having a fluid channel in which the conductance of the fluid is measured.

4. Spectral sensors

We take advantage of the developments in MEMS technologies to introduce new and improved methodology and engineering capabilities in the field of chemical and biochemical optical sensors for the analysis of a fluid. The proposed device has the advantages of size reduction and, therefore, increased availability, reduced consumption of chemical/biochemical sample, compatibility with other MEMS technologies, and integrability with computational circuitry on the chip.

Consequently, integrating MEMS and optical devices will give the added advantages of size reduction and integrability with the electrical circuitry. The integration and compatibility of sensors is very much in demand in the field of system on a chip. Here we extend CMOS technology to build an optical filter which can be used in a single chip microspectrometer. The chip contains an array of microspectrometer and photodetectors and the read out of their circuits.

By the nature of matter in the universe, most evident at the atomic and molecular level, it allows so much information to be deduced from its optical spectra. Because molecule and atoms can only emit or absorb photons with energies that correspond to certain allowed transition between quantum states, optical spectroscopy is one of the valuable tools of analytical chemistry (Schmidt, 2005). Optically based chemical and biological sensors are conveniently classified into five groups, according to the way light is modulated (Ellis, 2005). These light modulations are intensity, wavelength, polarization, phase, and time modulation. Here we focus on MEMS based sensors suitable for Intensity, wavelength, and time modulation.

4.1 Intensity modulation

As light passes through a material, its intensity attenuates as it interacts with the molecules, atoms, and impurities of the host material. The attenuation is an exponential function of the distance of its path length, x, traveled in the material. The absorption coefficient, α_λ, is defined relative to the concentration, M, and the cross section, S, of the absorbing molecules (Svanberg, 2001).

$$I_\lambda(x) = I_\lambda(0) . \exp(\alpha_\lambda.x) = I_\lambda(0) . \exp(-S.Mx/N) \qquad (4.1.1)$$

Where $I_\lambda(x)$ is the light intensity at distance x, $I_\lambda(0)$ the incident light intensity at x = 0, and N Avogado's number (6.022×10^{23} mol^{-1}).

Changes of the analyte concentration in the sample can alter the absorption coefficient α. An absorption based sensor measures these changes by the transmitted light intensity in terms of absorbance (A_λ) units:

$$A_\lambda = \log[I_\lambda(0)/ I_\lambda(x)] \qquad (4.1.2)$$

4.2 Wavelength modulation

Wavelength modulation can provide us with more information than just the intensity modulation. Several numbers of fixed wavelength sources are used simultaneously and their responses, intensity, are detected using photo detectors. Several sources that are modulated at different electrical frequencies can be used simultaneously in order to use a single photo detector. One of the wavelengths could serve as a reference channel for calibration.

Fluorescence occurs when an atom or a molecule makes a transition from a higher energy state to a lower one and emits lights. Excitation and subsequent emission can occur not only by photoluminescence but also by chemical reaction (chemiluminescence) or biological reaction (Bioluminescence). In resonance fluorescence, absorption and emission take place between the same two energy levels, and therefore the wavelength of the excitation and emission lights are the same. In non-resonant fluorescence, emission occurs either at higher wavelength than excitation wavelength (Stokes Fluorescence), or lower wavelength than excitation wavelength (anti- Stokes Fluorescence). The decay rate dN^*/dt of the fluorescence for a two level system is

$$dN^*/dt = -k_t . N^* \qquad (4.2.1)$$

where k_t is the total fluorescence rate, in sec^{-1}, and N^* is proportional to the number of electrons excited due to the fluorescent state in a time t. Hence

$$N^* = N_0^* . \exp(-k_t . t) \qquad (4.2.2)$$

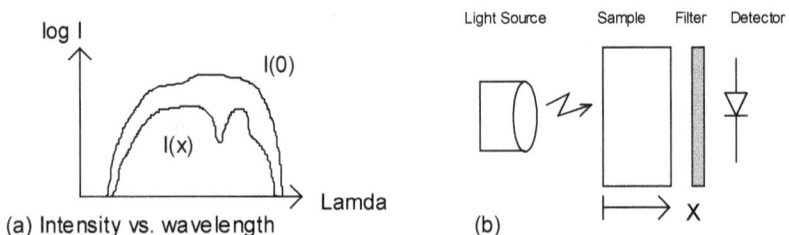

(a) Intensity vs. wavelength   (b)

Fig. 4.2. (a) Attenuation of the optical intensity as it travels along the x axis throught the matter versus the wavelength. (b) corresponding schematic for measurement.

4.3 Time modulation

Time modulation is essentially a subclass of intensity modulation. In time domain fluoremetry (TDF), a pulsed light source generates the photoluminescence. The fluorescence decay signal is measured as a function of time, and the decay curve determines the lifetime of the chemical sample. In time modulation base sensors measure the halftime of the sample.

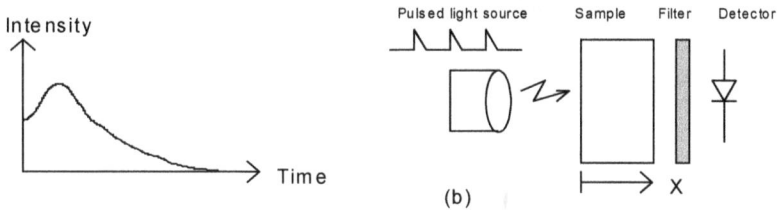

Fig. 4.3. (a) Fluorescence decay curve (b) corresponding schematic for measurement

5. MEMS based photo-sensors

An important part of any spectrometer, aside from the light source, is the optical filter and photo detectors. Recent engineering developments in the field of MEMS and microelectronics have shown that both of these devices can be produced in the micro level using existing technology (Hsu, 2008). Optical spectrometers can be produced using a tunable Fabry-Perot cavity (here simply called Fabry-Perot). The band-pass frequency range of the Fabry-Perot is a function of its cavity length (Patterson, 1997).

Fabry-Perot can be fabricated in the CMOS technology with photo-detectors integrated underneath it. In other words, Fabry-Perot is fabricated on top of a p-n diode in the CMOS technology. In this configuration, the p-n photo-detector is acting as a transducer that converts optical intensity of light that is passed through the Fabry-Perot to a proportional electrical signal. The existence of the Fabry-Perot in the optical path causes the photodiode to only respond to the light intensity of selected wavelength, which is set by the thickness of the Fabry-Perot cavity.

As illustrated in Fig. 5.1 below, the fabrication of Fabry-Perot and photodiode (FPPD), which starts with the fabrication of a p-n photo diode in a CMOS process technology, undergoes a post process in order to integrate a planer Fabry-Perot on top of the p-n photo diode. This process involves four steps. First, a portion of the top oxide layer immediately above the p-n diode is trimmed, by chemical itching, to reduce its effect on light

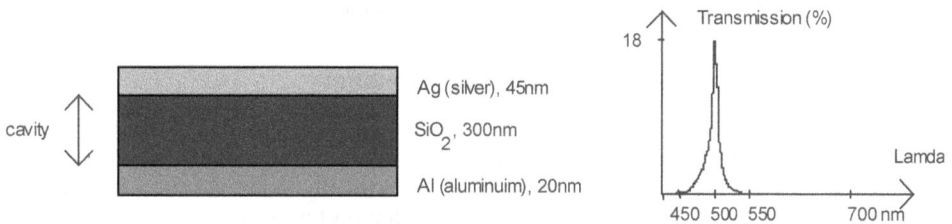

Fig 5.1. The Fabry-perot etalon with Al bottom Mirror

transmission. Second, a thin Aluminum layer is deposited, to form the lower mirror. Third, a layer of Silicon dioxide is added then etched to different sizes, using several masks. This way, each photodiode will have a different size of SiO_2 layer on top of it. Fourth, a thin layer of silver (Ag) is deposited on top of all oxide to form the top mirror layer (Tyree et al, 1994).

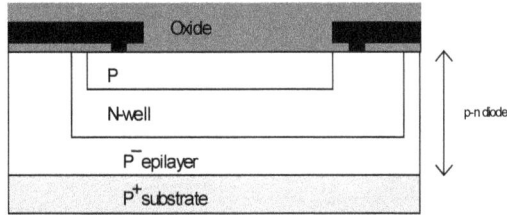

Fig. 5.2. Schematic structure for fabrication of a CMOS p-n photo diodes

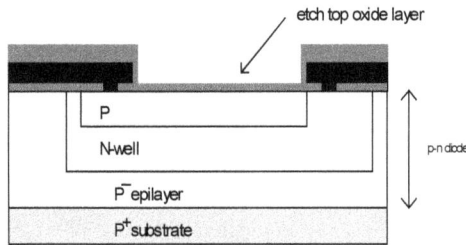

Fig. 5.3. Post CMOS process, 1st step, trimming the top oxide layer above the diode

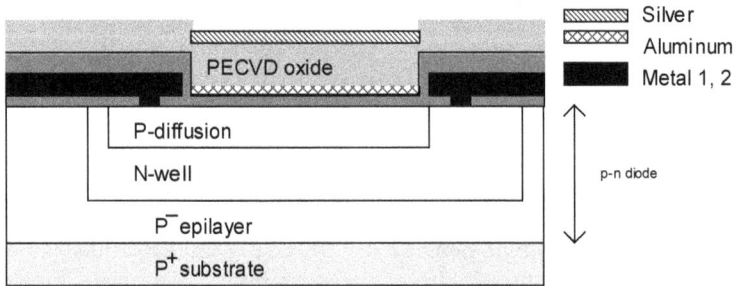

Fig. 5.4. Post CMOS process, Step 2nd, 3rd, and 4th. Depositing AL, PECVD oxide, and Silver, respectively, on top of p-n diodes to form Fabry-perot cavity filter

6. Optical micro-chemical and biochemical sensors

Optical sensors can be fabricated as shown in Fig. 6.1. A series of Fabry-Perot of different wavelength is fabricated in series, each having its own p-n photo-detectors, immediately underneath. These photodiodes are optically and electrically isolated from each other to reduce cross interference. A micro channel is fabricated on top of the series of Fabry-Perot photodetectors (FPFD) modules. Of course, FPFD modules can appear in any efficient configuration, such as a matrix format, under the flow channel. The entire structure of micro-channel and their FPFD modules can be fabricated in a twin parallel configuration, as shown in Fig. 6.2. In time modulation, this configuration can be used when one channel is empty and one channel is filled with chemical sample. In this situation, there are two received signals for each wavelength. One is the attenuated signal due to the sample, and the other one is a signal for cross-reference and evaluation of the intensity attenuation due to the chemical sample. This configuration can be also used in measurement of fluorescence. Two different dyes can be introduced in two channels in order to evaluate two different analyte concentrations.

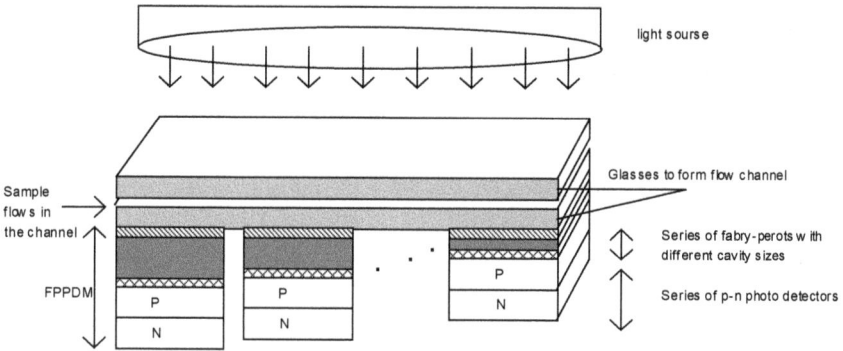

Fig. 6.1. Schematic structure of optical micro sensors using fabry-perot and p-n photo detectors

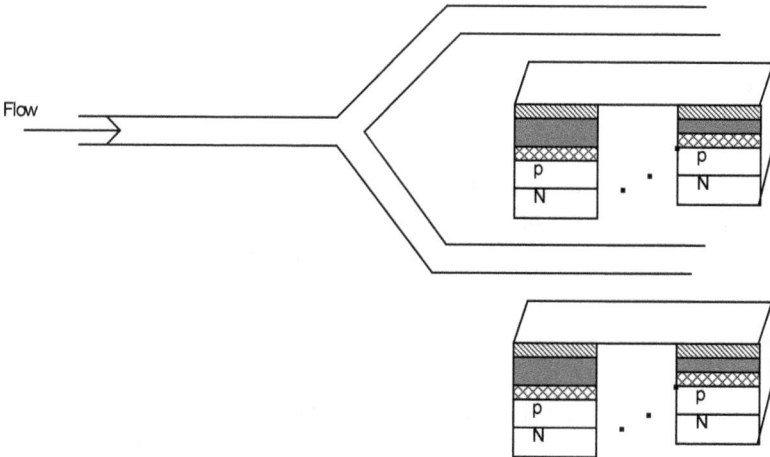

Fig. 6.2. Two parallel micro flow channels, each with its own FPFD module underneath

An array of FPPD is made of many individual FPPD that have different cavity thickness and therefore different range of pass band frequencies. The thickness of these oxide cavities is changed gradually in order to cover some desired range of the light spectrum. The array of FPPD can be formed in one or several columns, all entirely under the microchannel. Any light source that is transmitted through the micro channels will eventually reach these FPPD array under the channel. Each individual FPPD will react only to a small spectrum band of the light that is passed through its Fabry-Perot. Each individual FPPD is connected to the electronic circuit on the chip that will perform the signal conditioning and final post data processing.

7. Companion electronic circuitry

A block diagram of this circuitry is depicted in Fig. 7. All photodetector p-n diodes in the array of FPPD under the channel produce a current whose magnitude contains information related to light intensity. Furthermore this light intensity, which is absorbed by the photodiodes, depends on the content of the chemicals present in the micro-channel fluid. The main purpose of this electronic circuitry is to collect, condition, and interface these current signals to the post processing circuit. Since the information signals are in the form of diode currents, it is preferred to work with current mode (CM) electronic circuits.

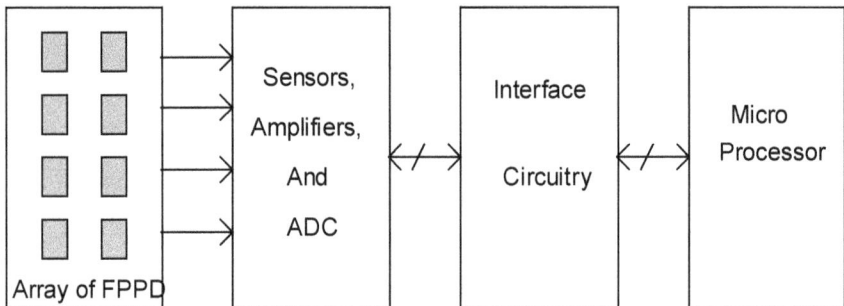

Fig. 7. Companion Electronic circuit connects array of FPPD to the microprocessor

8. Built-In-Self-Test

The Built-In-Self-Test (BIST) can interface with the sensors and other circuits under consideration. It can be built upon modifications of circuits and ideas available in the literature, such as the use of oscillations for mixed signal testing including the production line technique of using standard ring oscillator properties. The BIST is needed due to the fact that there are many interacting subsystems, and an error in one can perhaps drastically affect the operation of others.

BIST circuitry consists of a controller, a pattern generator and a multiple input signature analyzer. The Built-in Self-Test method allows core testing to be realized by commanding the core BIST controller to initiate self test and by knowing what the correct result should be. On-chip testing of embedded memories can be realized by either multiplexing their address and data lines to external SOC I/O pads or by using the core processor to apply enough read/write patterns of various types to ensure the integrity of the memory. This technique works best for small embedded memories. Some recommend providing embedded memories with their own BIST circuitry.

For BIST to be effective, there must be a means for on-chip test response measurement, on-chip test control for digital and analog test, and I/O isolation. There are three categories of measurements that can be distinguished: DC static measurements, AC dynamic measurements, and time domain measurements. The first of these, DC static measurements, includes the determination of the DC operating points, bias and DC offset voltages and DC gain. DC faults can be detected by a single set of steady state inputs. AC dynamic measurements measure the frequency response of the system under test. The input stimulus is usually a sine-wave form with variable frequency. Digital signal processing (DSP) techniques can be employed to perform harmonic spectral analysis. Time domain measurements derive slew rate, rise and delay times using pulse signals, ramps or triangular waveforms as the input stimuli of the circuit.

9. Smart signal processing

This stage consists of a mixed and intelligent DSP system that allows for the following functions to be performed.

- Analog-to-digital conversion: provides a signal interface between the sensor outputs (analog) and the signal processor inputs (digital).
- Determine fluid properties (physical and chemical): Neural and DSP algorithms as well as circuits can be used to carry out computations of fluid parameters such as dielectric constant, resistivity, spectrum, and chemical composition from the digitized sensor outputs.
- Detection and identification: The information obtained in step 2 above is fed to a microprocessor that can identify the chemical composition of the fluid and makes an intelligent decision in relation to the condition that is being monitored (water safe or not for drinking, dialysis needed or not, etc.). This can be readily programmed using look-up tables and threshold levels.
- Parameter selection and adjustment: These will be for various situations so as to include function selection to tell the sensor what to measure. In addition, the system must have the capability to compensate for deviations, detected by the built-in self test unit, of parameters such as amplifier gain, and micro-processor and neural circuit weight constants.

10. Summary

In this chapter we developed a general framework for the design and fabrication of a multi-sensor system on a chip, which includes intelligent signal processing, as well as a built-in self test and parameter adjustment units. Further, we outlined its architecture, and examined various types of sensors (fluid biosensors for measuring resistivity and dielectric constant, spectral sensors, MEMS based photo-sensors, and optical micro-chemical and biochemical sensors), and fabrication techniques, as well as develop a transistorized bridge fluid biosensor for monitoring changes in the dielectric constant of a fluid, which could be of use for in-home monitoring of kidney function of patients with renal failure.

11. Acknowledgments

This research was sponsored in part by the 2007 Wertheim Fellowship, US Naval Academy.

12. References

De Zuane, Handbook of Drinking Water Quality, Standards and Control,Van Nostrand Reinhold, New York, 1990.

Eggins, B. R., Biosensors: an Introduction, Wiley-Teubner, New York, 1996.

Ellis, A. M., Electronic and Photoelectron Spectroscopy: Fundamentals and Case Studies, Cambridge University Press, 2005.

Herwaarden, A. W. Van, P. M. Sarro, J. W. Gardner, and P. Bataillard, "Liquid and Gas Micro-calorimeters for (Bio)chemical Measurements," Sensors and Actuators, Vol. 43, 1994, pp. 24-30.

Hsu, T. R., MEMS and Microsystems: Design, manufacture, and Nanoscale Engineering, John Wiley, 2008.

Moskowitz, M., L. Sellami, R. Newcomb, and V. Rodellar, "Current Mode Realization of Ear-Type Multisensors," International Symposium on Circuits and Systems, ISCAS 2001, Sydney, Australia, volume 2, pp. 285-289.

Patterson, J. D., "Micro-Mechanical Voltage Tunable Fabry-Perot Filters Formed in (111) Silicon," National Aeronautics and Space Administration, Langley Research Center, 1997.

Scheller, F., and F. Schubert, Biosensors, Elsevier, Amsterdam, 1992.

Schmidt, W., Optical Spectroscopy in Chemistry and Life Sciences, Wiley-VCH, 2005.

Sellami, L., and R. W. Newcomb, "A Mosfet Bridge Fluid Biosensor," IEEE International Symposium on Circuits and Systems, May 30-June 2, 1999, Vol. V, pp. 140-143.

Singth, R. P., and D.R. Heldman, Introduction to Food Engineering,Academic Press, Inc., 1984.

Svanberg, S., Atomic and Molecular Spectroscopy: Basic Aspects and Practical Applications, Springer, 2001.

Turner, A. P. F., I. Karube, and G. S. Wilson, Editors, Biosensors, Fundamentals and Applications, Oxford University Press, Oxford, 1987.

Tyree, V., J.-I. Pi, C. Pina, W. Hansford, J. Marshall, M. Gaitan, M. Zaghloul, and D. Novotny, "Realizing Suspended Structures on Chips Manufactured by CMOS Foundry Processes through the MOSIS Service," MEMS Announcement, 41 pages, available fromXMOSIS@mosis-chip.isi.edu, 1994.

Van der Schoot and P. Berveld, "Use of Immobilized Enzymes in FET-Detectors," in Analytical Uses of Immobilized Biological Compounds for Detection, Medical and Industrial Uses, edited by G. G. Guilbault and M. Mascini, Reidel Publishing Co., Ultrecht, 1988, pp. 195-206.

Impedance Matching in VLSI Systems

Díaz Méndez J. Alejandro[1], López Delgadillo Edgar[2]
and Arroyo Huerta J. Erasmo[1]
[1]*National Institute for Astrophysics, Optics and Electronics*
[2]*Universidad Autónoma de Aguascalientes*
Mexico

1. Introduction

The continuous scaling process into submicrometric dimensions of silicon based devices has allowed the integration of a large number of systems in a single chip. Besides, the operating frequencies of such systems are higher and a large amount of information can be processed in a short period of time. On the other hand, while the core frequencies are increasing, higher data rates for off-chip interconnections become necessary, for example a processor that communicates with the memory in order to process information. Unfortunately, at high rates the signal wave length is comparable with the physical length of the interconnections, because of this, parasitic and transmission line effects have to be taken into account. As a consequence, the transmitted signal integrity is degraded resulting in communication errors (Thierauf S., 2004), (Brooks D., 2003).

It has been shown that, for modern off-chip communication systems, current mode signaling offers several advantages over voltage-mode at high data rates (Juan, 2007), but they need to be matched in impedance to the interconnection line. However, impedance matching requires termination resistors. Moreover, due to the large number of input/output circuits in a single chip, terminations have to be placed on-chip (Fan Y. & Smith J., 2003) so that the PCB area is not increased. One of the most important transmission line effects that degrades signal integrity in these signaling schemes is reflection loss. In this case, signal reflections traveling trough the line are present in either driver to receiver or receiver to driver directions. Unfortunately, it is difficult to achieve perfect matching of impedances due to the large process variations in the fabrication of interconnection lines and the different traces between them (Ramachandran N. et. al., 2003). Also, temperature variations and external effects are present inside and outside the chip. As a conclusion, impedance matching techniques must be developed in order to automatically adapt the impedance variations of the line.

In this chapter systems for on-die automatic impedance matching for off-chip signaling are described. In order to perform the automatic matching operation an algorithm that integrates the sign of the impedance matching error and the sign of the coupling branch current is implemented. The advantage of this algorithm is that it works without interfering with the driver operation. Computer simulations of layout extractions are presented. Also, a system of knowlegde- based impedance matching which avoids the calculation of a complex mathematical model is presented.

2. Signaling schemes

Electrical signaling schemes, which have become one of the most important topics in digital design and a hot topic in research, are the techniques used in the transmission of digital signals from one place to another through an interconnection (Dally & Poulton, 1998). Typical medias for the transmitted signals are on-chip, PCB and backplane interconnections as well as cable lines.

Electrical signaling schemes are classified into voltage mode and current mode signaling schemes depending on the signal carriers of the data through the interconnection. Besides, signaling schemes are also grouped into single ended, fully differential, pseudo differential and incremental signaling (Juan, 2007). In this section electrical signaling schemes are presented and the advantages of current mode over voltage mode are enlisted.

2.1 Voltage mode signaling

In Fig. 1 the model for a voltage mode signaling scheme is shown where the line driver is represented by a voltage source V_{DD} that corresponds to the value of the voltage swing. The resistance R_S represents the output impedance of the driver and the transitions between logic states is achieved by changing the position of the switch. Thus, these logic states, namely 1 and 0, are represented by two supply voltage levels. The circuit drives the output signal through the transmission line with characteristic impedance Z_O to the far end of it where a CMOS inverter compares the received voltage against a voltage reference derived from the power supply. Finally the voltage source V_N represents the power supply noise generated between the transmitter and the receiver at both ends of the line which, in fact, deteriorates the signal integrity at the far end.

An important property in voltage mode signaling is that due to the large swing of the signal the noise margins are also large. Even so, special care must be taken in the design of such circuits because of the swing dependent noise sources that are added to the data signals.

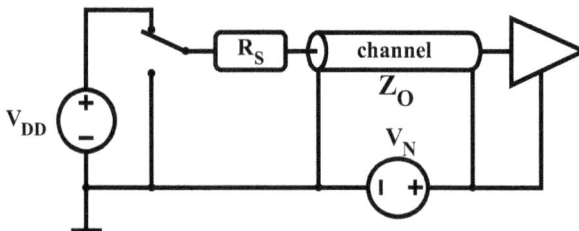

Fig. 1. Voltage mode signaling modeling.

Some realizations of circuits for voltage mode signaling are presented in (Ramachandran N. et. al., 2003),(Dehon et. al., 1993),(Deutschmann B. & Ostermann T., 2003), (Svensson C. & Yuan J., 1991), (Choy C. S. et. al., 1997), (Shin S. K. et. al., 2005) and (Balatsos A., 1998). The simplest voltage mode signaling circuit is shown in Fig. 2. It consists of an inverter stage at the near end of the channel where each transistor acts as a switch that directs the output node to the respective rail voltage (V_{DD} through M_1 and V_{SS} through M_2). Also, it is important to say that at any time one transistor of the circuit is inactive while the other is active. As a consequence the signal transmitted through the channel is the voltage at the output V_O of the inverter and can be computed with equation (1)for rising edge signal and (2) for falling edge, (Juan, 2007).

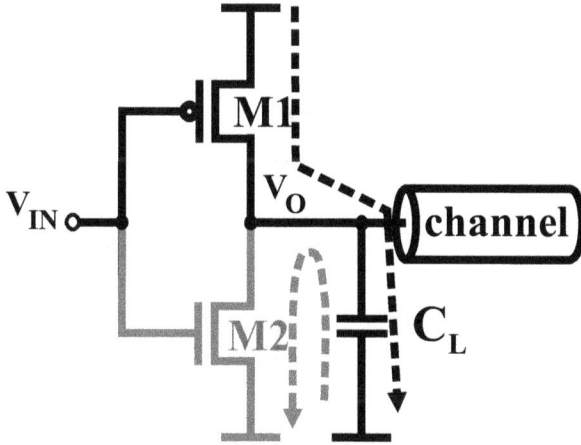

Fig. 2. Voltage mode signaling typical circuit.

$$C_L \frac{dv_o(t)}{dt} + \frac{v_o(t) - V_{DD}}{R_p} - C_L V_{OL} \delta(t) = 0 \tag{1}$$

$$C_L \frac{dv_o(t)}{dt} + \frac{v_o(t)}{R_n} - C_L V_{OH} \delta(t) = 0 \tag{2}$$

In equations (1) and (2) the constants R_n and R_p are the resistances of the NMOS and PMOS transistor channels when they are biased in the triode region. C_L is the load capacitance of the driver, V_{OL} and V_{OH} are the voltages that represent the logic states 0 and 1. Finally, the products $C_L V_{OL} \delta(t)$ and $C_L V_{OH} \delta(t)$ are the contribution of the initial voltage for the processes of charging and discharging respectively.

The power consumption for voltage mode signaling systems is shown in equation (3), where κ is the switching activity coefficient. It is clear that the dependence with the frequency represents a disadvantage for high frequency applications.

$$P \approx \kappa C_C V_{DD}^2 f \tag{3}$$

In the following subsections advantages and limitations of various voltage mode signalling schemes, such as single ended, fully differential, pseudo differential and incremental, are presented. This classification is obtained from (Juan, 2007).

2.1.1 Single ended signaling

In single ended signaling, only one conductor per channel is needed to carry the signal to the receiver side of the system. As shown in Fig. 3(a), the signal arriving to the far end of the line contains both the transmitted one and a noise component that is generated by the devices that are near to the conductor. This signal is compared against a reference signal V_{REF} that is generated locally in the receiver.

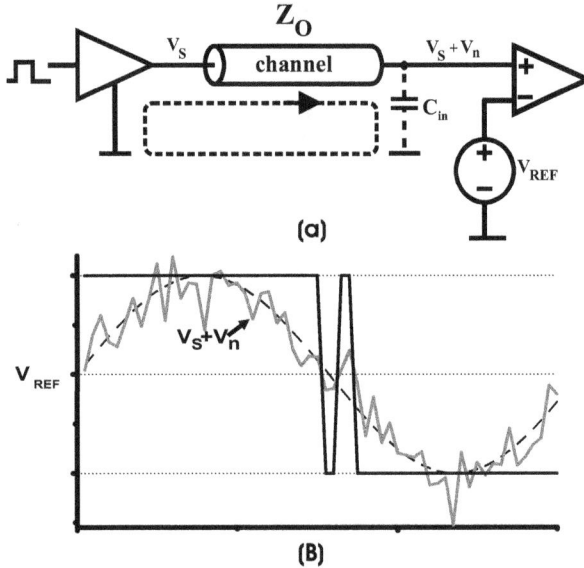

(a)

(B)

Fig. 3. (a) Single ended signaling in voltage mode; (b) input output voltage examples

In Fig. 3(b) the input and output voltage examples for a single ended voltage mode link are shown. Due to the single ended characteristic of the system the noise can not be rejected by the comparator and, in consequence, the output presents unwanted error symbols. That is why such systems are susceptible to coupled noise. In addition, in Fig. 3(a) the dotted line shows the path of the signal in the interconnection which goes from the driver to the receiver through the line and returns usually through the ground planes. The capacitor C_{in} is the input impedance of the comparator. This path represents a large area loop that results in high level electromagnetic emissions that affect devices located close to the channel.

2.1.2 Fully differential signaling

The fully differential signaling for voltage mode links is shown in Fig. 4. The main difference compared with the single ended scheme is that it uses two interconnections to carry the signal to the far end of the line. Another important characteristic is that conductors are so close to each other that the induced noise tends to be the same in each one. As a consequence the signals that are present at the line far ends are the transmitted positive voltage V_S^+ plus the induced potential V_n, i.e. $V_S^+ + V_n$ and, in the same way, the signal in the other polarity is $V_S^- + V_n$. At the receiver side the determination of the symbol is done by means of the voltage comparator configured in a differential way, because of this, the noise component is cancelled. As a conclusion, it can be said that fully differential configuration for signaling provides excellent common mode noise immunity.

By analyzing Fig. 4 the signals that charge the input capacitors C_{in+} and C_{in-} are the time varying differential currents that flow in opposite ways in each conductor. These currents

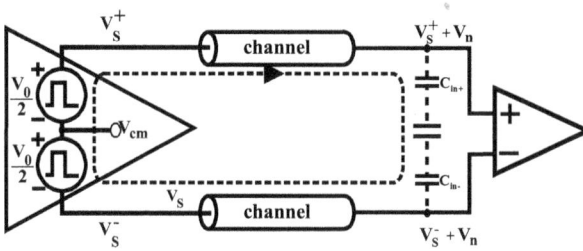

Fig. 4. Fully differential voltage mode signaling

form the closed loop shown with the dashed line in the Fig. 4. Compared with the one in single ended, the area occupied by this loop is small due to the proximity of the conductors. As a consequence, the electromagnetic coupling with other channels is small. Finally, the most important shortcoming of the differential signaling systems is the occupied area by the two lines.

2.1.3 Pseudo differential signaling

The pseudo differential signaling for voltage mode links is shown in Fig. 5, it is, in essence, a combination of the single ended and the differential signaling systems. In the pseudo differential scheme a single conductor is used as a reference for a group of signal paths. A common number for this group is four. As in fully differential signaling, the physical lines running from the transmitter to the receiver are so close between them that the induced noise V_{ni} is considered the same in all of them. In consequence, if differential comparators are considered as receivers then noise components can be eliminated.

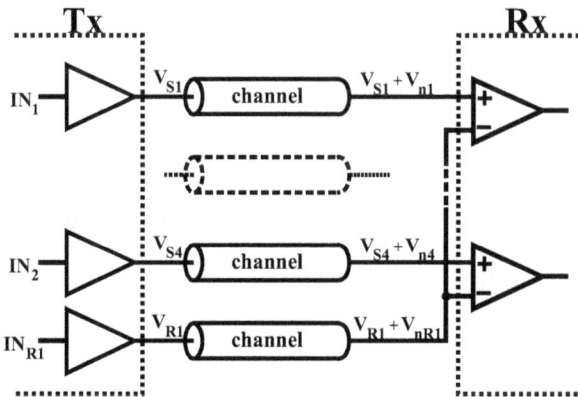

Fig. 5. Pseudo differential voltage mode signaling

It is clear that the main advantage of pseudo differential links is the reduced number of conductors that it needs. Unfortunately, the use of a single conductor as a reference signal also represents a drawback, because the area for the signal loops is increased. As a consequence, the channel inductance is larger compared to that in the fully differential approach. A solution for this drawback is presented in (Carusone A. et. al., 2001), where an incremental signaling approach with high signal integrity is presented.

2.2 Current mode signaling

A model for a current mode signaling system is shown in Fig. 6. In this case the line driver is represented by a current source I_S and the couple of switches that direct the current trough the line. At the far end of the link a resistor R_L is connected between the reference and signal lines and its purpose is not only to match the transmission line impedance but also to convert the current into voltage. This voltage is then changed to digital by the differential-mode comparator at the receiver side.

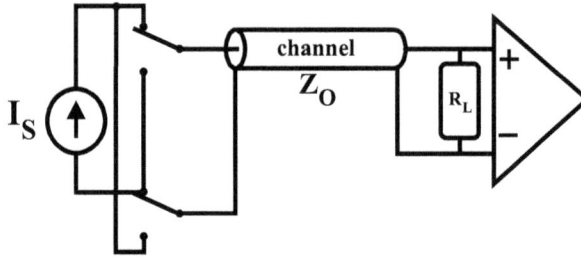

Fig. 6. Current mode signaling modeling.

In current mode signaling systems the symbols are represented by branch current signals. For the case of Fig. 6 when switches are in the upper side a current I_S flows through the line from transmitter to receiver and, in the ideal case, through the load resistor. In this case a voltage $V_L = I_S * R_L$ is present at the input of the comparator. In the opposite case, when switches are in the lower side, the current flows from receiver to driver, then the voltage drop is $V_L = -I_S * R_L$. In consequence, the total voltage swing at the far end of the link is $V_{sw} = 2 * I_S * R_L$.

One of the most basic drivers for current mode signaling systems is shown in Fig. 7 where the current I_S is directed to one branch or to the other just by switching the transistors M_1 and M_2. The control of this action is achieved by the digital data to be transmitted Dat and $\bar{D}at$. The resistances R are implemented to match the characteristic impedance Z_0 of the channel.

One of the most important advantages of current mode signaling is that the information is represented by branch currents and, due to the low impedance characteristic of the transmission media, the voltage swing for these systems is small even though the current signals are big. As a consequence, circuits can operate with a low voltage supply and current swings are not affected by the variations on the supply voltages (as opposed to voltage mode signaling). From the argument of swing invariance to supply changes, it can be concluded that current mode signaling has superior signal integrity compared with the voltage mode one.

An important issue in signaling systems is the propagation delay and is directly related to the rising and falling time of the signal. For a capacitive node, the rising(falling) time is shown in (4), (Juan, 2007), where I is the average current charging and discharging the node, C_n is the node capacitance and ΔV_n is the node voltage swing.

$$\Delta t = \frac{C_n \Delta V_n}{I} \qquad (4)$$

From equation (4) it can be inferred that if a small Δt is needed then the voltage swing ΔV_n must be minimized or the charging/discharging current must be big. Then, from equation

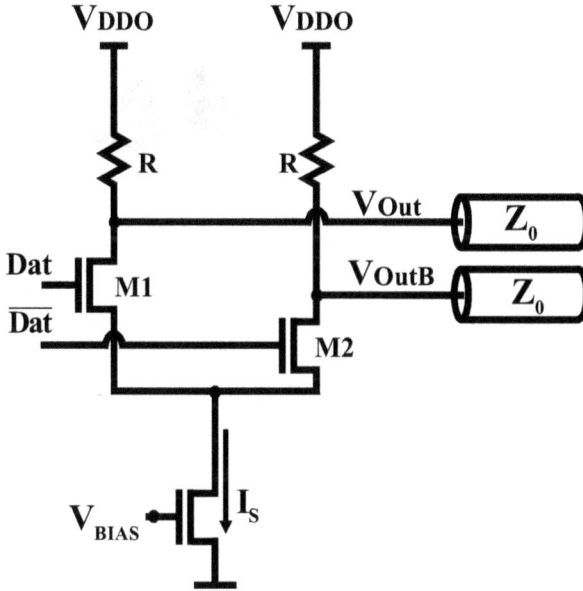

Fig. 7. Current mode signaling circuit.

(4),it can be concluded that current mode signalling systems have small propagation delay which make them suitable for high speed environments.

An important topic regarding electronic systems is power consumption. In particular, it is essential for signaling systems because they tend to consume big quantities of power. The power consumption in current mode circuits can be calculated by using equation (5), where it can be seen that there is not a dependence with frequency. Moreover, has only static power consumption which represents a benefit in high frequency applications.

$$P \approx I * V_{DD} \qquad (5)$$

In the following subsections two of the most important realizations for current mode signaling are presented. They are called unipolar and bipolar current mode signalling.

2.2.1 Unipolar current mode signaling

The symbol codification in unipolar current mode signaling (UCMS) is shown in Fig. 8, where a logical 1 is represented by the current I_1 and the logical 0 is represented by the absence of current. In this system the transmitter offset current is represented by I_{X0} and the possible values for the symbol are represented by the black area (it is clear that a zero current is easy to implement, that is why the black area corresponding to the symbol 0 is small). The receiver offset I_{r0} and sensitivity I_{rs} are also sketched in the scheme, they are near the reference current I_R.

In Fig. 9 the unipolar current mode signaling system is depicted. The UCMS block represents the driver which is, in this case, the one shown in Fig. 7. As stated before, the driver sinks a current I from one line each time, i. e. there is no signal flowing in both conductors at the

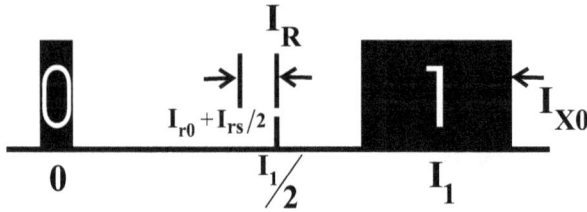

Fig. 8. Symbols in unipolar current mode signaling.

same time. At the far end of the lines, termination resistors R_T are placed in order to generate a differential voltage at the input of the receiver. This voltage is given by the difference between the positive and the negative inputs of the comparator as shown in equation (6).

$$\Delta V_{in} = |V_{in}^+ - V_{in}^-| = |V_{DD} - R_T I - V_{DD}| = R_T I \qquad (6)$$

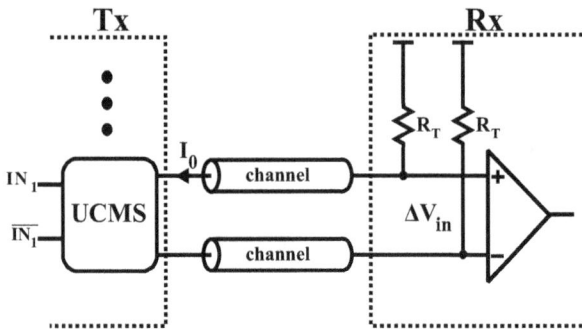

Fig. 9. Unipolar current mode signaling system.

As stated before, the variations in the voltage supply and in the ground sources do not have effect in the current flowing through the channel. Furthermore, due to the differential configuration at the receiver side, the common mode noise is eliminated. The disadvantage in unipolar signalling is that electromagnetic emission exists because only one conductor is carrying the current each time.

2.2.2 Bipolar current mode signaling

The symbol codification in bipolar current mode signaling is presented in Fig. 10. In this case the logical 1 is represented, as in unipolar signaling, by the current I_1. The difference is that the logical 0 is represented by the current $-I_1$.

By comparing Fig. 10 with Fig. 8 it can be appreciated that the allowed area for the logical 1 offset at the receiver has decreased. The reason for this is that the area of the offset for the logical 0 has to be increased because the current that represents it is now different from zero. The receiver offset I_{r0} and sensitivity I_{rs} are also sketched in the scheme, they are near the reference current I_R which is centered in zero amperes in bipolar signaling.

An example of bipolar current mode signaling systems is depicted in Fig. 11. It can be seen in the figure that current flows always in both interconnections but in opposite directions, as

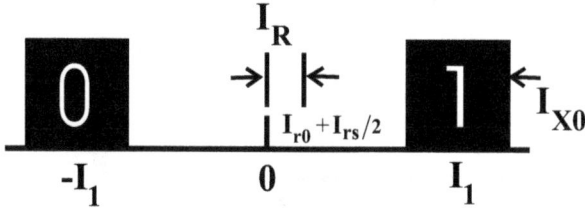

Fig. 10. Symbols in bipolar current mode signaling.

stated before in this section. This property allows these systems to have a low electromagnetic emission because field components are cancelled due to the opposite directions of the current. Another characteristic of bipolar signaling is that the load resistance R_T is placed between the interconnections in such a way that current signals generate a voltage which is compared by the differential mode comparator at the receiver.

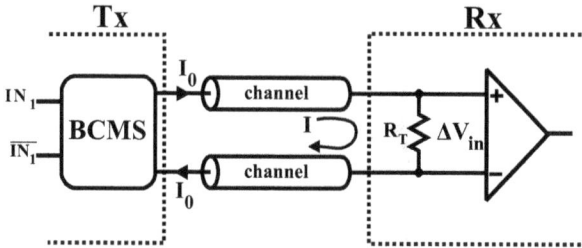

Fig. 11. Bipolar current mode signaling system.

Bipolar current mode signalling systems are also called low voltage differential signalling (LVDS) and a typical driver is shown in Fig. 12 where the both switches direct the signal to the corresponding conductor according to the data input IN_1. A disadvantage of LVDS links is that both switches current mode sources are implemented with transistors. Then, there are four transistors between the supply voltage and the ground, which is not so desirable in low supply voltage applications. Also common mode feedback circuits are needed increasing the driver size in a considerably amount.

When a N-bit parallel link is needed, a group of N bipolar current mode drivers are put together in a special array. This array is called current mode incremental signaling and specific details on this approach are presented in (Wang T. & Yuan F., 2007).

2.3 Specifications for signaling standards

The specifications for the signaling standards are essential in communication because they establish the voltage levels so that the driver and receiver agree with the logic high and low conditions.

An illustration of the specifications for digital signaling is shown in Fig. 13. The voltages V_H and V_L are the expected voltage levels for the logic values 1 and 0 respectively. In the transmitter side, the driver's goal is to have a high logic level that goes above a minimum voltage level, i.e. $V_{0H} \leq V_H$ and at the receiver side the accepted voltages must go above V_{IH}. Then the noise margin for the high logic level can be written as $NM_H = V_{0H} - V_{IH}$. In a similar way, the noise margin for a logic low is $NM_L = V_{IL} - V_{OL}$.

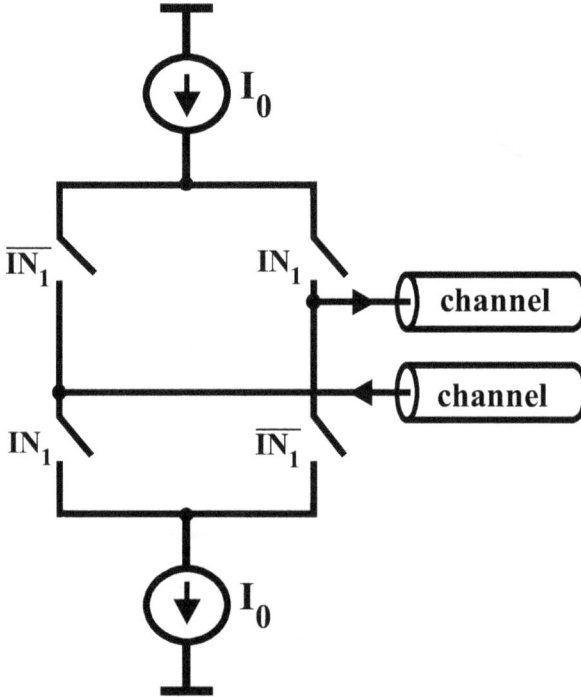

Fig. 12. Low voltage differential signaling driver.

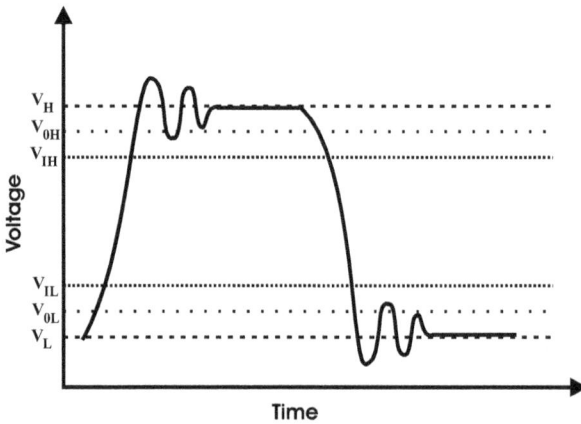

Fig. 13. Specifications for digital signaling.

In table 1 voltage specifications for some common standards are enlisted (Young B., 2001). Although some of them are current mode, the values are presented in voltage which represent the drop on the termination resistors. Another important issue is that the standards are in order from the lowest to the highest speed that can be achieved, then higher speeds are

Stan dard	V_{DDQ}	V_{0L}		V_{0H}		Termi nation	Driver
		Min	Max	Min	Max		
TTL	$5 \pm 10\%$		0.4	2.4		None	PP
LVTTL	$3.3 \pm 10\%$		0.4	2.4		None	PP
GTL			0.4			R_{system}	OD
HSTL	1.5 ± 0.1		0.4	$V_{DDQ} - 0.4$		1	PP
ECL	$-5.2 \mp 5\%$	-1.810	-1.620	-1.025	-0.880	$50\Omega(-2V)$	CM
PECL	$5 \pm 5\%$	3.190	3.380	3.975	4.120	50Ω	CM
LVPECL	$3.3 \pm 5\%$	1.490	1.680	2.275	2.420	50Ω	CM
LVDS		0.925			1.474	50Ω	CM

Table 1. Driver specifications for signaling standards

accomplished by systems with terminations, reduced voltage swings and with differential configuration.

3. Impedance matching techniques

As stated in previous sections, impedance matching techniques must be implemented in order to reduce return losses. It has been also shown that the fastest signaling standards implement termination resistors in order to match the interconnection impedance. Four of the most common termination techniques are shown in Fig. 14, (Brooks D., 2003). The first technique is the parallel termination (Fig. 14(a)) where a single resistor is connected either to ground or to V_{DD} and its value is equal to the characteristic impedance of the line. Although this is one of the most used methods, its disadvantage is that the current is always flowing through it, thus increasing the power consumption of the system.

The second termination technique is shown in Fig. 14(b), it is called Thevenin termination and consists of a couple of resistors, one connected to ground and the other to V_{DD}. The advantage of this scheme is that it provides pull up and pull down functions improving noise margins in some cases. The drawback of this system is that it is not easy to find the optimum values of the resistors in order to match the characteristic impedance of the line. The third technique is the AC termination and is depicted in Fig. 14(c). It is composed by a series connection of a resistor and a capacitor. Here the capacitance blocks the DC signals in order to reduce the power consumption but distortion can appear when high speed links are considered. Finally the series termination scheme is presented in Fig. 14(d). This is one of the most often used techniques, specially in voltage mode drivers.

Unfortunately the techniques presented in Fig. 14 are implemented with fixed devices and process, temperature and voltage supply variations are not taken into account for the design of such systems. In the following subsections some techniques are presented in which variable terminators are implemented to automatically adapt the impedance of the transmission line.

3.1 Automatic impedance matching control techniques

An important issue in automatic impedance matching is the control technique used in the adaptation process. It takes the reference signal which indicates the desired value of the impedance and the signal that represents the actual value of the impedance and process them in order to have the same value. The output of the circuit sets the value of the impedance that matches the interconnection.

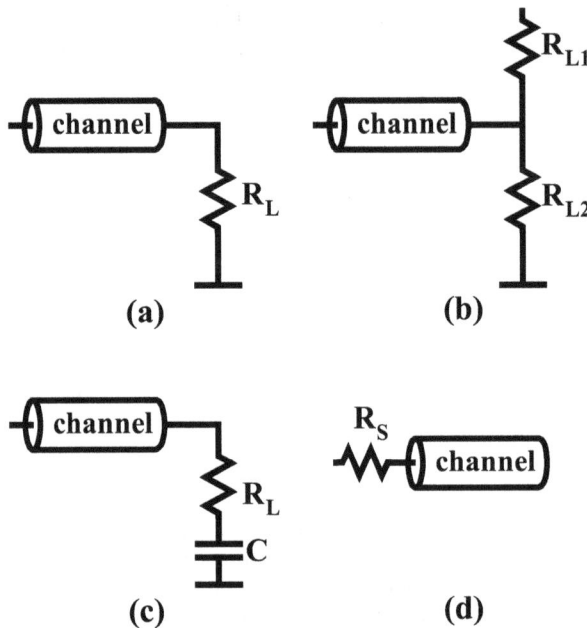

Fig. 14. Termination techniques. (a) Parallel, (b) thevenin, (c) AC and (d) Series.

One of the most common technique used to control the impedance in signaling systems is the one shown in Fig. 15. This technique represents the concept for the circuits presented in (Dehon et. al., 1993), (Koo K. et. al., 2001), (Koo K. et. al., 2006) and (Muljono H. et. al., 2003). It has in essence three stages, the first is a clocked comparator that decides if the impedance value is higher or lower than the reference one. The second stage is a counter which can be either binary or thermometer coded and its input is an $Up/\overline{Dn}$ signal that comes from the comparator. The last stage is a digital register that is used to hold the value of the counter when the matching condition of the impedance is fulfilled. The digital outputs of the register are used to control arrays of transistors in a pull up or pull down connection. The drawback of this technique is that switching noise in the supply lines can be generated due to the turning on and the turning off of the transistor array.

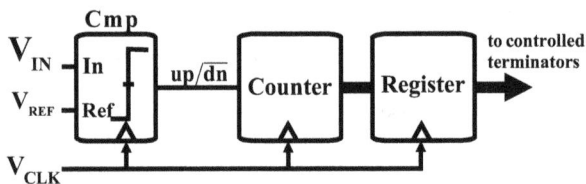

Fig. 15. Digital control for automatic impedance matching.

Another approach used in the control of automatic impedance matching is shown in Fig. 16 which is a simplified version of the one presented in (Ramachandran N. et. al., 2003). This circuit is designed to adapt directly the output impedance of an analog driver to the interconnection line by controlling a variable resistor at the output stage of the driver. The input signals to this control circuit are the input and the output of the line driver, they are processed by the first stage which is a peak to peak detector. The second stage is a differential difference amplifier (DDA) and its output is the analog control voltage. The drawback of this analog technique is that the driver speed is limited by the frequency response of the control circuits due to the direct signal measurement in the input and output ports.

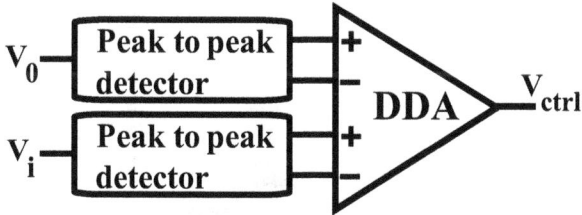

Fig. 16. DDA based analog control for automatic impedance matching.

A second analog approach for the control of an impedance matching system is shown in Fig. 17 which consists of a feedback amplifier. In this circuit the drawback of limited speed of the driver is eliminated by implementing an off-line matching of the impedance, i. e. a replica of the matching impedance is implemented in order to generate the reference signal and no measurements are made from the ports of the driver. The problem with this system is that variations in the interconnection impedance are not taken into account in the process of matching.

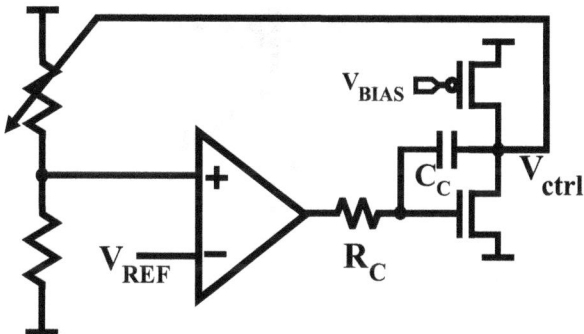

Fig. 17. Feedback amplifier based analog control for automatic impedance matching.

3.2 Reference signal generating circuits

In this section some of the most popular reference signal generating circuits are presented. Reference signals are an important subject in automatic impedance matching because they establish the value that the variable impedance must reach in order to fulfill the matching condition.

One of the most used circuits for reference signal generating is shown in Fig. 18, (Fan Y. & Smith J., 2003), (Koo K. et. al., 2006) and (Tae-Hyoung K. et. al., 2005). It consists of an off-chip precision resistor R_{REF} connected in series with a replica of the on-chip variable impedance used to match the interconnection R_v. The reference voltage V_{REF} is obtained from the node located between the resistors and its optimal value is the half rail voltage $V_{DD}/2$. An advantage of this technique is that the matching operation is independent from the driver data rate because measurements are not taken from the signal lines. The drawback is that an external resistor is needed which increases the area of the PCB. Furthermore, impedance variations of the off-chip interconnection are not taken into account since the reference is generated off line.

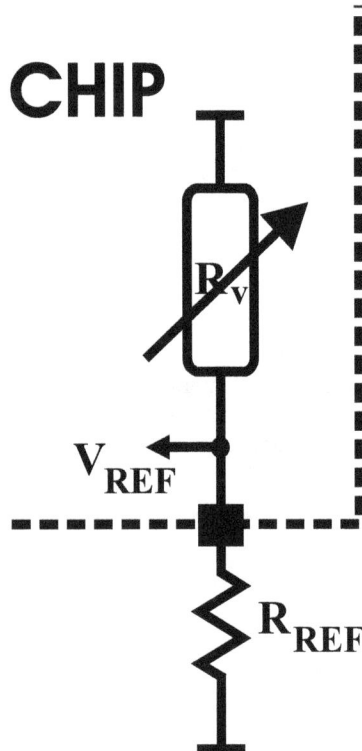

Fig. 18. Generation by dividing voltage.

An approach where current sources are implemented in order to generate voltage drops in a replica of the on chip variable impedance R_v and in an off-chip precision resistor R_{REF} is depicted in Fig. 19 (Dally & Poulton, 1998). Here, the two voltage references V_{REF} and V_{REF2} must have the same value in order to fulfill the matching condition of the impedance. As in the case of the circuit in Fig. 18, the off-chip resistor increases the area of the PCB which represents a disadvantage when high performance systems are needed. Also, variations in the interconnection impedance are not considered because references are generated in a circuit which is separated from the driver.

A modification of the circuit of Fig. 19 is shown in Fig. 20, (Koo K. et. al., 2001). In this case only the voltage drop from the on chip variable impedance R_v is considered as a voltage reference, avoiding the need of an off-chip resistor. In order to accomplish the automatic impedance matching operation, the reference voltage V_{REF} is compared against an internally generated voltage reference. The disadvantage of this technique is that the on-die process, voltage and temperature variations can move the internal reference away from its optimal

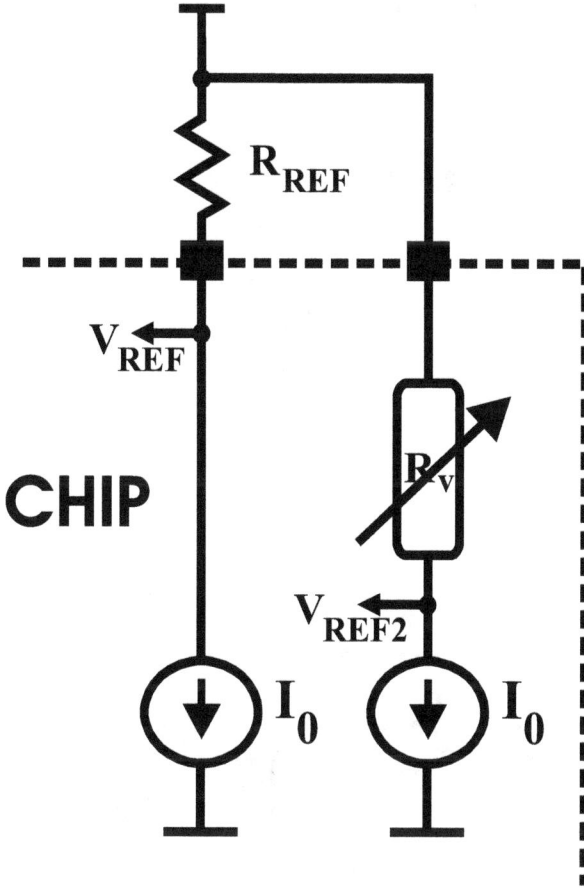

Fig. 19. Voltage drop reference generation.

value generating impedance errors. As in the case of the techniques presented before, the variations in the interconnection impedance are not considered.

In order to overcome the drawbacks related with the variations in the interconnection impedance, the technique shown in Fig. 21 has been implemented in (Dehon et. al., 1993) and (Dally & Poulton, 1998). In this case a voltage mode driver is implemented in order to drive an interconnection that is terminated with an open at the far end. The reference signal is taken from the node between the matching resistance and the interconnection channel and its shape is as shown at the bottom of the figure. This shape is composed by the signal arriving to the channel from the driver and the one reflected from the far end. In this case the matching condition is fulfilled when the middle part of the reference signal is the same as $V_{sw}/2$, where V_{sw} is the total swing of the transmitted signal. The drawback of this system is the difficulty in generating the correct timing signals in order to make the voltage comparisons in the correct time.

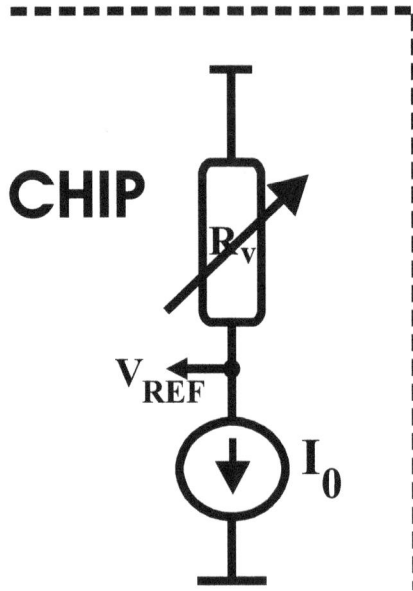

Fig. 20. On chip reference generation.

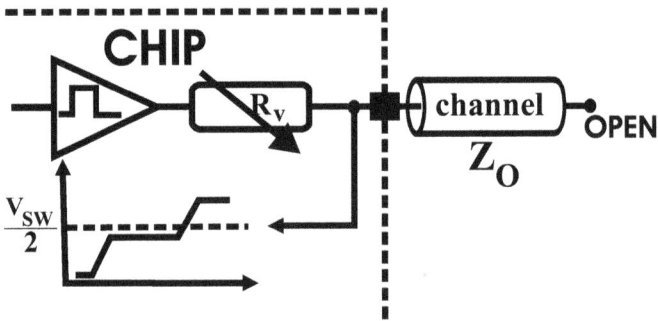

Fig. 21. Reference generation by signal reflection.

Finally, an approach for reference signal generation in current mode drivers is depicted in Fig. 22 which is an idea presented in (Munshi A. et. al., 1994). In this technique the current mode driver sinks or sources a current I_S from or to the interconnection which is in a parallel array with the matching impedance R_v. Then the matching condition in this scheme is fulfilled when the current I_v flowing though R_v is $I_v = I_0$, in other words, when $I_v = I_S/2$. The advantage of this method is that variations in the impedance of the interconnection are taken into account since reference signals are measured directly from the data link. The drawback is that measuring currents may modify the impedance branch.

In a similar way, reference signals for voltage mode interconnections can be obtained by considering the variable impedance and the interconnection line as a voltage divider. In this case, the reference voltage must be equal to $V_S/2$.

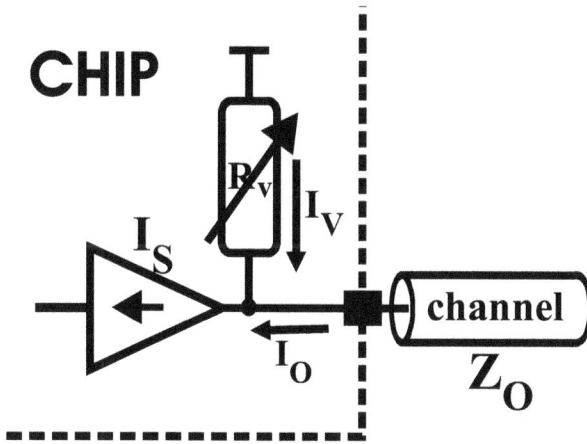

Fig. 22. Reference generation by current division.

4. Automatic impedance matching design based on the sign of the error

An automatic impedance matching based on an optimization algorithm that uses the sign of the error and the sign of the coupling branch current is proposed. A possible implementation of the system, simulation and experimental results are presented.

4.1 Mathematical approach

The mathematical formulation of the proposed method for impedance matching is based on Fig. 23, which is a modification of the system proposed in (Munshi A. et. al., 1994). In this system the output driver is modeled with a simple time variant current source and its output impedance is set to infinity. The current is divided between two branches, one is the transmission line with characteristic impedance Z_0 and the other is the matching branch Z_g which is used to avoid reflections in the line. This matching impedance is mainly composed by two elements, a current dependent voltage source V_g and a fixed impedance Z_P. The output voltage V_S is the voltage drop that results when the output current I_S flows through the parallel configuration of Z_g and Z_0. Then the system can be seen either by its current or by its voltage characteristics.

As shown in equation (7), the voltage dependent source V_g has a linear relationship with the coupling branch current I_g.

$$V_g = H_g I_g \tag{7}$$

where, the transimpedance H_g is a variable parameter with units of Ohms (Ω).

By analyzing the Fig 23 it can be deduced that the current I_g, flowing in the coupling branch, can be expressed as in the equation (8).

$$I_g = \frac{1}{Z_P}(V_S - V_g) \tag{8}$$

From equation (8), and making some mathematical manipulation, it is possible to prove that the matching impedance is given in terms of the fixed impedance Z_P and the transimpedance

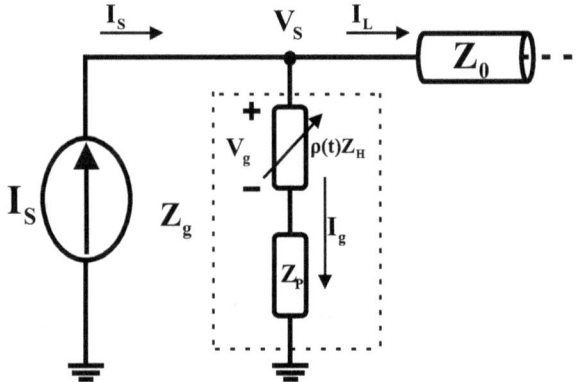

Fig. 23. Impedance Matching Synthesis Circuit

H_g. This is shown in the equation (9).

$$\frac{V_S}{I_g} = Z_g = Z_P + H_g \tag{9}$$

From equation (9) it can be inferred that H_g should be modified in order to achieve an impedance value $Z_g = Z_0$, which is the coupling condition. In consequence, H_g is defined as:

$$H_g(t) = \rho(t)Z_H \tag{10}$$

where, Z_H is constant and has units of Ohms(Ω) and $\rho(t)$ is an impedance matching coefficient and its optimal value is achieved when the impedance matching condition $Z_g = Z_L$ is fulfilled. It is clear that the fulfillment of the condition implies: $I_g = I_L = 1/2I_S$. From this expression, the impedance matching error e is defined as in equation (11).

$$e = \frac{1}{2}I_S - I_g \tag{11}$$

From equations (7), (10) and (11), the goal of the system is to dynamically adapt the coefficient $\rho(t)$ in such a way that the error e is minimized. A suitable technique to accomplish this goal is the LMS (Least Mean Square) (Carusone A. & Johns D., 2000), in which the criteria is to optimize temporal estimations of $E[e^2]$. Then, it is possible to express ρ as follows:

$$\rho(t) = -2\mu \int_{-\infty}^{t} e(u)\nabla_\rho e(u)du \tag{12}$$

where, μ is constant and establishes the speed of adaptation of the system, $\nabla_\rho e(u)$ represents the gradient of the error related with the parameter ρ . Also, for the equation 12 is considered that e^2 is a noisy estimation of $E[e^2]$.

Once established the conditions and the optimization method, it is necessary to find ρ as a function of the system parameters. Then, by substituting equation (8) in (11), the impedance matching error is given by:

$$e = \frac{1}{2}I_S - \frac{1}{Z_P}(V_S - V_g) \tag{13}$$

From equations (7),(10) and (13), it is possible to find the gradient of the error as shown in equation (14).

$$\nabla_\rho e = -\frac{1}{Z_P} Z_H I_g \tag{14}$$

Finally, from equations (14) and (12) the coefficient ρ as a function of the system parameters is presented in (15).

$$\rho(t) = \frac{2\mu Z_H}{Z_P} \int_{-\infty}^{t} e(u) I_g(u) du \tag{15}$$

It can be seen from (15) that for practical implementation the silicon area can be large due to the multiplication operation. In order to simplify the system, the SS-LMS (Sign-Sign LMS) algorithm, (Carusone A. & Johns D., 2000), is considered, where the sign function is applied to the error and to the matching branch signals. Thus, the multiplication results in a trivial operation. Consequently, the impedance matching coefficient become:

$$\rho(t) = \alpha \int_{-\infty}^{t} Sgn(e(u)) Sgn(I_g(u)) du \tag{16}$$

where, $\alpha = \frac{2\mu Z_i}{Z_P}$ is constant.

Based on equation (16), the block diagram of the proposed system for automatic impedance matching is depicted in Fig. 24. The system input is the driver voltage V_S and the reference for the error signal is the driver current divided by two. The block $\frac{1}{Z_p}$ represents the fixed impedance and the transimpedance block $H(\alpha)$ performs the dependent source V_g. The impedance optimization is made by the sign blocks together with the multiplication and integration blocks. The coefficient α specifies the speed of matching and the error level around the optimal impedance.

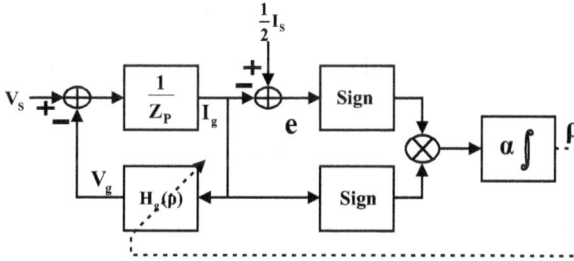

Fig. 24. Block Diagram for Impedance Matching.

4.2 Proposed implementation

One condition that must be established for the circuit implementation of the system of Fig. 24 is that the impedance Z_P is real and that its value is smaller than the load impedance Z_0. It means that Z_P must accomplish the following conditions: $0 < Z_P < Z_0$ and $Z_P = Re(Z_P)$. Also, the sign operation for the error can be expressed as:

$$Sgn(e) = Sgn\left(\frac{1}{2} I_S - I_g\right)$$
$$= Sgn\left(\frac{1}{2} Z_P I_S - Z_P I_g\right) \tag{17}$$

In the same way, the sign operation for I_g is:

$$Sgn(I_g) = Sgn(Z_P I_g) \tag{18}$$

As shown in (17) and (18), the sign operation can be implemented as a voltage level comparator. In this way, the inputs for (17) are the voltage across the fixed impedance Z_P and the reference voltage across the impedance with the same value as Z_P.

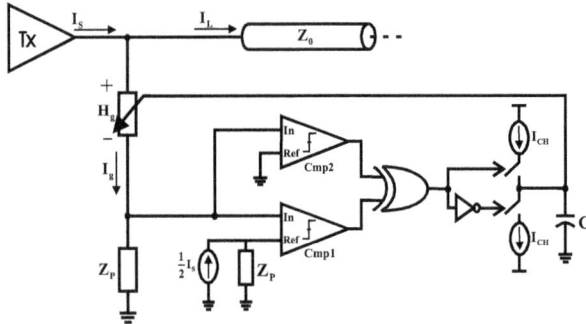

Fig. 25. Automatic Impedance Matching System Implementation.

The proposed circuit implementation for the automatic impedance matching system is shown in Fig. 25, where it can be seen that the error signal sign and the sign of the current in the coupling branch give as a result logic levels, then the multiplication is trivial and is implemented with a Xor logic gate, as shown in (19).

$$Sgn(e)Sgn(I_S) \Rightarrow Sgn(e) \oplus Sgn(I_S) \tag{19}$$

Another important operation for the system is the integration, which is implemented by means of a charge pump and a filter (Lopez et al., 2009). The current I_{CH} of the pump is directly related with the parameter α, as a consequence, the matching speed is established by this current. Finally, the current dependent voltage source $V_g = H_g I_g$ is implemented with a variable resistance, and its value is controlled by the voltage in the charge pump filter.

4.3 Proposed test vehicle

As stated before the mathematical representation of the system to be implemented is that shown in equation (20).

$$\rho(t) = \beta \int_{-\infty}^{t} Sgn\left(\frac{1}{2}I_S(u) - I_g(u)\right) Sgn(I_g(u)) du \tag{20}$$

Even though equation (20) is expressed in terms of branch currents, it is difficult to achieve a practical implementation for it. This is due to the fact that sensing a branch current is more complicated than sensing a node voltage. Therefore in the proposed implementation shown in Fig. 26, the inputs to the calibration circuit are voltages. The core of the scheme is the unipolar current mode differential driver (Dally & Poulton, 1998). This driver sinks the current I_S from its outputs depending on the logic state of its inputs Dat and $\overline{Dat}$. The terminators Z_g are programmable resistors that can be analog programmed via V_{CTRL} to match the interconnection impedance Z_0. The voltage reference V_{REF} is generated by sinking a current $\frac{I_S}{2}$ from a replica of the programmable resistance Z_g. The voltages V_{Out} of the driver and V_{REF} are the inputs for the impedance calibration circuit. This circuit generates the control voltage V_{CTRL} by implementing equation (20).

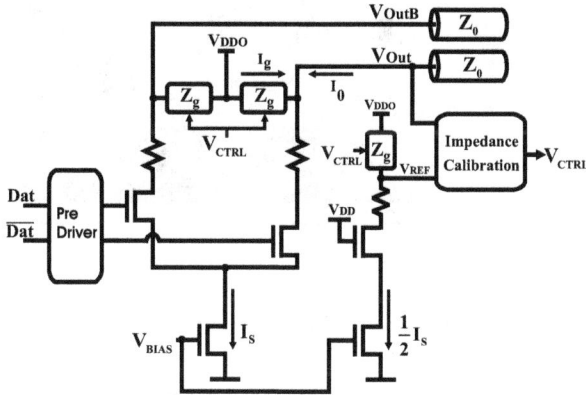

Fig. 26. Proposed Implementation.

By analyzing Fig. 26 one can find that the output voltage V_{Out} and the reference voltage V_{REF} are those described by equations (21) and (22), respectively.

$$V_{Out} = V_{DD0} - I_g Z_g. \tag{21}$$

$$V_{REF} = V_{DD0} - \frac{I_S}{2} Z_g. \tag{22}$$

Using equations (21) and (22) we can define the voltage mode error as follows:

$$e_V = V_{Out} - V_{REF} = Z_g \left(\frac{I_S}{2} - I_g \right). \tag{23}$$

Also, by assuming that $Z_g > 0$, it is inferred that the sign of the errors in voltage and in current mode are the same, this can be verified by equation (24), moreover, the function $Sgn(I_g)$ can be calculated directly from the input to the driver Dat. Consequently this shows that measuring voltages instead of currents is a good option to implement the SS-LMS technique without affecting the operation of the driver.

$$Sgn \left(\frac{I_S}{2} - I_g \right) = Sgn \left(Z_g \left(\frac{I_S}{2} - I_g \right) \right) \tag{24}$$

A print of the Mentor Graphics screen of the layout of the system is shown in Fig. 27 and it was designed in the $0.35 \mu m$ C35B4C3 AMS technology. The circuit enclosed by the dashed line corresponds to the current mode driver, the pre-driver, the programmable resistors and the voltage reference circuit. The SS-LMS based impedance matching algorithm is shown outside the dashed line.

In order to verify the performance of the impedance calibration circuit, the system was simulated with post layout extractions using Mentor Graphics tools. To test the circuit, different resistive loads (45Ω, 50Ω and 55Ω) were attached to the circuit. The signal rate

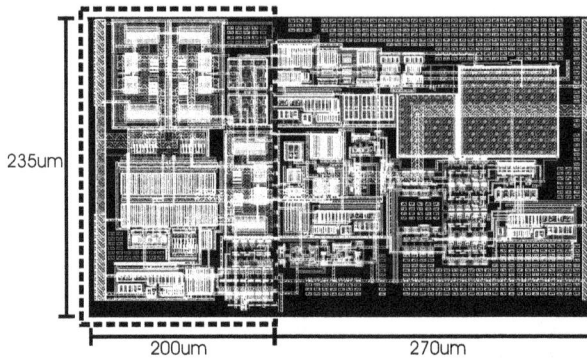

Fig. 27. System Layout.

of the data inputs to the system Dat and $\overline{Dat}$ is $1Gb/s$ and clock frequency for the impedance calibration circuit is established at $300MHz$.

In Fig. 28 the time domain signals for the output and the reference signal are shown. In this case, the load impedance for the system is set to 50Ω. As can be seen, the system adapts after some time.

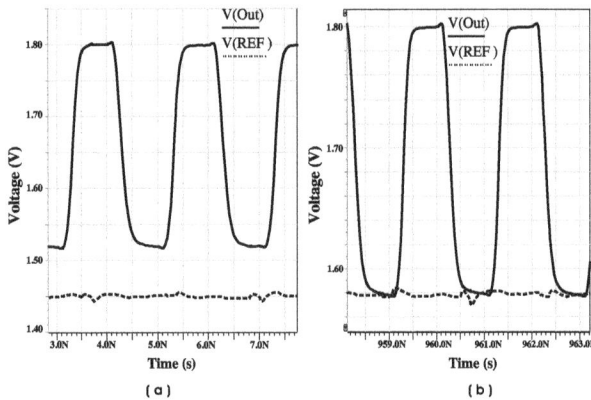

Fig. 28. Time domain analysis, (a) before and (b) after adaptation.

In Fig. 29, learning curves are shown for the 50Ω load impedance case (Fig. 29a) and for 45Ω and 55Ω (Fig. 29b). Those curves are normalized error signals in dB as a function of time and the adaptation time can be seen on them.

The last post layout simulations deal with worst case power and speed scenarios as well as temperature. The Mentor Graphics kit is used to perform this task. Only worst power and speed cases are presented because they result in the poorest performance compared with the others. Figs. 30a, Fig. 30b and Fig. 30c show the simulation results for worst power, worst speed and temperature variations (100 degree Celsius) respectively. As can be observed in the figure, the error always converges to a level bellow the $-30dB$.

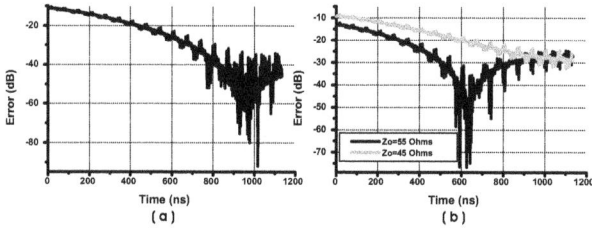

Fig. 29. Error for (a) 50 Ohms and (b) 45 and 55 Ohms loads.

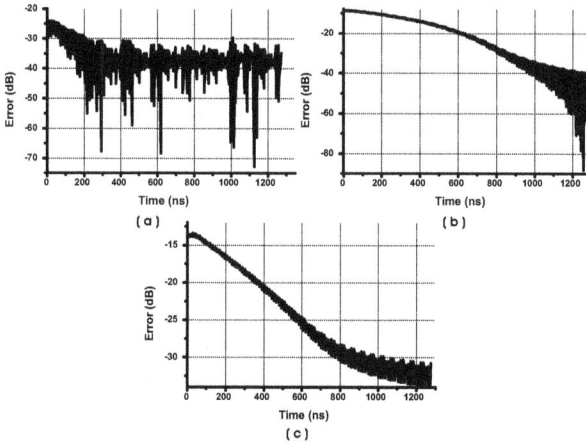

Fig. 30. Error in the (a) worst power, (b) worst speed and (c) temperature variation cases.

5. Knowledge-based impedace matching control design

Formulation of a complete mathematical model for impedance mismatch is a very complex process, since the parameters involved depend on many factors like process variations, length variations of the interconnection lines, temperature, etc. In this sense, knowledge based algorithms represent interesting alternatives which can be explored when looking for solutions to the impedance mismatch problem using adaptive schemes.

Fuzzy logic formalizes the treatment of vague knowledge, and approximates reasoning through inference rules (Zadeh, L. 1999). It establishes the mechanisms to generate practical solutions to problems where traditional methods, which may require precise mathematical models, may not be suitable. Because of this, fuzzy control represents a good alternative to solve the impedance mismatch problem through on-chip adaptive mechanisms (Arroyo et al., 2009).

Fig. 31 shows the general structure of the fuzzy controller used to adapt the system. The structure is simple and was designed in UMC 90nm CMOS technology. The membership functions are implemeted using differential pairs, while the multipliers are four-quadrant multipliers. The controller was designed to work in current-mode, therefore the sum operation is simply the sum of the currents in a node. Since the implementation of the divisor circuit is not trivial, a normalizer circuit can be used as an alternative.

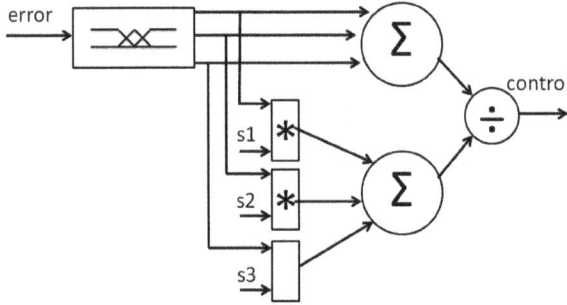

Fig. 31. General structure of the fuzzy controller

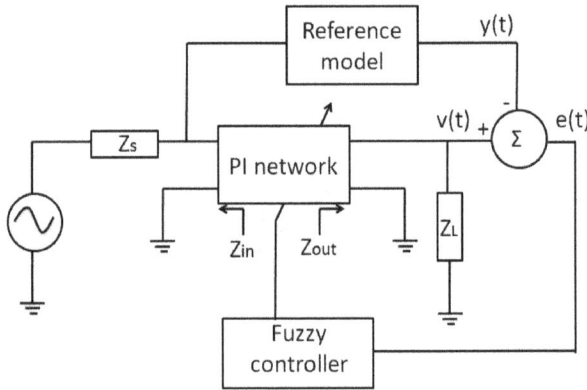

Fig. 32. Impedance matching system

Fig. 32 depicts a block diagram of the fuzzy control-based scheme for the impedance matching system. In order to allow the impedance matching process, a two port network with a standard π configuration is inserted between the source and the load. It is considered that both, the source impedance Z_S and the load impedance Z_L, allow complex values in the general case. The reference model is used to generate the reference signal $y(t)$, which is necessary to obtain the error. The error signal is used as the input to the fuzzy controller and is given by

$$e(t) = v(t)1 - y(t) \tag{25}$$

where $e(t)$ is the error, $v(t)$ is the current output of the system and $y(t)$ is the desired output. The output of the fuzzy controller is used to adaptively change the value of one of the capacitors of the π network. The system iterates until the impedance matching condition given by (2) is fulfilled, i.e.

$$Z_L = Z_{out}^* \tag{26}$$

where $(*)$ denotes the complex conjugate. It is clear that the fulfillment of this condition implies: $e(t) = 0$. Fig. 33 and Fig. 34 show the evolution in time of the absolute value of the adapted impedance and the normalized mean square error, respectively. As can be seen, the fuzzy controller adapts the impedance matching network, leading the system in every case to match the value of the load impedance.

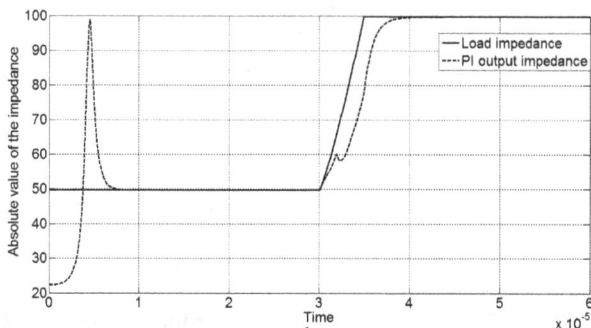

Fig. 33. Adaptation process of the impedance matching system

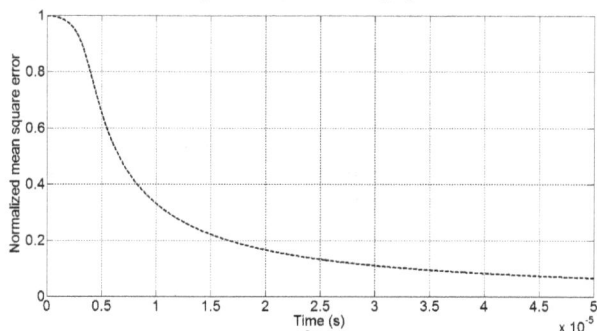

Fig. 34. Normalized mean square error

6. Conclusion

In this chapter systems for on-die automatic impedance matching for off-chip signaling were described. A review of different techniques for impedance matching was presented. Based on that, two algorithms were proposed and implemented in order to perform the automatic impedance matching control: the first one is based on the integration of the sign of the impedance matching error and the sign of the coupling branch current, the second one uses a fuzzy controller in the feedback path in order to adapt the impedance of the matching network. Advantages and performance of these algorithms were discussed and proved by presenting computer simulations of layout extractions.

7. References

Dally W. J. & Poulton J. W. (1998). *Digital Systems Engineering*, Cambridge University Press.

Juan F. (2007). *CMOS Current-Mode Circuits for Data Communications*, Springer.

DeHon A. et. al (1993). Automatic impedance control, *IEEE International Conference on Solid State Circuits*, pp. 164 - 165, Feb. 1993.

Ramachandran N. et. al. (2003), A 3.3-v cmos adaptive analog video line driver with low distortion performance, *IEEE Journal of Solid-State Circuits*, vol. 38, no. 6, pp. 1051 - 1058.

Deutschmann B. & Ostermann T. (2003). Cmos output drivers with reduced ground bounce and electromagnetic emission, *Proceedings of the 29th European Solid - State Circuits Conference, 2003*, ESSCIRC 03, pp. 537 - 540.

Svensson C. & Yuan J. (1991). High speed cmos chip to chip communication circuit, *IEEE International Sympoisum on Circuits and Systems*, vol. 4, pp. 2228 - 2231, June 1991.

Choy C. S. et. al. (1997). A low power-noise output driver with an adaptive characteristic applicable to a wide range of loading conditions, *IEEE Journal of Solid-State Circuits*, vol. 32, no. 6, pp. 913 - 917, June 1997.

Shin S. K. et. al. (2005). A slew rate-controlled output driver having a constant transition time over the variations of process, voltage and temperature, *Proceedings of the IEEE Custom Integrated Circuits Conference*, pp. 231 - 234.

Balatsos A. (1998), Clock buffer ic with dynamic impedance matching and skew compensation, *Masters of Applied Science Thesis*, Toronto, Canada.

Carusone A. et. al. (2001). Differential signaling with a reduced number of signal paths, *IEEE Transactions on Circuits, and Systems II*, vol. 48, no. 3, pp. 294 - 300, Mar. 2001.

Wang T. & Yuan F. (2007), A new current-mode incremental signaling scheme with applications to gb/s parallel links, *IEEE Transactions on Circuits and Systems I: Regular Papers*, vol. 54, no. 2, pp. 255 - 267, Feb. 2007.

Young B. (2001). *Digital Signal Integrity Modeling and Simulation with Interconnects and Packages*, Prentice Hall PTR.

Brooks D. (2003), *Signal Integrity Issues and Printed Circuit Board Design*, Prentice Hall PTR.

Koo K. et. al. (2001). A new impedance control circuit for usb2.0 transceiver, *Proceedings of 27th European Solid State Circuits Conference*, pp. 237 - 240, Sept. 2001.

Koo K. et. al. (2006). A versatile i/o with robust impedance calibration for various memory interfaces, *IEEE International Symposium on Circuits and Systems*, pp. 1003 - 1006, May 2006.

Muljono H. et. al. (2003). A 400-mt/s 6.4-gb/s multiprocessor bus interface, *IEEE Journal of Solid-State Circuits*, vol. 38, no. 11, pp. 1846 - 1856, Nov. 2003.

Fan Y. & Smith J. (2003). On-die termination resistors with analog impedance control for standard cmos technology, *IEEE Journal of Solid-State Circuits*, vol. 38, no. 2, pp. 361 - 364, Feb. 2003.

Tae-Hyoung K. et. al. (2005). A 1.2v multi gb/s/pin memory interface circuits with high linearity and low mismatch, *IEEE International Sympoisum on Circuits and Systems*, vol. 2, pp. 1847 - 1850, May 2005.

Munshi A. et. al. (1994). Adaptive impedance matching, *IEEE International Symposium on Circuits and Systems*, vol. 2, pp. 69 - 72.

Carusone A. & Johns D. (2000). Analog adaptive filters: Past and present, *IEE Proceedings Circuits, Devices and Systems*, vol. 174, no. 1, pp.82 - 90, Feb. 2000.

Zadeh, L. (1965). *Fuzzy sets*, Information and Control, page numbers (338-353), Academic Press.

Arroyo-Huerta, E.; Díaz Méndez, A.; Ramírez-Cortés, J.M. & Sánchez-García, J.C. (2009). An adaptive impedance matching approach based on fuzzy control, *52nd IEEE Midwest Symposium on Circuits and Systems*, ISBN: 978-1-4244-4479-3, Cancún, México, August 2009.

Thierauf S.(2004). A High-Speed Circuit Board Signal Integrity, *Artech House, Inc.*, 2004

E. Lopez-Delgadillo, J.A. Diaz-Mendez, M.A. Garcia-Andrade, M.E. Magana, F. Maloberti (2009). A Self Tuning System for On-Die Terminators in Current Mode Off-Chip Signaling, *52nd IEEE Midwest Symposium on Circuits and Systems*, Cancún, México, August 2009.

3

Carbon Nanotube- and Graphene Based Devices, Circuits and Sensors for VLSI Design

Rafael Vargas-Bernal and Gabriel Herrera-Pérez
Instituto Tecnológico Superior de Irapuato (ITESI),
México

1. Introduction

With the reduction in power consumption and size chip, the electronic industry has been searching novel strategies to overcome these constraints with an optimal performance. Carbon nanotubes (CNTs) due to their extremely desirable electrical and thermal properties have been considered for their applicability in VLSI Design. CNTs are defined as sheets of graphene rolled up as hollow cylinders. They can basically be classified into two groups: single-walled (SWNTs) and multi-walled (MWNTs) as shown in Figure 1. SWNTs have one shell or wall and whose diameter ranging from 0.4 to 4 nm, while MWNTs contain several concentric shells and their diameter ranging from several nanometers to tens of nanometers.

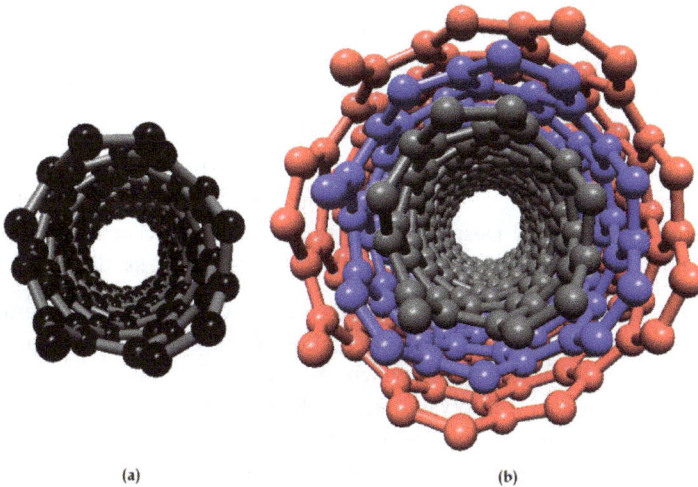

(a)   (b)

Fig. 1. Types of carbon nanotubes: single-walled nanotube (SWNT) and multiple-walled nanotube (MWNT).

The electrical properties the SWNTs can be either of metallic or semiconducting materials depending on their chirality, that is, the direction in which they get rolled up. However, MWNTs are always metallic materials. The main applications of carbon nanotubes in electronics are biochemical sensors, data storage, RF applications, logic circuits and/or semiconductor materials (Xu et al., 2008). Nowadays, graphene nanoribbons (GNRs) or carbon nanotubes unrolled are presented as attractive candidate for next-generation of integrated circuit applications derived of the anomalous quantum Hall effects and massless Dirac electronic behavior (Lu & Lieber, 2007).

The main objective of this review related with carbon nanotubes and graphene nanoribbons is assessing the current status in VLSI design and provides a vision of the future requirements for electrical subsystems based on carbon nanotubes: technology, products and applications. This chapter presents a comprehensive study of the applicability of carbon nanotubes and graphene nanoribbons as base materials, with special emphasis into the advantages and limitations, in the design of elements for VLSI design such as interconnects, electronic devices such field-effect transistors, diodes and supercapacitors; optoelectronic devices such as solar cells and organic light-emitting diodes; electronic circuits such as logic gates, and digital modulators; and bio/chemical sensors such as biosensors and gas sensors.

2. Electrical properties of carbon nanotubes

One promising direction for the VLSI Design is the use of carbon nanotubes as the active part of the device, circuit or sensor. Carbon nanotubes (CNTs) are macromolecular one-dimensional systems with unique physical and chemical properties (Zhou et al., 2007). Such properties are derived of that all chemical bonds are satisfied and they are very strong, which also leads to total mechanical, thermal and chemical stability (Baughman et al., 2002). The electronic structure and electrical properties of CNTs are derived from those of a layer of graphite (graphene sheet). The specific electrical properties of the carbon nanotubes are obtained as result of their particular band structure and the hexagonal shape of its first Brillouin zone. CNTs can carry out high electrical current densities at low electron energies. When high electron energies are used, this quantity of energy destroys the CNT structure, which is not desirable from any point of view (Mamalis et al., 2004; Terrones, 2003, 2004).

This section analyzes the electrical characteristics of carbon nanotubes and graphene nanoribbons through their physical structure with the aim of presenting the attractive interest for using them in VLSI Design. The advantages and drawbacks of the use of CNTs and graphene nanoribbons as active part of an electrical device are studied.

Among physical variables of the carbon nanotube related with the electrical performance are diameter, chirality, length, position, and orientation. Each graphene sheet is wrapped in accordance with a pair of indices (n, m), which represents the number of unit vectors along two directions in the honeycomb crystal lattice of graphene. If $m = 0$, the nanotubes are called zigzag nanotubes, if $n = m$, the nanotubes are called armchair nanotubes and otherwise, they are called chiral nanotubes (see Figure 2) (Hayden & Nielsch, 2011; Hetch et al., 2007; Marulanda, 2010).

Two physical properties of the graphene modify its electrical properties: symmetry and electronic structure. There are three types of electrical behavior as shown in Figure 3: 1) if $n = m$, the nanotube is metallic; 2) if n-m is equal to $3j$, where j is a positive integer ("$3j$" rule), then the nanotube is semiconducting with a very small band gap, and 3) otherwise, the nanotube is a moderate semiconducting. The $3j$ rule has exceptions due to the curvature effects in carbon nanotubes with small diameter, which can influence in the electrical properties. A metallic carbon nanotube can present semiconducting behavior and vice versa (Avouris, 2002).

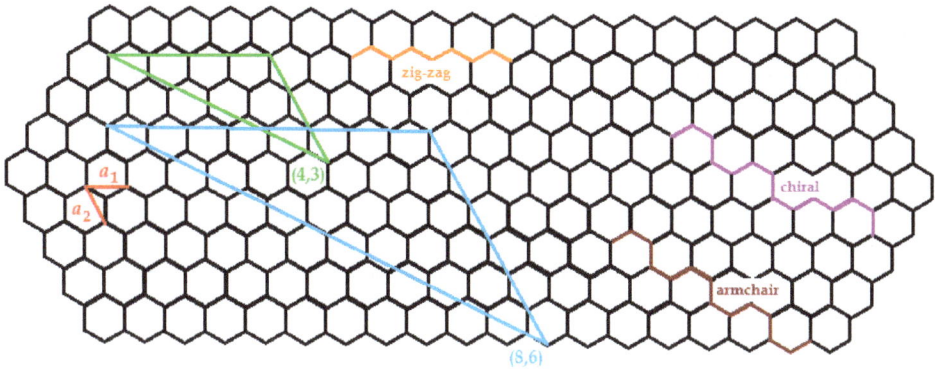

Fig. 2. Classification of carbon nanotubes by chiral indices: zig-zag, chiral, and armchair.

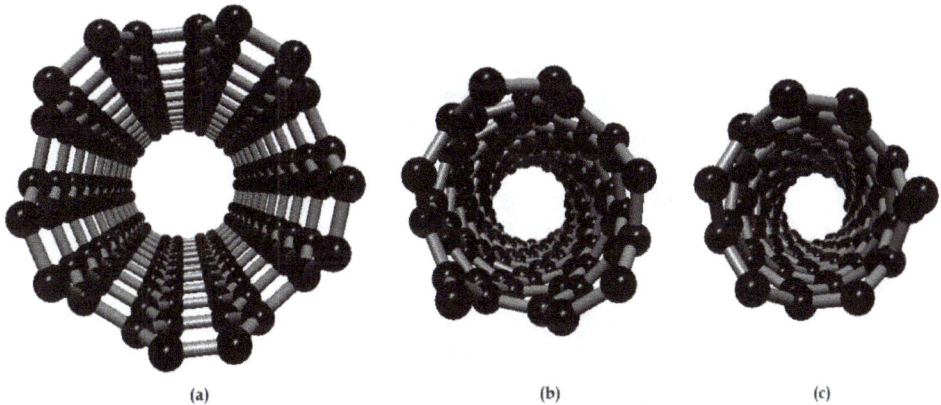

Fig. 3. Classification of carbon nanotubes by electrical properties: (a) metallic nanotube, (b) semiconducting nanotube, and (c) moderate semiconducting nanotube.

The interaction among electrons in an one-dimensional conductor such as a carbon nanotube can be modeled as a Tomonaga-Luttinger liquid, since electronic properties are derived of the collective excitations of charge and spin waves with a bosonic nature, that is, mass-less current flow (Danilchenko et al., 2010). Carbon nanotubes show two different electrical behaviors depending of the range of temperature: ballistic current transport at room temperature and Coulomb blockade phenomena at low temperatures. Ballistic transport is presented when the effective distance between contacts, where voltage is applied, is shorter than the mean free path. Coulomb blockade occurs when electrons hop on to and off from a single atom between two contacts due to a high contact electrical resistance (Hierold, 2008; Léonard, 2009).

In particular, metallic carbon nanotubes allow that very large electrical currents can be used to design high speed nanoscale electronic devices due to its wide band gap. Metallic multiwall CNT can carry a current density on the order of 10^8 A/cm^2 and have the capacity of dissipated power of 1.82 mW (Shacham-Diamand et al., 2009). Individual carbon nanotubes can be considered as quasi-one-dimensional (1D) conductors. Multi-walled nanotubes (MWNTs) are considered two-dimensional (2D) conductors due to their coaxial distribution of SWNTs with intertube spacing of ~ 3.4 Å. Metallic carbon nanotubes present high dielectric constant, while semiconducting carbon nanotubes have low dielectric constant (Joachim et al., 2000; Kang et al., 2007; Krompiewski, 2005).

One of the most promising applications of the electrical properties of carbon nanotubes is the use of them in the fabrication of electronic devices. Special interest is given to the use of soft and ductile matrices to portable, light, and flexible electronics. In the design of electronic devices, the precise and tunable control of the electronic properties is essential to the high performance VLSI circuits. During the synthesis of carbon nanotubes, both metallic and semiconducting carbon nanotubes are obtained (Kanungo et al., 2010), forming sets of carbon nanotubes called bundles. A bundle containing tens to hundreds of tubes is denominated a rope; in this structure, the carbon nanotubes are separated ~ 3.2 Å forming a close-packed triangular lattice where the diameters are almost identical (see Figure 4) (Hou et al., 2008).

Fig. 4. An ideal bundle of carbon nanotubes.

A bundle of carbon nanotubes is formed by van der Waals interactions among neighboring nanotubes. It is waited that cooperative effects among nanotubes be originated in a bundle (Kim et al., 2010). The presence of multiple carbon nanotubes can substantially reduce the electrical resistance to the electrical current carried out by them, if this is compared with the electrical resistance of an individual nanotube. It is true, only when the bundles have direct physical contact, non electrical, with any material in the device with the aim of reducing the temperature generated by the current carried out. The electrical transport in a bundle has interesting electrical properties such as single electron transport (Coulomb blockade allow us control the number of electrons in the electrical conduction one by one) and metallic resistivity (increased with the temperature). Additionally, the electronic transport in a bundle is modified by the direction and magnitude of the applied electrical field and the electrostatic screening produced by the carbon nanotubes surrounding to a specific carbon nanotube, as shown in Figure 5. Such electrostatic screening leads to a tunable switching behavior which is induced by electric field perpendicular or transverse to the bundle axis. In the case of semiconducting nanotubes, the applied electrical field produces band gap closure; while for metallic nanotubes, it produces a band gap opening. In this way, only for metallic nanotubes it is possible to modulate the conductivity of the bundle through of the applied field and splitting of the valance and conduction bands thanks to the symmetry breaking of the electrostatic screening between adjacent nanotubes due to a weak electrical interaction presented in the intertube region between them. It is necessary to remember that the level of electrostatic screening inversely determines the electrostatic field and Coulomb potential of the ions in the nanotubes. For semiconducting nanotubes, the band gap is reduced thanks to the increase of size of valence and conduction bands generated by the Stark effect derived of the applied electrical field to the nanotubes (Haruehanroengra & Wang, 2007).

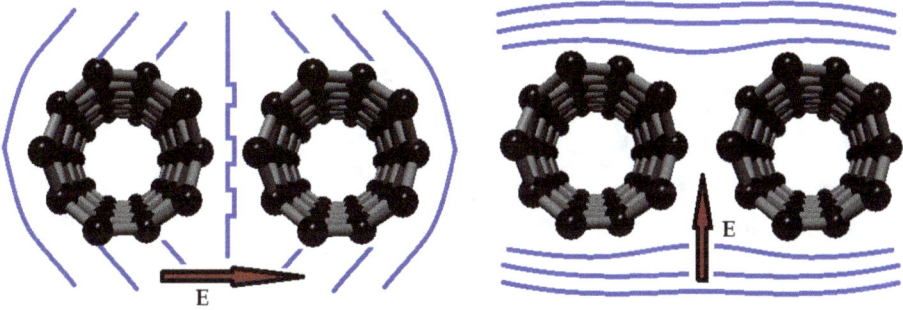

Fig. 5. Electrical field applied in a bundle of carbon nanotubes. Red arrow indicates the direction of the field.

Arrays of carbon nanotubes have electrical properties which can be controlled by means of its length, diameter, and chirality (Jain et al., 2011). A uniformity of the properties can be achieved when performance characteristics such as high yield, reproducibility, sensitivity, and specificity are guaranteed. This is obtained through synthesis procedures, dispersion procedures, and deposition processes whose quality allows us the integration of the carbon nanotubes with the same physical properties before and after of the dispersion of bundles (Hong et al., 2010).

Due to the presence of bundles of nanotubes, it is necessary the development of methods which allow us to separate nanotubes for extending their use in electronic applications. Several methods to separate bundles based on monovalent side wall functionalization have been developed even with the aim of improving solubility, purification and exfoliation. Unfortunately, these methods can lead to disrupts π transitions, generate changes in electrical resistance, and can even produce the tube fragmentation due to the formation of impurity states near the Fermi level. New strategies based on the use of mixtures of metallic and semiconducting nanotubes are producing high mobility semiconducting combinations without laborious separation requirements to use all carbon nanotubes obtained during the synthesis. The use of divalent functionalizations which produce impurity states far away from the Fermi level, can even lead to generate high performance semiconducting inks of low cost which can be applied in printable VLSI electronics. In addition, divalent functionalization offers a different strategy to control the electrical properties slightly taking into account tube type, size, and chirality. Adequate addends used in the functionalization allow us to transform metallic nanotubes into semiconducting nanotubes (Javey, 2008).

In 2004, graphene arose as a product of exfoliation of graphite, with the form of a two-dimensional sheet of sp^2-hybridized carbon (Novoselov et al., 2004). In the same manner that carbon nanotubes, it has unique electrical, mechanical and thermal properties. Such properties have been exploited in the development of energy-storage materials, transparent conducting electrodes (Alkire et al., 2009; Hu et al., 2007), field-effect transistors, digital and analog integrated circuits, integrated circuit interconnects, solar cells, ultracapacitors, and electrochemical sensors such as single molecule gas detectors and biosensors. High electron mobility at room temperature, low electrical resistivity, and symmetry of carrier mobilities between electrons and holes, are the electrical properties attractive to apply graphene in the design of electronic devices of high-performance. A similar classification to the carbon nanotubes with respect to the electrical behavior of the graphene is illustrated in Figure 6.

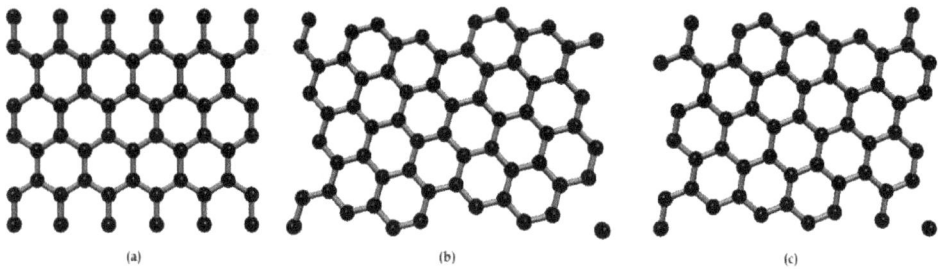

(a)   (b)   (c)

Fig. 6. Classification of graphene by electrical properties: (a) metallic graphene, (b) semiconducting graphene, and (c) moderate semiconducting graphene.

Graphene nanoribbons (GNRs) can be defined as rectangles made from graphene sheets with widths going from a few nanometers to tens of nanometers and lengths from nanometers to micrometers. They are considered as quasi-1D nanomaterials and can have metallic (zigzag) or semiconducting (armchair (AGNR) or zigzag (ZGNR)) behavior depending of its chirality and orientation. Both types are denoted in accordance with the number of chains either, armchair or zigzag, found in its width. High electrical and thermal conductivity, low noise and bidimensional structure are properties which can be useful to produce integrated circuit interconnects with GNRs. The size of GNRs allows us to control the band gap of the material to be electrically manipulated in an electronic device generating a wide versatility of design (Ferry et al., 2009; Guildi & Martín, 2010).

GNRs possess a richer energy band structure than the graphene, since an external electric field can be used to tune a specific bandgap (Chen, X. et al., 2011). Semiconducting AGNRs have electrical behavior as semiconductor material of indistinct manner with respect to the carbon chain position, and metallic AGNRs present both metallic as semiconducting behavior which is related with the change of chain associated with the "$3j$ rule" in carbon chain position (Law et al., 2004; Philip-Wong, 2011).

Graphene has interesting electrical properties such as electron-hole symmetric band structure, high carrier mobilities, ballistic transport, and absence of band gap (Reddy et al., 2011). But also, some disadvantages associated with its use in field-effect transistors such as lack of gate control, high off-state leakage current and saturation not controlled by drain voltage. Different methodologies are being developed to overcome, adapt to, and even use these electrical characteristics for its application in electronic devices. The use of graphene as electronic material resides in the reduction of energy consumption, linear energy dispersion, carriers with zero mass, linear current-voltage characteristic, high Fermi velocities, very low channel electrical resistance, mobilities and saturation velocities for a high current-carrying capability (6 orders higher than copper), low density of states, and the increase of frequency operation of the devices based on these qualities (Geim & Novoselov, 2007). Depending of the bias voltage, the sheet of graphene can present electrical resistance in the range of Kilo-ohms to ohms for low voltages and high voltages, respectively. OFF-state leakage currents in field-effect transistors based on graphene are detrimental for digital circuits, but these are very useful to analog circuits where ON-state modulating small voltages and current signals are a common case. For applications as high performance RF circuits, the graphene offers an alternative material given that its cut-off frequency is very high, and it has high compatibility with VLSI systems based on silicon. Due to its nature structurally malleable, the electrical properties of the graphene can be favorably modified by mechanical strain and stress (Geim, 2009). Graphene can also be used in interconnects and optoelectronics.

3. Carbon nanotube interconnects

The interconnects distribute a large quantity of signals used for the diverse elements of a VLSI design such as clock signals, power, or ground in an integrated circuit (IC), and also to various circuits on a chip. Local, intermediate and global interconnects are the levels of operation of such interconnections. The use of Cu as material for interconnects represents a current paradigm for high-performance integrated circuits due to that line dimensions, and grain size become comparable to the bulk mean free path (MFP) of electrons (~ 40 nm). In addition, higher RC delays reduce the operation speed of ICs. When a new proposal in VLSI design is done, the main characteristic must be the compatibility with current IC manufacturing. The two most promising potential candidates that can be used as material for interconnects are optical and carbon-nanotube (CNT) based interconnects (Cho et al., 2008; Koo et al., 2007; Kreupl et al., 2002).

This section provides a summary of the novel challenges that are being realized in nanometer-scale on-chip interconnects. Special topics associated with the operational effects such as performance and reliability are analyzed, with the aim of identifying the electrical characteristics that can be obtained in resistivity, interconnect delay, and current-carrying capability. Finally, the prospective applications of GNRs for interconnects are discussed.

Carbon nanotubes can be integrated into multilevel interconnects to meet emerging needs: delay, lifetime, parasitic resistance, inductive effects, bandwidth density, electromigration (Hosseini & Shabro, 2010), energy efficiency, power dissipation, and lowering temperature of the interconnection. Additionally, the use of carbon nanotubes makes possible the development of three-dimensional hyper-integration architectures with a high performance: versatility, scalability, adaptability, high-density interconnects, and a reduced number of defects, (Ahn et al., 2006; Bakir & Meindl, 2009; Papanikolau et al., 2011; Shacham-Diamand et al., 2009; Xie et al., 2010; Zhou & Wang, 2011).

Electrical transport in MWNTs presents three different cases. When a MWNT operates at conditions of low energy (thermal and electrical), electrical current in carried by the outermost shell of it. At intermediate energy, only metallic shells contribute to the electrical transport of current. Finally, at high energy all shells of the MWNT carry electrical current. In this manner, is complicated to adjust the operation of MWNTs with the aim of that these can be used in interconnections of VLSI systems (Srivastava, 2004, 2009; Tan et al., 2008).

Since inherently carbon nanotubes can provide high electrical current density, numerous applications, including interconnects for VLSI design, have been suggested as a novel way of reducing physical spaces with an optimal performance. The electrical resistance for CNTs, as large as 1 μm with perfect contacts, is about of 6.45 KΩ. This value is high to be used in interconnects, therefore, carbon nanotubes are placed in parallel in large numbers (a bundle) with the aim of reducing the total electrical resistance. A CNT bundle is generally a mixture of single-walled CNTs, multi-walled CNTs or single-walled CNTs and multi-walled CNTs. CNTs bundle offers a promising alternative to place metallic contacts and vias at the local level for VLSI circuits, with the advantage that they can be grown with low or high indexes when lower or higher current densities are needed, respectively. In particular, the length-to-diameter ratio of the CNT interconnects have significant implications for the design of on-chip capacitors and inductors (Nojeh & Ivanov, 2010).

Due to the very high frequencies used to carry signals in the integrated circuits, the ballistic transport presented by carbon nanotubes and graphene allow us to design advanced interconnect networks (see Figure 7). Since metallic carbon nanotubes are almost insensitive to the disorder, they are considered as perfect 1D electrical conductors. In a similar way, tube-tube connections, junctions and even tube-metal contacts that also are used in the interconnection of VLSI systems must work reliably with minimal electrical losses in the contact points. By nature, nanotube-metal interface presents a tunneling barrier. The research associated with these phenomena has searched solutions based on fabrication methods to modulate the electrical characteristics of the interface (Li, J. et al, 2003).

Tube-tube junctions involve physical contact, with small structural deformation, between two tubes and these are not chemically bonded. This type of junction is found in interconnections between ropes, MWNTs, and crossed-over tubes (Andriotis et al., 2001, 2002). The electrical transport is realized by means of tunneling transport between tubes, producing an alteration in electrical transport of the individual tubes involved in the junction, due to the weak electrical coupling. When specific junctions called "X", "Y" or "T" are been used in the connections, these have proved to be stable and therefore, these can be useful when it is required to joint multi-terminal electronic devices by means of carbon nanotubes or in the case of wiring interconnection (Chen & Wang, 2009; Li, H. et al., 2008, 2009).

Plating a hollow structure inside used as via or trench in circuits VLSI by means of carbon nanotubes have not been reasonable due to high hydrophobicity of graphene sheets, which hinder the entry of solvent and dissolved species (Shacham-Diamand et al., 2009). This problem is emphasized for carbon nanotubes with diameters less than 50 nm.

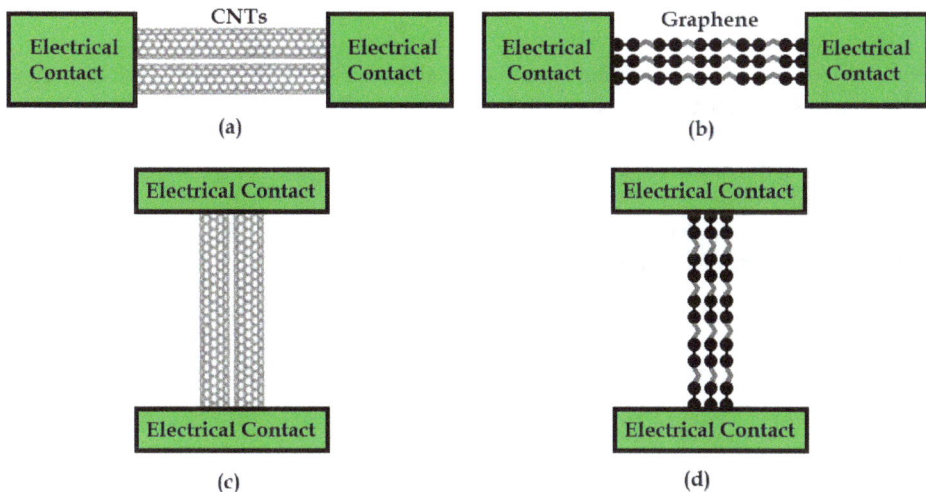

Fig. 7. Electrical connections in VLSI circuits: (a) interconnects based on carbon nanotubes, (b) interconnects based on graphene, (c) vias based on carbon nanotubes, and (d) vias based on graphene.

The graphene presents a higher conductance with respect to Cu for interconnects in the range of nanometers. Among the properties exploited of the graphene for interconnects are: high carrier mobility at room temperature, thermal conductivity, higher mechanical strength, reduced capacitance coupling between adjacent wires, width-dependent transport gap, temperature coefficient, and ballistic transport. When line widths of the graphene nanoribbons are reduced below 8 nm, the resistivity of GNRs is insignificant. Additionally, the use of graphene in interconnects extends the life of high performance for silicon-based integrated circuit technology. In thermal characteristics, the graphene interconnects allow us to cool heat flux, to remove hot-spots, and to spread lateral heat (Goel, 2007).

Additionally to the electrical properties, the mechanical and thermal properties of CNTs and graphene nanoribbons must be taking into account in the design of interconnects. Mechanical properties such as strength, stability and minimal elastic deformation can be achieved thanks to its topology and low density. By another side, carbon nanotubes exhibit good thermal conductivity and high thermal stability, which are necessary to support high current densities (Giustiniani et al., 2011).

Within of the novelties to come in this sector are the scaling of the ordinary interconnects by means of an accurate and reproducible patterning of nanoscale structures based on carbon nanotubes and/or graphene nanoribbons. The use of self-assembly is more and more feasible given the advancements in the development of supermolecular networks. These changes will allow the perfect alignment and optimal charge transport among the elements interconnected in a VLSI system. Additionally, these techniques increase the yield given place to a massive fabrication and lower costs, which are essential in a VLSI system.

4. Carbon nanotube based devices

As active part of electronic devices, the CNTs have been used to control their electrical properties. In this manner, carbon nanotubes can implement electronic devices such as diodes, transistors, Schottky rectifiers (Behman et al., 2008), supercapacitors (Chen, P.-C. et al, 2009), solar cells (Jia et al., 2011; Nogueira et al., 2007), and organic light-emitting diodes, by combining semiconductor and metallic behaviors (Terrones, 2003, 2004; Tseng et al., 2004). Different strategies and topologies have been proposed with the aim of improving their performance. Transistors and Schottky rectifiers can be obtained by means of metallic-semiconducting junctions (Hur et al., 2004).

This section analyses the performance characteristics, topologies, and applications of the electronic devices fabricated by means of carbon nanotubes with emphasis to VLSI Design. It is explained as the choice of material is critical for a successful application with high performance in electronic devices such as field-effect transistors, *p-n* diodes, supercapacitors, solar cells, and organic-light-emitting diodes.

The development of the carbon nanotube field-effect transistors (CNFETs) was due to the historical motivation of reducing or make insignificant short-channel effects, and to improve performance of transistors in these length scales (Burke, 2004). The use of semiconducting carbon nanotubes is strategic given that metallic nanotubes cannot be fully switched off. The main advantages of this type of transistors are: ballistic electron transport over its lengths (Hasan et al., 2006), higher current density, lower power consumption with respect to silicon versions, and faster operation speed (Burghard et al., 2009). There are four main topologies to design CNT field-effect transistors: 1) back-gated CNTFETs, 2) top-gated CNTFETs, 3) wrap-around gate CNTFETs, and 4) suspended CNTFETs, as shown in Figure 8.

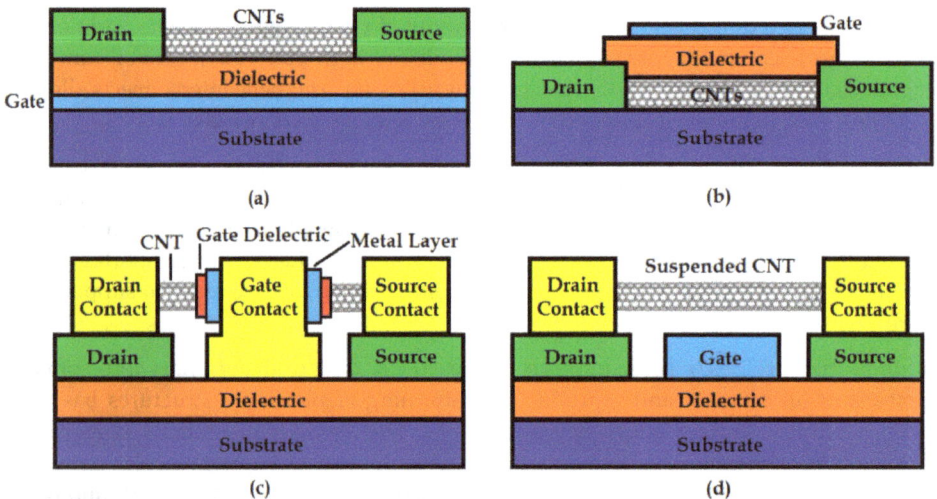

Fig. 8. Cross sections of different geometries of carbon nanotube field-effect transistors: (a) back-gated CNTFETs, (b) top-gated CNTFETs, (c) wrap-around gate CNTFETs, and (d) suspended CNTFETs.

In the case of back-gated CNTFETs, the main disadvantages found for its use are a poor contact between the gate dielectric and CNT, difficult switching between ON and OFF states when low-voltages are applied, and a Schottky barrier between CNTs and drain and source regions. In the case of top-gated CNTFETs, these offer several advantages over back-gated CNTFETs, but it fabrication process is more complicated (Singh et al., 2004). In wrap-around gate CNTFETs, the entire circumference of the nanotube is gated and therefore, electrical performance is enormously improved, reducing leakage current and increases the device ON/OFF ratio. Finally, in the case of suspended CNTFETs is searched the reduction of the contact between the substrate and gate oxide, and therefore, it decreases scattering at the CNT-substrate interface with the drawback of limiting its use in applications where high ON/OFF ratio are required (Kocabas et al., 2005, 2006).

The CNTFETs can be classified in two types: 1) n-type CNTFETs, when electrons are majority carriers for positive gate voltages, and 2) p-type CNTFETs, when holes are majority carriers for negative gate voltages. An ohmic contact is found when a current-voltage relationship is linear and symmetric (electrons and holes are transported in the same time), while a Schottky-barrier is presented when current-voltage relationship is non-linear and asymmetric (a unique type of electrical carrier is transported) (Lin, A. et al., 2009).

Four electrical transport regimes can be found in transistors based on carbon nanotubes, which are distinguished in accordance with the length of the nanotube compared with their mean free path, and by the type of contact between the nanotubes and the source/drain metals: 1) *ohmic-contact ballistic,* when charge injection is realized by the source and drain contacts into the carbon nanotubes and vice versa, producing a high current flow; 2) *ohmic-contact diffusive,* when bidirectional charge transport suffers scattering between source and drain contacts and carbon nanotubes with a limited current flow; 3) *Schottky-barrier ballistic,* when the gate voltage controls the thickness of the barrier and drain voltage can lower the barrier producing bidirectional high current flow: in ON-state, electrons tunneling from the source, and in OFF-state, holes tunneling form the drain; and 4) *Schottky-barrier diffusive,* when the combination of gate and drain voltages reduces the Schottky barrier and the charge transport suffers scattering producing a reduced current flow (Appenzeller et al., 2005; Cao et al., 2007).

With the introduction of graphene as active material for electronic devices, new field-effect transistors were introduced, namely these are called GFETs. A GFET uses as active material, graphene, for ballistic transport of carriers. As it was illustrated for carbon nanotube, also can be built four types of GFETs: 1) back-gated GFETs, 2) top-gated GFETs, 3) wrap-around gate GFETs, and 4) suspended GFETs. Last two topologies are not available now, but these will be fabricated in a pair of years. Back-gated GFETs present large parasitic capacitances and poor gate control. However, when smooth edges of the graphene nanoribbons are achieved, ON/OFF ratios as high as 10^6 are obtained, which is attractive for digital applications. Top-gated GFETs are the preferred option for analogical practical applications. In wrap-around gate GFETs, the entire rectangle of the graphene nanoribbon will be gated (see Figure 9).

Nowadays, carbon nanotube-based field-effect transistors (FETs) have operating characteristics that are comparable with those devices based on silicon. The active part in field-effect transistors is the electrical channel established by means of the carbon nanotube in the substrate connecting source and drain terminals. SWNTs have been the ideal candidates as semiconducting materials due to that them can be doped to address the type of conductivity either n-type or p-type and, in this way, to manipulate the level of electrical conduction. The carbon nanotube based FETs can achieve high gain (> 10), a large on-off ratio (>10^5), and room-temperature operation (Lefenfeld, et al., 2003).

Carbon nanotubes and graphene nanoribbons are very sensitive to their environments including charges, vacuum levels, and environment chemical components, due to their ultrasmall diameters and large surface-to-volume ratios. Carrier mobility in carbon nanotubes is very susceptible to charge fluctuations derived of the defects located at the ambient surrounding the CNTs and graphene nanoribbons. The mobility fluctuation is the dominant $1/f$ noise mechanism for the narrow channel carbon nanotubes operating in strong inversion region with a small source-drain bias.

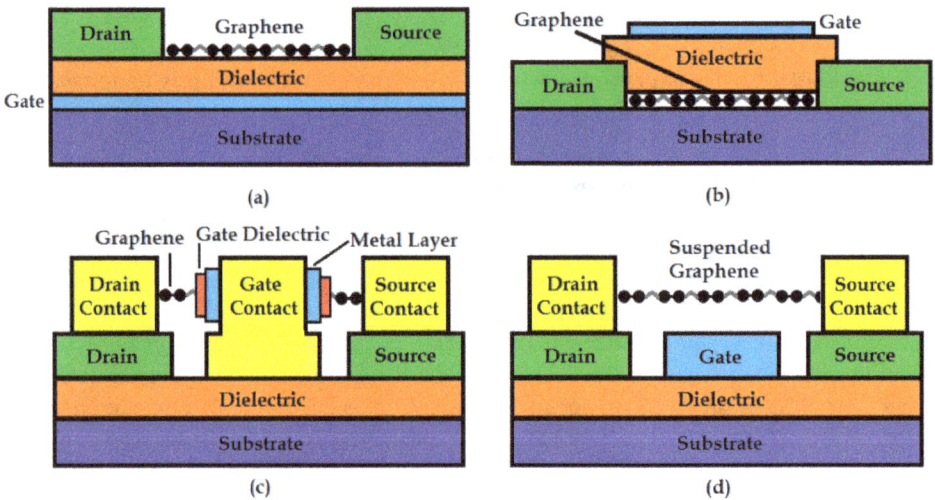

Fig. 9. Cross sections of different geometries of graphene field-effect transistors: (a) back-gated GFETs, (b) top-gated GFETs, (c) wrap-around gate GFETs, and (d) suspended GFETs.

At ambient temperature, semiconducting SWNTs generally show unipolar p-type behavior. By doping with potassium, the unipolar p-type behavior can be switched to unipolar n-type behavior. p-n Diodes can be designed by covering one-half of the gate of a single channel field-effect transistor with polymers such as polyethylenimine (PEI) and poly(methyl methacrylate) (PMMA) (Mallick et al., 2010; Zhou, Y. et al., 2004).

Those field-effect transistors that have been fabricated with functionalized nanotubes exhibit high electron mobilities, high on-current, and very high on/off ratios which are necessary in high-speed transistors, single- and few-electron memories, and chemical/biochemical sensors. Studies on scaling resistivity are being realized with the aim of identifying the influence of device parameters in the on/off ratio (Sangwan et al., 2010).

A supercapacitor is an electrochemical capacitor with relatively high energy density of small size and lightweight (hundreds of times greater than those of electrolytic capacitors).

Carbon nanotubes together with ceramic materials can be used to design supercapacitors by means of heterogeneous films. The use of ceramic materials allows increasing their electrical energy accumulated as voltage, while carbon nanotubes offer the properties of flexibility and transparence. Among the optimized properties are specific capacitance, power density, energy density, and long operation cycles. Supercapacitors require electrodes with large surface area, which can be obtained by means of sets of carbon nanotubes operating as electrical conductive networks (Lekakou et al., 2011). These electrodes must be capable of supporting high power and energy density, with reduced internal electrical resistance and produced with lower cost.

Flexible electronics is now a reality thanks to the successful development of the organic electronics working to low-temperature (Lin, C.-T. et al., 2011). Devices such as organic thin film transistors (OTFTs), large-area displays (Wang, C., 2009), solar cells (Rowell et al., 2006), organic light-emitting diodes (OLEDs), and sensors can be implemented based on carbon nanotubes. The electrical properties improved with the use of carbon nanotubes are transistor on-off ratio, threshold voltage, and transistor transconductance. Additionally to the electrical properties, this type of devices can be fabricated to low-cost. Carbon nanotubes can be used to fabricate transparent conductive thin films (Facchetti & Marks, 2010; Ginley, 2010) which are exploited as hole-injection electrodes for organic light-emitting diodes (OLEDs) either for rigid glass or flexible substrates (Wang, 2010; Wiederrecht, 2010; Zhang et al., 2006). The incorporation of CNTs in polymer matrices used to design OLEDs allow changing electrical characteristics of the polymer due to that the CNTs operate as doping materials. Carbon nanotubes introduce additional energy levels or forming carrier traps in the host polymers, therefore, the CNTs facilitate and block the transport of charge carrier and improving the performance at specific dopant concentrations. Such concentrations must be controlled by percolation and functionalization of the carbon nanotubes with the polymer.

The integration of hybrid materials forming heterojunctions has allowed improving the efficiency of solar cells by means of the reduction of internal resistance, which is directly associated with the fill factor, transport and separation of charges that are useful for an optimal performance. Additionally, the use of carbon nanotubes provides the possibility of tailoring the electrical and structural properties to increase the optical efficiency of the light applied to the solar cell. Two great operative advantages of carbon nanotubes are being exploited in organic photovoltaics: higher electrical charge transport properties and elevated number of exciton dissociation centers (Nismy et al., 2010). Such dissociation makes that holes are transported by a hopping mechanism and the electrons are transferred through the nanotube. The ballistic transport of the electrons in carbon nanotubes produces very high carrier mobility in the active layer. Through a well-distributed percolation and careful functionalization of carbon nanotubes it is possible increase the charge transported thanks to the multiple transfer pathways among nanotubes. If carbon nanotubes are used as transparent electrodes in solar cells, then they collect electrical charge carriers (Hatakeyama et al., 2010, Liu, X et al., 2005).

The main strategy to come is the use of multiple nanotubes operating in parallel either individually or forming well-defined bundles with the aim of controlling the on-current in a wide range of electrical current going from micro-amperes to mili-amperes. In this manner, it is very useful to develop methodologies to produce arrays of nanotubes with well-characterized characteristics with the aim of obtaining high-performance applications.

5. Carbon nanotube based circuits

Carbon nanotubes can be exploited as molecular device elements and molecular wires. Each device element is based on a suspended, crossed nanotube geometry that leads to bistable, electrostatically switchable ON/OFF states. Such device elements can be addressed by means of control elements to manipulate large arrays (10^{12} elements or more) using carbon nanotube interconnects (Ishikawa et al., 2009; Tulevski et al., 2007).

This section discusses circuits based on carbon nanotubes that have been proposed in the last decade for VLSI Design. The necessary steps to leading to the carbon nanotubes based circuits toward integrated circuits are analyzed in detail. Different realizations of analog and digital circuits are studied, which can be used for integrated circuits in VLSI Design. The advantages and limitations of the performance of such circuits in their analog and digital versions are summarized.

The electrical properties of carbon nanotubes are making possible the complete design of VLSI systems under a unique active material (Hosseini & Shabro, 2010). Semiconducting carbon nanotubes can be used to build transistors, devices and circuits, while metallic carbon nanotubes are used to build interconnects and vias. Circuits such as ring oscillators (Pesetski et al., 2008), inverter pair (Nouchi et al., 2008), NOR gate, nonvolatile random access memory, etc. can be designed with field-effect transistors based on carbon nanotubes. Nowadays, simulation software has shown that CNT transistor circuits can operate at upper GHz frequencies (Vasileska & Goodnick, 2010).

The integration of multiple field-effect transistors can be realized to build digital logic circuits. In circuits where back-gated transistors are used, the same gate voltage is applied to all transistors associated with the circuit. Therefore, to increase the potentially of such circuits different strategies are being developed with the aim of applying different voltages to the gates in each transistor associated with the circuit. In digital circuits, the transistors must have electrical characteristics that can favor high performance such as: high gain, high ON/OFF ratio, excellent capacitive coupling between the gate and nanotube, and room-temperature operation (Cao et al, 2006; Jamaa, 2011). Until now, one-, two-, and three-transistor circuits have showed digital logic operations, giving place to logic inverters, NOR gates, static random-access memory cells, and AC ring oscillators (Wang, C., 2008).

Logic inverters are logical devices with one input and one output (see Figure 10 (a)). A logic inverter converts a logical "0" into a logical "1", and vice versa (Bachtold et al., 2001). Therefore, an inverter circuit operates as a basic logic gate to swap between two logical voltage levels "0" and "1". NOR gates are logical devices with two or more inputs and one output. A NOR gate of two inputs operates as follows: an output "1" is obtained when both inputs are "0", and an output "0" is achieved when one or both inputs are "1". A static random-access memory (SRAM) cell is built as a latch by feeding the output of two serial logical inverters together, that is, a bistable circuit generated to store each bit (Rueckes et al., 2000). Each cell has three different states: standby (the circuit is idle, both logic inverters are blocked to be used), reading (the data has been requested, first logic inverter is used) and writing (updating the contents, second logic inverter is used). An AC ring oscillator produces an oscillating AC voltage signal by means of the connection of three logical inverters in a ring, that is, the output of the last inverter is fed to the input of the first inverter. Such circuit has not statically stable solution, since the output voltage of each inverter oscillates as a function of time.

Digital circuits based on carbon nanotubes depends of the diameter of them, because it is directly proportional to the Schottky barrier height formed by the carbon nanotube and metal contacts of the source and drain terminals (Andriotis et al., 2006, 2007, 2008; Javey & Kong, 2009). In addition, larger diameters reduce the I_{ON}/I_{OFF} ratio and voltage swing, which is the key to achieve very high speed operation and high definition of the output signal, respectively. In the same way, diameters in the range of 1 to 1.5 nm have the highest performance in current drive, which allow us to reduce the delay and increase the short circuit power that are used during switching (Cao & Rogers, 2008). Given the demand of driving large capacitive loads, the carbon nanotube based transistors must be designed with complex architectures to support high current densities. This last implies the use of efficient methodologies for controlled dispersion of carbon nanotubes or the design of bundles with high-uniformity in diameter, chirality, and orientation. The first logic inverter based on graphene (Traversi et al., 2009) was operated to low power consumption and presents inability to the direct connection in cascade configuration due to the different output logic voltage levels.

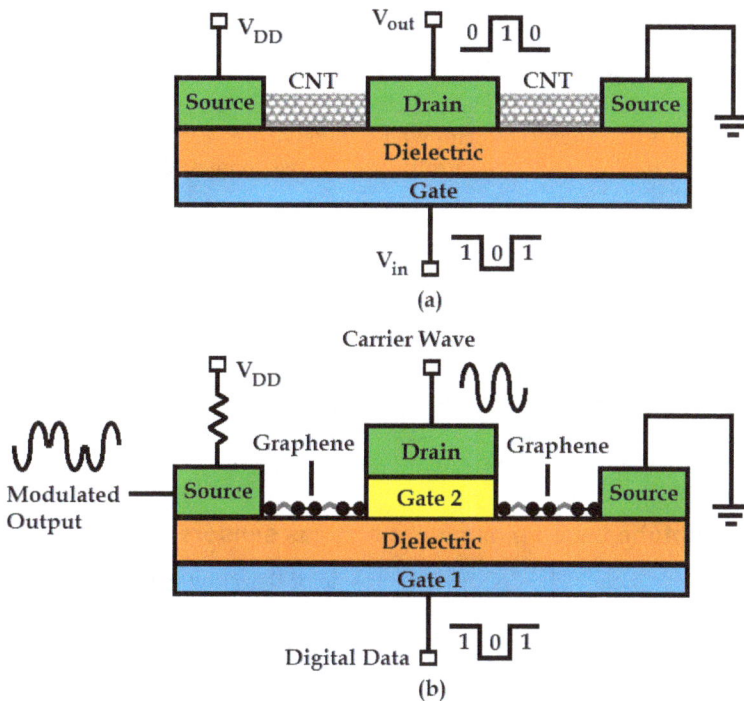

Fig. 10. VLSI circuits: (a) Carbon nanotube-based logic inverter, and (b) graphene-based digital modulator.

CNTs can be used to yield radio-frequency analog electronic devices such as narrow band amplifiers operating in the VHF frequency band with power gains as high as 14 dB. Examples of advanced analog circuits based on carbon nanotubes are resonant antennas, fixed RF amplifiers, RF mixers, and audio amplifiers. Hundred of devices, interconnected into desired planar layouts on commercial substrates are possible; thereby such systems can achieve complex functionality (Kocabas et al., 2008).

In RF applications, GFETs achieve high carrier mobility and saturation velocity (Lin et al., 2011). Mixers are RF circuits that are used to create new frequencies from two electrical signals applied to it. These signals have different frequency and when they are applied to the mixer, then are obtained two new signals corresponding to the sum and difference of the original frequencies. They are used to shift signals from one frequency range to another, for its transmission in RF systems such as radio transmitters. The radio frequency mixer based on graphene can produce frequencies up to 10 GHz, therefore, secure applications such as cell phones and military communications are feasible (Lin et al., 2011). Any limitations can be found for the use to full scale of graphene in VLSI circuits: different ohmic contact between materials, poor adhesion between metals and oxides, and high vulnerability to damage in the integration processes.

Among the main characteristics that graphene offers for VLSI Design are flexible, transparent material, and it operates to room temperature. Thanks to their electrical properties, the graphene is an ideal material to build more energy-efficient computers and other nanoelectronic devices. Nowadays, it is necessary to develop methods that allow us to separate graphene nanoribbons by a thin nonconductive material. Among the proposals that have been made are the use of one-atom-thick sheets of alloys of boron and nitrogen whose electrical behavior is nonconductive, and whose physical appearance is similar to graphene. The contents of such alloy must be controlled due to the geometrical arrangements that can be obtained.

Due to the ambipolarity (conduction of holes and electrons with equal efficiency), it is possible to design electronic devices (Vaillancourt et al., 2008; Xu et al., 2008). In Figure 10 (b), a digital modulator for communications circuits based on graphene is illustrated. This circuit is based on a graphene transistor including two gates: gate 1 controls the magnitude of current flowing through the transistor, and gate 2 controls the polarity of this current. The electrical operation of this circuit is similar to an electronic inverter, where gate 1 delivers a digital data stream as input, and it modulates such signal with the carrier wave applied to the drain to mix both signals, given place to a modulated signal.

6. Carbon nanotube based biosensors and gas sensors

Chemical sensors include a class of devices capable of detecting gas molecules or chemical signals in biological cells. Significant progress has been achieved in the detection of explosives, nerve agents, toxic gases and nontoxic gases due to the threat of terrorism and the need for homeland security. The biosensors and gas sensors based on one-dimensional nanostructures are very attractive, because they present high sensitivity and fastest response to the surrounding environment, thanks to their reduced dimensions and large surface-to-volume ratio (Sinha et al., 2006; Star et al., 2004; Wong et al., 2010).

Carbon nanotubes are promising candidates for designing gas sensors due to their excellent chemical and superficial properties derived of their chemical composition and high-aspect-ratio between its length and diameter, respectively. Levels as low as ppt (parts per trillion) or ppb (parts per billion) can be detected in comparison with their predecessors based on microsystems (MEMS) which could detect only ppm (parts per million). The basic structures used to design gas sensors are based on chemoresistors and FETs with one-dimensional nanostructures. An excellent biosensor or gas sensor is obtained when an appropriate control of the chemical and physical variables associated with the detection is presented. Therefore, the use of one-dimensional nanostructures improves the sensitivity, selectivity, stability, and response time (Balasubramanian & Burghard, 2006; Rivas et al., 2009).

This section analyses the different proposes of carbon nanotube based biosensors and carbon nanotube based gas sensors that were published in the last decade. This review discusses various design methodologies for CNT-based biosensors and CNT-based gas sensors as well as their application for the detection of specific biomolecules and gases. Recent developments associated with the topologies to design CNT-based chemiresistors and CNT-based field-effect transistors are highlighted.

Carbon nanotubes and graphene are technologically attractive to develop sensors due to four great characteristics: 1) each atom in its structure is physically accessible under any environment condition; 2) any perturbation in atoms can be electrically measured; 3) structural stability; 4) superior sensing performance at room temperature; and 5) tunable electrical properties (Wong et al., 2010). These characteristics have allowed the development of chemical, molecular and biological sensors (Oliveira & Mascaro, 2011; Wang, 2009).

Traditionally, pristine high-quality nanotubes are functionalized with functional groups to produce chemical or biochemical coatings (Wang, J., 2005; Zourob, 2010) or sites where very high sensitivity and selectivity to specific gases or to biochemical species is presented. Such gases or biochemical species can be detected by means of a change in electrical resistivity or capacitance presented in individual carbon nanotubes or bundles of them with the presence of this species. The change presented can be an increase or a reduction with respect to the value of the electrical parameter without the chemical or biochemical specie before mentioned (Chen, P.-C. et al., 2010). The chemical and biochemical sensors based on carbon nanotubes have even achieved sensitivities in the order of parts per billion to parts per million for specific gases or biochemical species depending of the molecule size and physicochemical properties (Bradley et al., 2003). Therefore, in any occasions it is necessary to add catalysts to improve the chemical activity during the chemical or biochemical detection (Cao & Rogers, 2009). In particular, the functionalization required by the biosensors regularly needs to favor the biocompatibility with the biological environment and realize the monitoring of information related with biological events and processes (Gruner, 2006; Ishikawa et al., 2009, 2010). In this manner, the biological species must be not affected by the biochemical interaction between the biosensor and the associated biological subject (Dong et al., 2008; Jia et al., 2008).

The basic construction blocks to design chemical or biochemical sensors based on carbon nanotubes can be divided into two different configurations: two-terminal CNT devices or three-terminal transistor-like structures. In the case of two-terminals devices, these can be modeled by an electrical resistor or an electrical capacitor (see Figure 11). In the case of three-terminals, they are modeled by a bipolar junction transistor (BJT) or a metal-oxide-semiconductor field effect transistor (MOSFET), the latter being the most common for VLSI systems (Zhao et al., 2008).

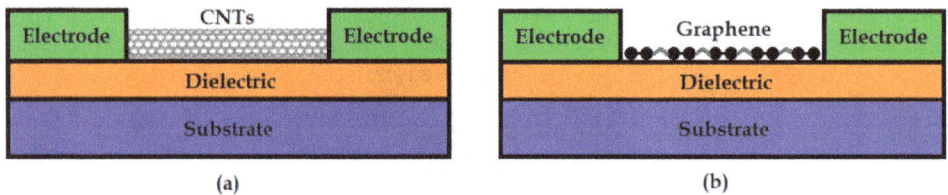

(a)   (b)

Fig. 11. Cross section of resistive gas sensors and biosensors: (a) sensors based on carbon nanotubes, and (b) sensors based on graphene.

The use of complex morphologies and structures based on composites containing carbon nanotubes and polymers in the design of gas sensors, has allowed the detection of polar and nonpolar gases making use of the change of dielectric constant to enhance sensitivity to minute quantities of gas molecules (Jesse et al., 2006; Mahar et al., 2007).

Graphene is exploited due to its inexhaustible structural defects and functional groups. These are advantageous in electroanalysis and electrocatalysis for electrochemical applications such as gas sensors and biosensors. Physisorbed ambient impurities by graphene such as water and oxygen can produce an effect similar to hole-doping and therefore a behavior similar to a p-type material (Traversi et al., 2009). Then, the graphene can be exploited as a sensing material for the design of chemical and/or biochemical sensors. When graphene is doped, well-identified localized states are added and band gap is introduced to the electrical properties generating an interesting alternative to design sensors (Barrios-Vargas et al., 2011).

The main changes to be realized in the optimization of performance of gas sensors are the search of methods which allow us to synthesize identical and reproducible CNTs will give place to gas sensors with high quality and high performance, independently of the type of chemical functionalization required for the detection.

7. Conclusions

In accordance with the review proposed here, CNTs are very attractive as base material to the design of components for VLSI Design. Chemical modifications of CNTs allow to the designer improve the selectivity of the electrical properties for the different applications. In the future, the use of hybrid materials where carbon nanotubes are involved will be a priority, given that the use of composite materials to design electronic devices, circuits and sensors requires multiple physical and chemical properties that a unique material cannot provide by itself. In the search for reducing electrical resistance presented by carbon nanotubes, different strategies have been developed to improve the efficiency of interconnection between devices based on carbon nanotubes and metallic electrodes used to lead the electrical bias to them. The implementation of digital and analog circuits with CNFETs or graphene nanoribbons will produce a great advance toward VLSI design using nanoelectronics. Still, hurdles remain as it was described in each section of the chapter.

8. Acknowledgement

This work was supported by ITESI, CONACYT under thematic network RedNyN agreement I0110/229/09, and PROMEP agreement 92434. The first author thank to his wife and son for their time and patience to realize this study.

9. References

Ahn, J.H.; Kim, H.-S.; Lee, K.J.; Jeon, S.; Kang, S.J.; Sun, Y.; Nuzzo, R.G. & Rogers, J.A. (2006), Heterogeneous Three-Dimensional Electronics by Use of Printed Semiconductor Nanomaterials, *Science*, Vol. 314, pp. 1754-1757, 15 December 2006.

Alkire, R.C.; Kolb, D. M.; Lipkowski, J. & Ross, P.N. (2009), *Advances in Electrochemical Science and Engineering Vol. 11 Chemically Modified Electrodes*, Wiley-VCH, ISBN 978-3-527-31420-1, Federal Republic of Germany.

Andriotis, A.N.; Menon, M.; Srivastava, D. & Chernozatonskii, L. (2001), Rectification Properties of Carbon Nanotube "Y-Junctions", *Physical Review Letters*, Vol. 87, No. 6, pp. 066802(4).

Andriotis, A.N.; Menon, M.; Srivastava, D. & Chernozatonskii, L. (2002), Transport Properties of Single-Wall Carbon Nanotube Y Junctions, *Physical Review B*, Vol. 65, No. 16, pp. 165416(13).

Andriotis, A.N. & Menon, M. (2006), Are Electrical Switching and Rectification Inherent Properties of Carbon Nanotube Y Junctions?, *Applied Physics Letters*, Vol. 89, No. 13, pp. 132116(3).

Andriotis, A.N. & Menon, M. (2007), Structural and Conducting Properties of Metal Carbon-Nanotube Contacts: Extended Molecule Approximation, *Physical Review B*, Vol. 76, No. 4, pp. 045412(7).

Andriotis, A.; Menon, M. & Gibson, H. (2008), Realistic Nanotube-Metal Contact Configuration for Molecular Electronics Applications, *IEEE Sensors Journal*, Vol. 8, No. 6, pp. 910-913.

Appenzeller, J.; Lin, Y.M.; Knoch, J.; Chen, Z. & Avouris, P. (2005), Comparing Carbon Nanotube Transistors – The Ideal Choice: A Novel Tunneling Device Design, *IEEE Transactions on Electron Devices*, Vol. 52, No. 12, pp. 2568-2576.

Avouris, P. (2002), Molecular Electronics with Carbon Nanotubes, *Accounts of Chemical Research*, Vol. 35, No. 12, pp. 1026-1034.

Bachtold, A.; Hadley, P; Nakanishi, T. & Dekker, C. (2001), Logic Circuits with Carbon Nanotube Transistors, *Science*, Vol. 294, No. 5545, pp. 1317-1320.

Bakir, M.S. & Meindl, J.D. (2009), *Integrated Interconnect Technologies for 3D Nanoelectronic Systems*, Artech House, ISBN 978-1-59693-246-3, United States of America.

Balasubramanian, K. & Burghard, M. (2006), Biosensors based on Carbon Nanotubes, *Analytical and Bioanalytical Chemistry*, Vol. 385, No. 3, pp. 452-468.

Barrios-Vargas, J.E. & Naumis, G.G. (2011), Doped Graphene: The Interplay between Localization and Frustration due to the Underlying Triangular Symmetry, *Journal of Physics: Condensed Matter*, Vol. 23, No. 37, pp. 375501(4).

Baughman, R.H.; Zakhidov, A.A. & de Heer, W.A. (2002), Carbon Nanotubes – The Route Toward Applications, *Science*, Vol. 297, No. 5582, pp. 787-792.

Behnam, A.; Johnson, J.L.; Choi, Y.; Ertosun, M.G.; Okyay, A.K.; Kapur, P.; Saraswat, K.C. & Ural, A. (2008), Experimental Characterization of Single-Walled Carbon Nanotube

Film-Si Schottky Contacts using Metal-Semiconductor-Metal Structures, *Applied Physics Letters*, Vol. 92, No. 24, pp. 243116(3).

Bradley, K.; Gabriel, J.-C.P.; Briman, M.; Star, A. and Grüner, G. (2003), Charge Transfer from Ammonia Physisorbed on Nanotubes, *Physical Review Letters*, Vol. 91, No. 21, pp. 218301(4).

Burghard, M.; Klauk, H. & Kern, K. (2009), Carbon-based Field-Effect Transistors for Nanoelectronics, *Advanced Materials*, Vol. 21, Nos. 25-26, pp. 2586-2600.

Burke, P.J. (2004), AC Performance on Nanoelectronics: Towards a Ballistic THz Nanotube Transistor, *Solid-State Electronics*, Vol. 48, Nos. 10-11, pp. 1981-1986.

Cao, Q.; Xia, M.-G.; Shim, M. & Rogers, J.A. (2006), Bilayer Organic-Inorganic Gate Dielectrics for High-Performance, Low-Voltage, Single-Walled Carbon Nanotube Thin-Film Transistors, Complementary Logic Gates, and p-n Diodes on Plastic Substrates, *Advanced Functional Materials*, Vol. 16, No. 18, pp. 2355-2362.

Cao, Q.; Xia, M.; Kocabas, C.; Shim, M.; Rogers, J.A. & Rotkin, S.V. (2007), Gate Capacitance Coupling of Single-Walled Carbon Nanotube Thin-Film Transistors, *Applied Physics Letters*, Vol. 90, No. 2, pp. 023516(3).

Cao, Q. & Rogers, J.A. (2008), Random Networks and Aligned Arrays of Single-Walled Carbon Nanotubes for Electronic Device Applications, *Nano Research*, Vol. 1, No. 4, pp. 259-272.

Cao, Q. & Rogers, J.A. (2009), Ultrathin Films of Single-Walled Carbon Nanotubes for Electronics and Sensors: A Review of Fundamental and Applied Aspects, *Advanced Materials*, Vol. 21, No. 1, pp. 29-53.

Chen, C. & Zhang, Y. (2009), *Nanowelded Carbon Nanotubes: From Field-Effect Transistors to Solar Microcells*, Springer, ISBN 978-3-642-01498-7, Germany.

Chen, P.-C.; Shen, G.; Sukcharoenchoke, S. & Zhou, C. (2009), Flexible and Transparent Supercapacitor based on In_2O_3 Nanowire/Carbon Nanotube Heterogeneous Films, *Applied Physics Letters*, Vol. 94, No. 4, pp. 043113(3).

Chen, P.-C.; Sukcharoenchoke, S.; Ryu, K.; de Arco, L.G.; Badmaev, A.; Wang, C. & Zhou, C. (2010), 2,4,6-Trinitrotoluene (TNT) Chemical Sensing based on Aligned Single-Walled Carbon Nanotubes and ZnO Nanowires, *Advanced Materials*, Vol. 22, No. 17, pp. 1900-1904.

Chen, X.; Wang, H.; Wan, H.; Song, K & Zhou, G. (2011), Semiconducting States and Transport in Metallic Armchair-Edged Graphene Nanoribbons, *Journal of Physics: Condensed Matter*, Vol. 23, No. 31, pp. 315304(8).

Cho, H.; Koo, K.-H.; Kapur, P. & Saraswat, K. C. (2008), Performance Comparisons between Cu/Low-κ, Carbon-Nanotube, and Optics for Future on-Chip Interconnects, *IEEE Electron Device Letters*, Vol. 29, No. 1, pp. 122-124.

Danilchenko, B. A.; Shpinar, L.I.; Tripachko, N.A.; Voitsihovska, E.A.; Zelensky, S.E. & Sundqvist, B. (2010), High Temperature Luttinger Liquid Conductivity in Carbon Nanotube Bundles, *Applied Physics Letters*, Vol. 97, No. 7, pp. 072106(3).

Dong, X.; Lau, C.M.; Lohani, A.; Mhaisalkar, S.G.; Kasim, J.; Shen, Z.; Ho, X.; Rogers, J.A. & Li, L.-J. (2008), Electrical Detection of Femtomolar DNA via Gold-Nanoparticle Enhancement in Carbon-Nanotube-Network Field-Effect Transistors, *Advanced Materials*, Vol. 20, No. 12, pp. 2389-2393.

Facchetti, A. & Marks, T.J. (2010), *Transparent Electronics: From Synthesis to Applications*, First Edition, John Wiley & Sons, ISBN 978-0-470-99077-3, Great Britain.

Ferry, D.K.; Goodnick, S.M. & Bird, J. (2009), *Transport in Nanostructures*, Second Edition, Cambridge University Press, United States of America.

Geim, A.K. & Novoselov, K.S. (2007), The Rise of Graphene, *Nature Materials*, Vol. 6, No. 3, pp. 183-191.

Geim, A.K. (2009), Graphene: Status and Prospects, *Science*, Vol. 324, No. 5934, pp. 1530-1534.

Ginley, D. S. (2010), *Handbook of Transparent Conductors*, Springer, ISBN 978-1-4419-1637-2, United States of America.

Giustiniani, A.; Tucci, V. & Zamboni, W. (2011), Carbon Nanotubes Bundled Interconnects: Design Hints based on Frequency- and Time-Domain Crosstalk Analyses, *IEEE Transactions on Electron Devices*, Vol. 58, No. 8, pp. 2702-2711.

Goel, A. K. (2007), *High-Speed VLSI Interconnections*, Second Edition, John Wiley & Sons, ISBN 978-0-471-78046-5, United States of America.

Gruner, G. (2006), Carbon Nanotube Transistors for Biosensing Applications, *Analytical and Bioanalytical Chemistry*, Vol. 384, No. 2, pp. 322-335.

Guildi, D.M. & Martín, N. (2010), *Carbon Nanotubes and Related Structures: Synthesis, Characterization, Functionalization, and Applications*, Wiley-VCH, ISBN 978-3-527-32406-4, Federal Republic of Germany.

Haruehanroengra, S. & Wang, W. (2007), Analyzing Conductance of Mixed Carbon Nanotube Bundles for Interconnect Applications, *IEEE Electron Device Letters*, Vol. 28, No. 8, pp. 756-759.

Hasan, S.; Salahuddin, S.; Vaidyanathan, M. & Alam, M.A. (2006), High-Frequency Performance Projections for Ballistic Carbon-Nanotube Transistors, *IEEE Transactions on Nanotechnology*, Vol. 5, No. 1, pp. 14-22.

Hatakeyama, R.; Li, Y.F.; Kato, T.Y. & Kaneko, T. (2010), Infrared Photovoltaic Solar Cells based on C_{60} Fullerene Encapsulated Single-Walled Carbon Nanotubes, *Applied Physics Letters*, Vol. 97, No. 3, pp. 013104(3).

Hayden, O. & Nielsch, K. (2011), *Molecular- and Nano-Tubes*, Springer, ISBN 978-1-4419-9442-4, United States of America.

Hecht, D.S.; Hu, L. & Grüner, G. (2007), Electronic Properties of Carbon Nanotube/Fabric Composites, *Current Applied Physics*, Vol. 7, No. 1, pp. 60-63.

Hierold, C. (2008), *Advanced Micro & Nanosystems Vol. 8 Carbon Nanotube Devices: Properties, Modeling, Integration and Applications*, Wiley-VCH, ISBN 978-3-527-31720-2, Federal Republic of Germany.

Hong, S.W.; Banks, T. & Rogers, J.A. (2010), Improved Density in Aligned of Single-Walled Carbon Nanotubes by Sequential Chemical Vapor Deposition on Quartz, *Advanced Materials*, Vol. 22, No. 16, pp. 1826-1830.

Hosseini, A. & Shabro V. (2010), Thermally-aware Modeling and Performance Evaluation for Single-Walled Carbon Nanotube-based Interconnects for Future High Performance Integrated Circuits, *Microelectronic Engineering*, Vol. 87, No. 10, pp. 1955-1962.

Hou, P.-X.; Liu, C. & Cheng, H.-M. (2008), Purification of Carbon Nanotubes, *Carbon*, Vol. 46, No. 15, pp. 2003-2025.

Hu, L.; Gruner, G.; Gong, J.; Kim, C.-J. "CJ" & Hornbostel, B. (2007), Electrowetting Devices with Transparent Single-Walled Carbon Nanotube Electrodes, *Applied Physics Letters*, Vol. 90, No. 9, pp. 093124(3).

Hur, S.-H.; Khang, D.-Y.; Kocabas, C. & Rogers, J.A. (2004), Nanotransfer Printing by Use of Noncovalent Surface Forces: Applications to Thin-Film Transistors that Use Single-Walled Carbon Nanotube Networks and Semiconducting Polymers, *Applied Physics Letters*, Vol. 85, No. 23, pp. 5730-5732.

Ishikawa, F.N.; Stauffer, B.; Caron, D.A. & Zhou, C. (2009), Rapid and Label-Free Cell Detection by Metal-Cluster-decorated Carbon Nanotube Biosensors, *Biosensors and Bioelectronics*, Vol. 24, No. 10, pp. 2967-2972.

Ishikawa, F.N.; Curreli, M.; Olson, C.A.; Liao, H.-I.; Sun, R.; Roberts, R.W.; Cote, R.J.; Thompson, M.E. & Zhou, C. (2010), Importance of Controlling Nanotube Density for Highly Sensitive and Reliable Biosensors Functional in Physiological Conditions, *ACS Nano*, Vol. 4, No. 11, pp. 6914-6922.

Jain, D.; Rouhi, N.; Rutherglen, C.; Densmore, C. G.; Doorn, S. K. & Burke, P.J. (2011), Effect of Source, Surfactant, and Deposition Process on Electronic Properties of Nanotube Arrays, *Journal of Nanomaterials*, Vol. 2011, Article ID 174268(7).

Jamaa, M.H.B. (2011), *Regular Nanofabrics in Emerging Technologies: Design and Fabrication Methods for Nanoscale Digital Circuits*, Springer, ISBN 978-94-007-0649-1, United States of America.

Javey, A. (2008), The 2008 Kavli Prize in Nanoscience: Carbon Nanotubes, *ACS Nano*, Vol. 2, No. 7, pp. 1329-1335.

Javey, A. & Kong, J. (2009), *Carbon Nanotube Electronics*, Springer, ISBN 978-0-387-36833-7, United States of America.

Jia, J.; Guan, W.; Sim, M.; Li, Y. & Li, H. (2008), Carbon Nanotubes based Glucose Needle-Type Biosensor, *Sensors*, Vol. 8, No. 3, pp. 1712-1718.

Jia, Y.; Cao, A.; Bai, X.; Li, Z.; Zhang, L.; Guo, N.; Wei, J.; Wang, K.; Zhu, H.; Wu, D. & Ajayan, P.M. (2011), Achieving High Efficiency Silicon-Carbon Nanotube Heterojunction Solar Cells by Acid Doping, *Nano Letters*, Vol. 11, No. 5, pp. 1901-1905.

Joachim, C.; Gimzewski, J.K. & Aviram, A. (2000), Electronics using Hybrid-Molecular and Mono-Molecular Devices, *Nature*, Vol. 408, No. 6812, pp. 541-548.

Kang, S.J.; Kocabas, C.; Kim, H.-S.; Cao, Q.; Meitl, M.A.; Khang, D.-Y. & Rogers, J. A. (2007), Printed Multilayer Superstructures of Aligned Single-Walled Carbon Nanotubes for Electronic Applications, *Nano Letters*, Vol. 7, No. 11, pp. 3343-3348.

Kanungo, M.; Malliaras, G.G. & Blanchet, G.B. (2010), High Performance Organic Transistors: Percolation Arrays of Nanotubes Functionalized with an Electron Deficient Olefin, *Applied Physics Letters*, Vol. 97, No. 5, pp. 053304(3).

Kim, G.; Bernholc, J. & Kwon, Y.-K. (2010), Band Gap Control of Small Bundles of Carbon Nanotubes using Applied Electric Fields: A Density Functional Theory Study, *Applied Physics Letters*, Vol. 97, No. 6, pp. 063113(3).

Kocabas, C.; Hur, S.-H.; Gaur, A.; Meitl, M.A.; Shim, M. & Rogers, J.A. (2005), Guided Growth of Large-Scale, Horizontally Aligned Arrays of Single-Walled Carbon Nanotubes and Their Use in Thin-Film Transistors, *Small*, Vol. 1, No. 11, pp. 1110-1116.

Kocabas, C.; Shim, M. & Rogers, J.A. (2006), Spatially Selective Guided Growth of High-Coverage Arrays and Random Networks of Single-Walled Carbon Nanotubes and Their Integration into Electronic Devices, *Journal of the American Chemical Society*, Vol. 128, No. 14, pp. 4540-4541.

Kocabas, C.; Kim, H.-S.; Banks, T.; Rogers, J.A.; Pesetski, A.A.; Baumgardner, J.E.; Krishnaswamy, S.V. & Zhang, H. (2008), Radio Frequency Analog Electronics based on Carbon Nanotube Transistors, *PNAS*, Vol. 105, No. 5, pp. 1405-1409.

Koo, K.-H.; Cho, H.; Kapur, P. & Saraswat, K.C. (2007), Performance Comparisons between Carbon Nanotubes, Optical, and Cu for Future High-Performance On-Chip Interconnect Applications, *IEEE Transactions on Electron Devices*, Vol. 54, No. 12, pp. 3206-3215.

Kreupl, F.; Graham, A.P.; Duesberg, G.S.; Steinhögl, W.; Liebau, M.; Unger, E. & Hönlein, W. (2002), Carbon Nanotubes in Interconnect Applications, *Microelectronic Engineering*, Vol. 64, Nos. 1-4, pp. 399-408.

Krompiewski, S. (2005), Spin-Polarized Transport through Carbon Nanotubes, *Physica Status Solidi B: Basic Solid State Physics*, Vol. 242, No. 2, pp. 226-233.

Law, M.; Goldberg, J. & Yang, P. (2004), Semiconductor Nanowires and Nanotubes, *Annual Review of Materials Research*, Vol. 34, pp. 83-122.

Lefenfeld, M.; Blanchet, G. & Rogers, J.A. (2003), High-Performance Contacts in Plastic Transistors and Logic Gates That Use Printed Electrodes of DNNSA-PANI Doped with Single-Walled Carbon Nanotubes, *Advanced Materials*, Vol. 15, No. 14, pp. 1188-1191.

Lekakou, C.; Moudam, O.; Markoulidis, F.; Andrews, T.; Watts, J.F. & Reed, G. T. (2011), Carbon-based Fibrous EDLC Capacitors and Supercapacitors, *Journal of Nanotechnology*, Vol. 2011, Article ID 409382(8).

Léonard, F. (2009), *The Physics of Carbon Nanotube Devices*, William Andrew, ISBN 978-0-8155-1573-9, United States of America.

Li, H.; Yin, W.-Y.; Banerjee, K. & Mao, J.-F. (2008), Circuit Modeling and Performance Analysis of Multi-Walled Carbon Nanotube Interconnects, *IEEE Transactions on Electron Devices*, Vol. 55, No. 6, pp. 1328-1337.

Li, H.; Xu, C.; Srivastava, N. & Banerjee, K. (2009), Carbon Nanomaterials for Next-Generation Interconnects and Passives: Physics, Status, and Prospects, *IEEE Transactions on Electron Devices*, Vol. 56, No. 9, pp. 1799-1821.

Li, J.; Ye, Q.; Cassell, A.; Ng, H.T.; Stevens, Ramsey; Han, Jie & Meyyappan, M. (2003), Bottom-Up Approach for Carbon Nanotube Interconnects, *Applied Physics Letters*, Vol. 82, No. 15, pp. 1566791(3).

Lin, A.; Patil, N.; Ryu, K.; Badmaev, A.; De Arco, L.G.; Zhou, C.; Mitra, S. & Wong, H.-S.P. (2009), Threshold Voltage and On-Off Ratio Tuning for Multiple-Tube Carbon Nanotube FETs, *IEEE Transactions on Nanotechnology*, Vol. 8, No. 1, pp. 4-9.

Lin, C.-T.; Hsu, C.-H.; Lee, C.-H. & Wu, W.-J. (2011), Inkjet-Printed Organic Field-Effect Transistor by Using Composite Semiconductor Material of Carbon Nanoparticles and Poly(3-Hexylthiophene), *Journal of Nanotechnology*, Vol. 2011, Article ID 142890(7).

Lin, Y.M.; Valdes-Garcia, A.; Han, S.-J.; Farmer, D.B.; Meric, I.; Sun, Y.; Wu, Y.; Dimitrakopoulos, C.; Grill, A.; Avouris, P. & Jenkins, K.A. (2011), Wafer-Scale Graphene Integrated Circuit, *Science*, Vol. 332, No. 6035, pp. 1294-1297.

Liu, X.; Luo, Z.; Han, S.; Tang, T.; Zhang, D. & Zhou, C. (2005), Band Engineering of Carbon Nanotube Field-Effect Transistors via Selected Area Chemical Gating, *Applied Physics Letters*, Vol. 86, No. 24, pp. 243501(3).

Lu, W. & Lieber, C.M. (2007), Nanoelectronics from the Bottom Up, *Nature Materials*, Vol. 6, No. 11, pp. 841-850.

Mahar, B.; Laslau, C.; Yip, R. & Sun, Y. (2007), Development of Carbon Nanotube-based Sensors – A Review, *IEEE Sensors Journal*, Vol. 7, No. 2, pp. 266-284.

Mallick, G.; Griep, M.H.; Ayajan, P.M. & Karna, S.P. (2010), Alternating Current-to-Direct Current Power Conversion by Single-Wall Carbon Nanotube Diodes, *Applied Physics Letters*, Vol. 96, No. 23, pp. 233109(3).

Mamalis, A.G.; Vogtländer, L.O.G. & Markopoulos, A. (2004), Nanotechnology and Nanostructured Materials: Trends in Carbon Nanotubes, *Precision Engineering*, Vol. 28, No. 1, pp. 16-30.

Marulanda, J.M. (2010), *Carbon Nanotubes*, In-Tech, ISBN 978-953-307-054-4, Croatia.

Nismi, N.A.; Adikaari, A.A.D.T. & Silva, S.R.P. (2010), Functionalized Multiwall Carbon Nanotubes incorporated Polymer/Fullerene Hybrid Photovoltaics, *Applied Physics Letters*, Vol. 97, No. 3, pp. 033105(3).

Nogueira, A. F.; Lomba, B.S.; Soto-Oviedo, M.A.; Correia, C.R.D.; Corio, P.; Furtado, C.A. & Hümmelgen, I.A. (2007), Polymer Solar Cells using Single-Wall Carbon Nanotubes Modified with Thiophene Pedant Groups, *Journal of Physical Chemistry C: Nanomaterials, Interfaces and Hard Matter*, Vol. 111, No. 49, pp. 18431-18438.

Nojeh, A. & Ivanov, A. (2010), Wireless Interconnect and the Potential for Carbon Nanotubes, *IEEE Design & Test of Computers*, Vol. 27, No. 4, pp. 44-52.

Nouchi, R.; Tomita, H.; Ogura, A.; Shiraishi, M. & Kataura, H. (2008), Logic Circuits using Solution-Processed Single-Walled Carbon Nanotube Transistors", *Applied Physics Letters*, Vol. 92, No. 25, pp. 253507(3).

Novoselov, K.S.; Geim, A.K.; Morozov, S.V.; Jiang, D.; Zhang, Y.; Dubonos, S.V.; Grigorieva, I.V. & Firsov, A.A. (2004), Electric Field Effect in Atomically Thin Carbon Films, *Science*, Vol. 306, No. 5696, pp. 666-669.

Oliveira, A.C. & Mascaro, L.H. (2011), Evaluation of Acetylcholinesterase Biosensor based on Carbon Nanotube Paste in the Determination of Chlorphenvinphos, *International Journal of Analytical Chemistry*, Vol. 2011, Article ID 974216 (6).

Papanikolaou, A.; Soudris, D. & Radojcic, R. (2011), *Three Dimensional System Integration: IC Stacking Process and Design*, Springer, ISBN 978-1-4419-0961-9, United States of America.

Pesetski, A.A.; Baumgardner, J. E.; Krishnaswamy, S.V.; Zhang, H.; Adam, J. D.; Kocabas, C.; Banks, T. & Rogers, J.A. (2008), A 500 MHz Carbon Nanotube Transistor Oscillator, *Applied Physics Letters*, Vol. 93, No. 12, pp. 123506(2).

Philip-Wong, H.-S. & Akinwande, D. (2011), *Carbon Nanotube and Graphene Device Physics*, Cambridge University Press, ISBN 978-0-521-51905-2, United Kingdom.

Reddy, D.; Register, L.F.; Carpenter, G.D. & Banerjee, S.K. (2011), Graphene Field-Effect Transistors, *Journal of Physics D: Applied Physics*, Vol. 44, No. 31, pp. 313001(20).

Rivas, G.A.; Rubianes, M.D.; Pedano, M.L.; Ferreyra, N.F.; Luque, G. & Miscoria, S.A. (2009), *Carbon Nanotubes: A New Alternative for Electrochemical Sensors*, Nova Science Publishers, ISBN 978-1-607-41314-1, United States of America.

Rowell, M.W.; Topinka, M.A.; McGehee, M.D.; Prall, H.-J.; Dennler, G.; Sariciftci, N.S.; Hu, L. & Gruner, G. (2006), Organic Solar Cells with Carbon Nanotube Network Electrodes, *Applied Physics Letters*, Vol. 88, No. 23, pp. 233506(3).

Rueckes, T.; Kim, K.; Joselevich, E.; Tseng, G.Y.; Cheung, C.-L. & Lieber, C.M. (2000) Carbon Nanotube-based Nonvolatile Random Access Memory for Molecular Computing, *Science*, Vol. 289, pp. 94-97.

Sangwan, V.K.; Behnam, A.; Ballarotto, V.W.; Fuhrer, M.S.; Ural, A. & Williams, E.D. (2010), Optimizing Transistor Performance of Percolating Carbon Nanotube Networks, *Applied Physics Letters*, Vol. 97, No. 4, pp. 043111(3).

Schaman-Diamand, Y.; Osaka, T.; Datta, M. & Ohba, T. (2009), *Advanced Nanoscale ULSI Interconnects: Fundamentals and Applications*, Springer, ISBN 978-0-387-95867-5, United States of America.

Singh, D.V.; Jenkins, K.A.; Appenzeller, J.; Neumayer, D.; Grill, A. & Wong, H.-S. (2004), Frequency Response of Top-Gated Carbon Nanotube Field-Effect Transistors, *IEEE Transactions on Nanotechnology*, Vol. 3, No. 3, pp. 383-387.

Sinha, N.; Ma, J. & Yeow, J.T.W. (2006), Carbon Nanotube-based Sensors, *Journal of Nanoscience and Nanotechnology*, Vol. 6, No. 3, pp. 573-590.

Srivastava, N. & Banerjee, K. (2004), Interconnect Challenges for Nanoscale Electronic Circuits, *TMS Journal of Materials (JOM)*, Vol. 56, No. 10, pp. 30-31.

Srivastava, N.; Li, H.; Kreupl, F. & Banerjee, K. (2009), On the Applicability of Single-Walled Carbon Nanotubes as VLSI Interconnects, *IEEE Transactions on Nanotechnology*, Vol. 8, No. 4, pp. 542-559.

Star, A.; Han, T.-R.; Joshi, V.; & Stetter, J.R. (2004) Sensing with Nafion Coated Carbon Nanotube Field-Effect Transistors, *Electroanalysis*, Vol. 16, Nos. 1-2, pp. 108-112.

Tan, C.S.; Gutmann, R.J. & Reif, L.R. (2008), *Wafer Level 3-D ICs Process Technology*, Springer, ISBN 978-0-387-76532-7, United States of America.

Terrones, M. (2003), Science and Technology of the Twenty-First Century: Synthesis, Properties, and Applications of Carbon Nanotubes, *Annual Review of Materials Research*, Vol. 33, pp. 419-501.

Terrones, M. (2004), Carbon Nanotubes: Synthesis and Properties, Electronic Devices and Other Emerging Applications, *International Materials Reviews*, Vol. 49, No. 6, pp. 325-377.

Traversi, F.; Russo, V. & Sordan, R. (2009), Integrated Complementary Graphene Inverter, *Applied Physics Letters*, Vol. 34, No. 22, pp. 223312(3).

Tseng, Y.-C.; Xuan, P.; Javey, A.; Malloy, R.; Wang, Q.; Bokor, J. & Dai, H. (2004), Monolithic Integration of Carbon Nanotube Devices with Silicon MOS Technology, *Nano Letters*, Vol. 4, No. 1, pp. 123-127.

Tulevski, G.S.; Hannon, J.; Afzali, A.; Chen, Z.; Avouris, P. & Kagan, C.R. (2007), Chemically Assisted Directed Assembly of Carbon Nanotubes for the Fabrication of Large-Scale Device Arrays, *Journal of the American Chemical Society*, Vol. 129, No. 39, pp. 11964-11968.

Vaillancourt, J.; Zhang, H.; Vasinajindakaw, P.; Xia, H.; Lu, X.; Han, X.; Janzen, D.C.; Shin, W.-S.; Jones, C.S.; Stroder, M.; Chen, M.Y.; Subbaraman, H.; Chen, R.T.; Berger, U. & Renn, M. (2008), All Ink-Jet-Printed Carbon Nanotube Thin-Film Transistor on a Polyimide Substrate with an Ultrahigh Operating Frequency of Over 5 GHz, *Applied Physics Letters*, Vol. 93, No. 24, pp. 243301(3).

Vasileska, D. & Goodnick, S.M. (2011), *Nano-Electronic Devices: Semiclassical and Quantum Transport Modeling*, Springer, ISBN 978-1-4419-8839-3, United States of America.

Wang, C.; Ryu, K.; Badmaev, A.; Patil, N.; Lin, A.; Mitra, S.; Wong, H.-S.P. and Zhou, C. (2008), Device Study, Chemical Doping, and Logic Circuits based on Transferred Aligned Single-Walled Carbon Nanotubes, *Applied Physics Letters*, Vol. 93, No. 3, pp. 033101(3).

Wang, C.; Zhang, J.; Ryu, K.; Badmaev, A.; De Arco, L.G. & Zhou, C. (2009), Wafer-Scale Fabrication of Separated Carbon Nanotube Thin-Film Transistors for Display Applications, *Nano Letters*, Vol. 9, No. 12, pp. 4285-4291.

Wang, J. (2005), Carbon-Nanotube based Electrochemical Biosensors: A Review, *Electroanalysis*, Vol. 17, No. 1, pp. 7-14.

Wang, Y. & Yeow, J.T.W. (2009), A Review of Carbon Nanotubes-based Gas Sensors, *Journal of Sensors*, Vol. 2009, Article ID 493904 (24).

Wang, Z.M. & Neogi, A. (2010), *Nanoscale Photonics and Optoelectronics*, Springer, ISBN 978-1-4419-7233-0, United States of America.

Wiederrecht, G. (2010), *Handbook of Nanoscale Optics and Electronics*, Academic Press, ISBN 978-0-12-375178-2, Spain.

Wong, C.P.; Moon, K.-S. & Li, Y. (2010), *Nano-Bio-Electronic, Photonic and MEMS Packaging*, Springer, ISBN 978-1-4419-0039-5, United States of America.

Xie, Y.; Cong, J. & Sapatnekar, S. (2010), *Three-Dimensional Integrated Circuit Design: EDA, Design and Microarchitectures*, Springer, ISBN 978-1-4419-0783-7, United States of America.

Xu, G.; Liu, F.; Han, S.; Ryu, K.; Badmaev, A.; Lei, B.; Zhou, C. & Wang, K.L. (2008), Low-Frequency Noise in Top-Gated Ambipolar Carbon Nanotube Field Effect Transistors, *Applied Physics Letters*, Vol. 92, No. 22, pp. 223114(3).

Xu, Y.; Srivastava, A. & Sharma, A.K. (2010), Emerging Carbon Nanotube Electronic Circuits, Modeling, and Performance, *VLSI Design*, Vol. 2010, Article ID 864165 (8).

Zhang, D.; Ryu, K.; Liu, X.; Polikarpov, E.; Ly, J.; Tompson, M.E. & Zhou, C. (2006), Transparent, Conductive, and Flexible Carbon Nanotube Films and Their Application in Organic Light-Emitting Diodes, *Nano Letters*, Vol. 6, No. 9, pp. 1880-1886.

Zhao, Y.-L.; Hu, L.; Stoddart, J.F. & Grüner, G. (2008), Pyrenecyclodextrin-Decorated Single-Walled Carbon Nanotube Field-Effect Transistors as Chemical Sensors, *Advanced Materials*, Vol. 20, No. 10, pp. 1910-1915.

Zhou, C.; Kumar, A. & Ryu, K. (2007), Small Wonder: The Exciting World of Carbon Nanotubes, *IEEE Nanotechnology Magazine*, Vol. 1, No. 1, pp. 13-17.

Zhou, W. & Wang, Z.L. (2011), *Three-Dimensional Nanoarchitectures: Designing Next-Generation Devices*, Springer, ISBN 978-1-4419-9821-7, United States of America.

Zhou, Y.; Gaur, A.; Hur, S.-H.; Kocabas, C., Meitl, M.A.; Shim, M. & Rogers, J. A. (2004), p-Channel, n-Channel Thin Film Transistors and p-n Diodes based on Single Wall Carbon Nanotube Networks, *Nano Letters*, Vol. 4, No. 10, pp. 2031-2035.

Zourob, M. (2010), *Recognition Receptors in Biosensors*, Springer, ISBN 978-1-4419-0918-3, United States of America.

4

Three-Dimensional Integrated Circuits Design for Thousand-Core Processors: From Aspect of Thermal Management

Chiao-Ling Lung[1,2], Jui-Hung Chien[2], Yung-Fa Chou[2],
Ding-Ming Kwai[2] and Shih-Chieh Chang[1]
[1]Department of Computer Science, National Tsing Hua University
[2]Information and Communications Research Laboratories,
Industrial Technology Research Institute
Taiwan

1. Introduction

As the performance of a processing system is to be significantly enhanced, on-chip many-core architecture plays an indispensable role. Since there are fast growing numbers of transistors on the chips, two-dimensional topologies face challenges of significant increases in interconnection delay and power consumption (Hennessy & Patterson, 2007; Kurd et al., 2001). Explorations of a suitable three-dimensional integrated circuit (3D IC) with through-silicon via (TSV) to realize a large number of processing units and highly dense interconnects certainly attracts a lot of attention. However, the combination of processors, memories, and/or sensors in a stacked die leads to the cooling problem in a tottering situation (Tiwari et al., 1998). One solution to overcome the obstacles and continue the performance scaling while still is to integrate on chip many cores and their communication network (Beigne, 2008; Yu & Baas, 2006). Through concerted processors, routers, and links, the network-on-chip (NoC) provides the advantages of low power dissipation and abundance of connectivity. Moreover, because of the widespread uses of radio frequency (RF), micro-electro-mechanical systems (MEMS) (Lu, 2009), and various sensors in mobile applications, proposals of three-dimensional integrated circuit (3D IC) with through silicon via (TSV) implementations in a layered architecture have been reported (Lee, 1992; Tsai & Kang, 2000). For interconnection scalability from layer to layer, 3D fabrics are a necessity. Consequently, a thermal solution which has a high heat removing rate seems unavoidable. Since there are fast growing numbers of transistors on the chips, two-dimensional topologies face challenges of significant increases in wire delay and power consumption. The two factors are often regarded as the primary limitations for current processor architectures (Hennessy & Patterson, 2007; Kurd et al., 2001; Tiwari et al., 1998).

On the other hand, the high packing density of the stacked dies also hampers the heat dissipation of the NoC system. Thermal issues arise from increasing dynamic power losses which in turn raise the temperature. Thermal and power constraints are of great concern with 3D IC since die stacking can dramatically increase power density, if hotspots overlap each other, and additional dies are farther away from the heat sink.

Thermal-aware floorplanning is the key in which the inter-layer interconnection plays a role more than just signal transmission or power delivery. Figure 1 depicts the usage of thermal TSV to alleviate the heat accumulation, which is brought from that used in printed circuit boards (PCBs) (Lee et al., 1992). For 3D ICs, the problems of high power/thermal density can be more serious than that in the planar form. Thus, the thermal TSVs become essential for heat dissipation. Of particular interest is the design of an efficient heat transferring path. Some recent works discussed the placement of thermal TSVs. However, not only the routing but also the floorplan may need to be changed substantially after the thermal TSVs are inserted (Tsai & Kang, 2000). This leads to long iterations. Further, as the circuit complexity is increased, to insert the thermal TSVs without largely changing the floorplan is an important technology to be developed (Tsui et al., 2003). In order to keep the original routing and floorplan as much as possible, the temperature-driven design should be brought in early phases of the design procedure.

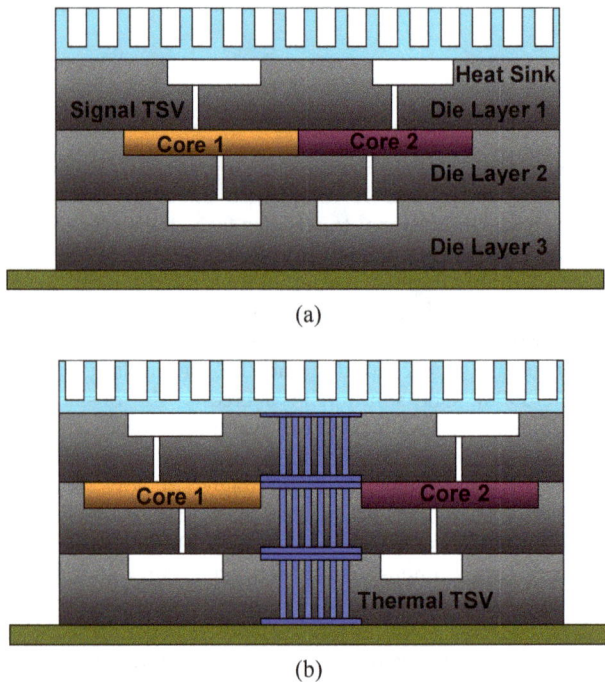

(a)

(b)

Fig. 1. 3D IC implementations of a multiprocessor system-on-chip (MP-SoC) with (a) a traditional structure and (b) with the insertion of thermal ridges.

2. Design and theoretical analysis of on-chip thermal ridge

2.1 Theoretical analysis

The thermal TSVs are intended to be placed in the inter-CG whitespace, which is called a thermal ridge. In this section, we derive analytical expressions for some key parameters.

2.1.1 Analytical model of the thermal ridge

At the transient state, the heat conduction can be described by the following equation

$$\frac{\partial}{\partial x}k_{xx}\frac{\partial T}{\partial x}+\frac{\partial}{\partial y}k_{yy}\frac{\partial T}{\partial y}+\frac{\partial}{\partial z}k_{zz}\frac{\partial T}{\partial z}+g=\rho C\frac{\partial T}{\partial \theta} \tag{1}$$

where T is the temperature, g is the heat generation rate in W/cm^2, ρ is the density of the material, C is the thermal capacity of the material, θ is time, and k is the thermal conductivity of the material. This fundamental thermal conduction equation describes that the temperature transmitting through the thermal volume depends on time θ and directional thermal conductivities k_{xx}, k_{yy}, and k_{zz} (Chieh et al., 2010; Lung et al., 2010). The boundary conditions of the top and bottom surfaces of the chip are *adiabatic* and those of the surrounding surfaces are *convective*.

For dissipating the heat into the substrate homogeneously, the inter-core-group thermal ridges are aligned orthogonally in column and in row. The temperature prediction of the many-core system is performed by utilizing CFD-RC which is commercial thermal and fluidic temperature simulation software. However, in order to illustrate the physical phenomenon more intuitively, a simplified one-dimensional conduction equation without taking the transient into consideration is utilized.

$$\frac{\partial}{\partial x}k_{xx}\frac{\partial T}{\partial x}=-g \tag{2}$$

The heat removing rate of the thermal ridge is assumed to be q. Let us consider two CGs. The temperature distribution between CG1 and CG2 can be expressed by

$$T=T_1+\frac{T_2-T_1}{w}x-\frac{q}{2k_s}[w-x]x \tag{3}$$

where T_1 and T_2 are the temperatures of CG1 and CG2, respectively, q is the heat conducted to the ambient environment by the thermal ridge, k_s is the equivalent thermal conductivity of the thermal ridge, and w is the width of the thermal ridge. Since T denotes the temperature at the location x, examining the mid-point $T_{1/2}$ by substituting x with $w/2$ into (3), we have

$$w=\left(\frac{8k_s}{q}\right)^{1/2}\left[\frac{T_1+T_2}{2}-T_{1/2}\right]^{1/2} \tag{4}$$

From (4), it is easy to see that if the mid-point temperature $T_{1/2}$ is targeted to be lower, w needs to be larger.

2.1.2 Effective thermal conductivity of the thermal ridge

The equivalent thermal conductivity k_{szz} of a thermal ridge is decided by the density of the thermal TSVs in the thermal ridge (Chieh et al., 2010; Lung et al., 2010). To determine k_{szz}, the effective thermal conductivity should be taken into account and described as the following equation:

$$k_{szz} = d \cdot k_{emb} + (1-d)k_{sub} \tag{5}$$

where k_{emb} is the equivalent thermal conductivity of the thermal TSVs, k_{sub} is the thermal conductivity of the silicon substrate, d is the percent contribution of the thermal TSVs in the thermal ridge. Since the orientation of the thermal TSV is longitudinal along the z direction, this effective thermal conductivity cannot be applied to the lateral heat transfer computation. For x and y directional heat transfer, the thermal conductivity should be applied by the following equation.

$$k_{sxx} = \left(1 - \sqrt{m}\right)k_{sub} + \cfrac{\sqrt{m}}{\cfrac{1-\sqrt{m}}{k_{sub}} + \cfrac{\sqrt{m}}{k_{emb}}} = k_{syy} \tag{6}$$

where m is the percent contribution of the metal lines for thermal conduction in the silicon substrate. In general, the vertical thermal conductivity k_{szz} is much larger than the lateral thermal conductivities k_{sxx} and k_{syy}. By (5) and (6), we can clearly figure out that k_{sxx} is around 10 W/mK and k_{szz} is around 120 W/mK. Thus, the heat flows through the thermal ridge almost dissipates by the heat sink instead of transferring laterally. By substituting the equivalent k_s and the temperature values of T_1, T_2 and $T_{1/2}$ into (3), we obtained that the widths of the thermal ridge should be 200 μm ~ 400 μm.

2.2 Design parameters and assumptions

Here, we focus on a mesh-connected NoC with 1,024 cores. A globally asynchronous, locally synchronous (GALS) digital-signal processor (DSP) design is adopted (Tran et al., 2009a, 2009b; Truong et al., 2008). Each DSP, constituting a tile, is composed of a core with an on-chip oscillator for its own clocking and a switch with associated buffers, as shown in Figure 2. The tile allows repetitive, mirrored layout, occupying an area of 0.168 mm² (410 μm × 410 μm) (Tran et al., 2009a, 2009b). Consider a simple power map with two major sources in the tile. One is attributed to the computation and the other to the communication. Correspondingly, the average power consumption at the active status is broken down to 17.6 mW and 1.1 mW, respectively (Tran et al., 2009a, 2009b).

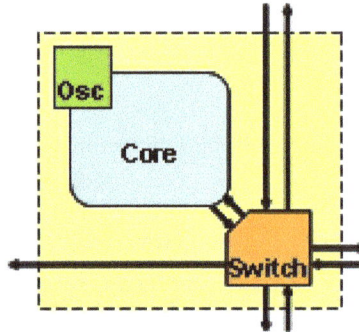

Fig. 2. The DSP element for a GALS many-core system.

The cores are arranged as a 32×32 square mesh. Since the international technology roadmap for semiconductor (ITRS) predicts that the maximum chip size will maintain similar dimensions, we assume 20 mm $\times$ 20 mm as our upper bound. Under such a constraint, the remaining area not occupied by the tiles is the input/output and peripheral circuits. The total power consumption of the chip is around 20 W, which leads to the average power density of 5 W/cm². Since ITRS also predicts the power density is reasonable up to the level of 100 W/cm², the power density assumed in this chapter is a probable value (Brunschwiler et al., 2009; Xu et al., 2004).

In this chapter, we assumed that there are three layers of the die stack and the many-core NoC is sandwiched in the middle. As mentioned earlier, a commercial tool based on finite element method (FEM) is used. The three-dimensional model of the NoC is created with the widely used package model, in a fashion similar to that shown in Figure 1. However, the heat sink is not modelled and analyzed in our case. Instead, it is simplified to a heat loss, and a proper heat transfer coefficient is applied to the boundary condition on the top surface where the heat sink would have been located originally.

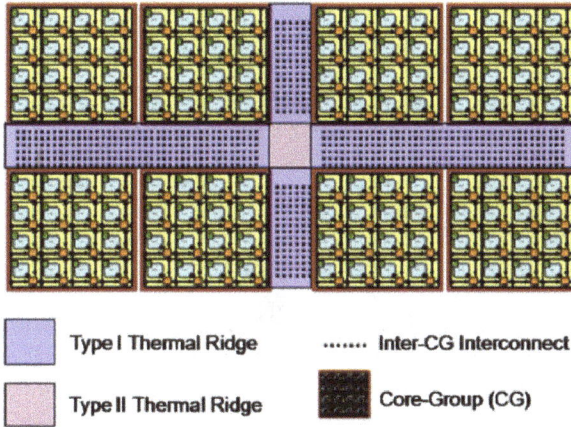

Fig. 3. Insertion of type I and type II thermal ridges into the NoC.

First, the 1,024 cores are divided into 8×8 CGs, each CG consisting of 4×4 cores. As shown in Figure 3, thermal ridges are inserted between the hottest CGs. By the locations where they are inserted, the thermal ridges can be categorized into two types. The type-I thermal ridge has a low density of thermal TSVs and the type-II thermal ridge has a high density of thermal TSVs. This is because the type-I thermal ridge is located between two CGs in which their routing dominates the most of the silicon area, even after the expansion to gain more whitespace. On the other hand, the type-II thermal ridge lies in the intersectional area having no wires passing through, and therefore, a large quantity of thermal TSVs can be planted.

The physical effect of the thermal ridge can be illustrated by using the electrical lumped model as shown in Figure 4. By the duality between electrical and thermal models, the temperature T is substituted by a voltage V, the power P is substituted by a current I, and the thermal resistance R by definition is proportional to the reciprocal of thermal conductivity k_s. The availability of the thermal ridge can be modelled by the equivalent circuits as follows.

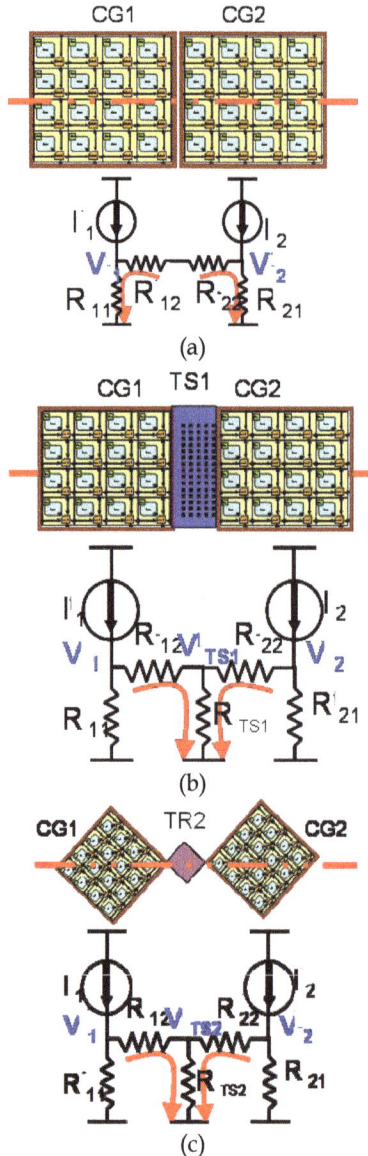

Fig. 4. Resistive thermal models of two adjacent CGs inserted with (a) no thermal ridge, (b) a type-I thermal ridge, and (c) a type-II thermal ridge.

Figure 4(a) shows the case when there is no thermal ridge between CG1 and CG2. It is clear in the schematic that no extra conduction path has been added to the ground. Since the vertical thermal resistance R_{11} (R_{21}) is much larger than the lateral thermal resistance R_{12} (R_{22}), the voltage V_1 (V_2) keeps at a high value. Figure 4(b) shows the case when a type-I thermal ridge is inserted between CG1 and CG2. Another conduction path is added through the thermal resistance R_{TS1}. As aforementioned, R_{TS1} is inversely proportional to k_s. As long as k_s is much larger than the thermal conductivity k_{sub} of the silicon substrate, R_{TS1} is much smaller than R_{11} (R_{21}); the current I_1 (I_2) goes mostly through R_{TS1}, rather than R_{11} (R_{21}). In addition, by voltage division, V_{TS1} is obviously lower than V_1 (or V_2). In other words, the temperature of the type-I thermal ridge is definitely lower than the temperature of CG1 or CG2. Figure 4(c) shows the case when a type-II thermal ridge is inserted at the intersectional area between the CGs to remove more heat. The value of R_{TS2} depends on that of k_s. Since the thermal TSVs are densely planted on the type-II thermal ridge, R_{TS2} is much smaller than R_{11} (or R_{21}). Compared with CG1 and CG2, the type-II thermal ridge, which has a lower temperature, is designed to be an on-chip heat sink.

2.2.1 Rotation of the hotspots

To verify the feasibility of the proposed scheme for thermal-aware floorplanning, we obtain the temperature distribution of the basic CG first. There are 4 × 4 cores within a CG as shown in Figure 5. The cores are homogenous, with the hotspot near the lower right corner. It is clear that since the hotspot is not located at the center of the core, when assembled into the CG, the temperature distribution is asymmetric.

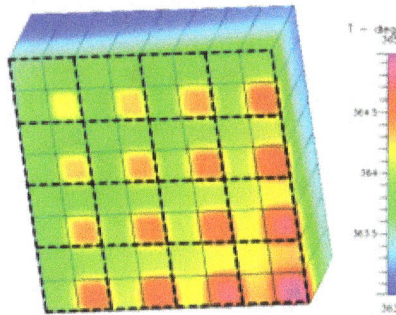

Fig. 5. Temperature distribution of the 16-core CG.

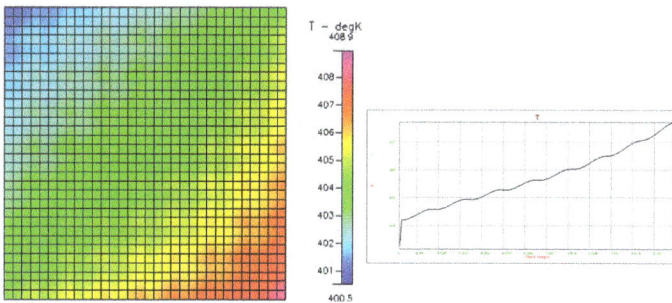

Fig. 6. Temperature distribution of the 1,024-core NoC with the same orientation of each core.

However, the situation becomes worse, when 64 such CGs are put together to construct the 1,024-core NoC. Figure 6 shows a typical layout in which the orientation of each core is kept the same as in the Figure 5, with the hotspot near the lower right corner. Apparently, the design maintains regularity in connectivity with the same routing distance between cores, but unfortunately, it is not thermal-aware. The temperature distribution is still asymmetric and the maximum temperature of the whole chip now rises up to 408.9 K which requires a heat sink. The lack of symmetry leads to that the heat sink cannot be placed at a simple orientation with equal heat dissipation ability.

Let us define the temperature non-uniformity as follows:

$$U = \frac{\Delta T}{\Delta x} \qquad (7)$$

where ΔT is temperature difference and Δx is distance between any two points on the single core. Hence, it represents the slope of the temperature gradient per unit length. Clearly, the bigger the value of U, the more severe the temperature difference between neighboring cores. In the case of Figure 6 the maximum U is around 4.1 K/cm the averaged U is around 3.1 K/cm.

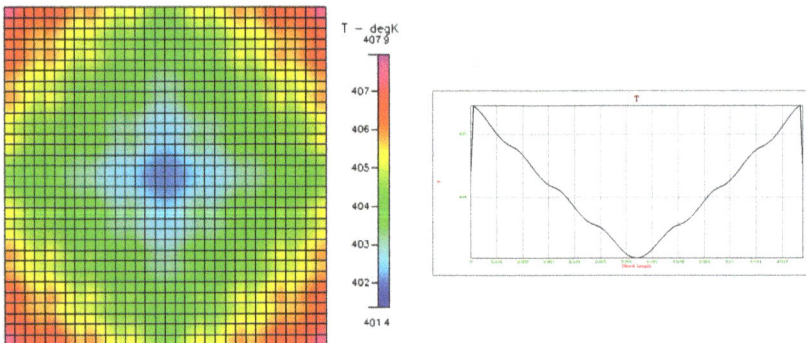

Fig. 7. Temperature distribution of the 1,024-core NoC with the orientation of every quarter of CGs rotated 90 degree.

To mitigate the non-uniformity, we may try to rotate either the cores in the CG or the CGs so as to align the temperature profile symmetrically (Xu et al., 2006). Figure 7 shows the latter approach by dividing the CGs into four quadrants, keeping the orientation of the second quadrant, and rotating the other three quadrants of the CGs to the upper left, upper right, and lower left corners, respectively.

To compare with those attained in Figure 6, the maximum temperature decreases 1 K, but the averaged temperature non-uniformity increases to 3.8 K/cm. If we rotate the cores in the CG in a similar fashion and then assemble such CGs, the result is not much different and hence is not shown here. This illustrates the fact that the rotation of the hotspots cannot reduce the maximum temperature effectively.

(a)

(b)

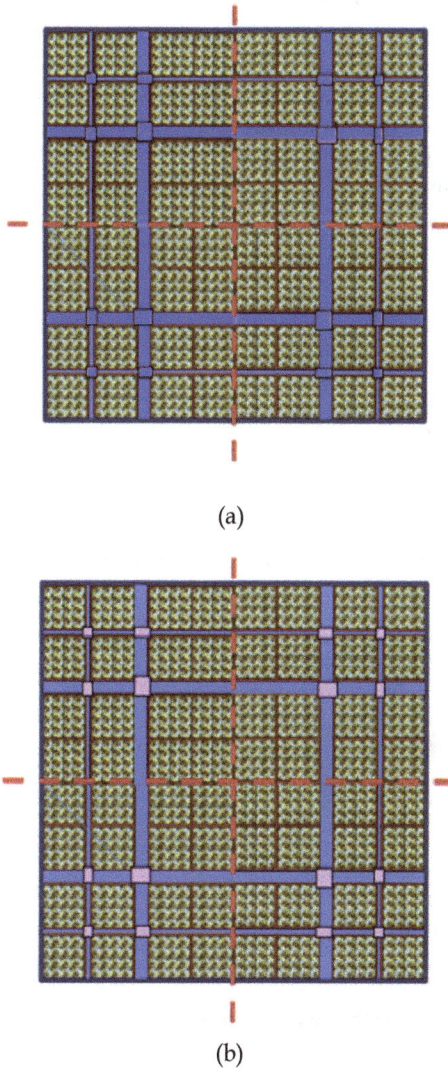

Fig. 8. The insertion places of thermal ridges. (a) Type I only. (b) Type I and Type II.

2.2.2 Insertion of the thermal ridges

The primary objective of the thermal ridges is to reduce the maximum temperature and the temperature non-uniformity at the same time. The thermal ridges are introduced into the design, with the required extra space under the constraint of manufacturing cost. In our case, at most 20% of the chip area is allowed for the thermal ridges and their locations are depicted in the Figure 8. Straits with widths of 400 μm and 200 μm are created by expanding the routing distances between CGs.

2.3 Simulation results of the proposed scheme

First, the type-I thermal ridges are inserted into the straits, except for their intersectional areas as shown in Figure 8(a). The resulting temperature distribution is shown in Figure 9. The maximum temperature is 373.4 K, which occurs in the center of the chip. To compare with the previous solutions, the maximum temperature significantly decreases 35 K by using the thermal ridges. The temperature difference at the center of the chip is about 32 K. Also, the thermal map changes a lot, since the thermal ridges are distributed in the suburb areas.

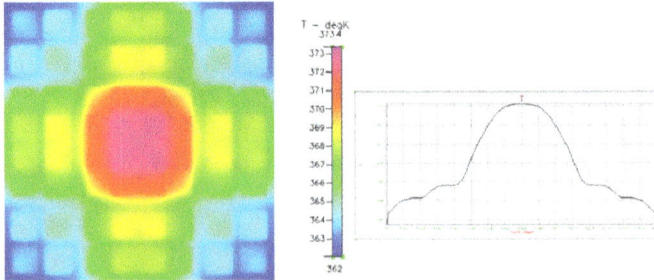

Fig. 9. The temperature distribution of the 1024-core NoC with type I thermal ridge.

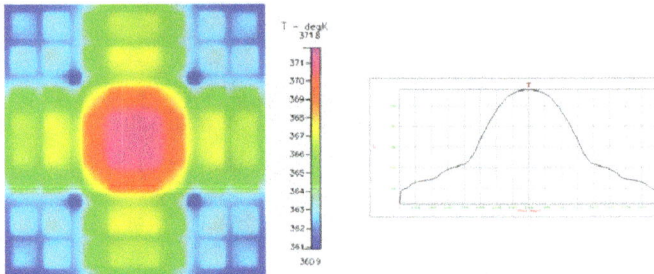

Fig. 10. Temperature distribution of the 1,024-core NoC with type-I and type-II thermal ridges.

Furthermore, the design affects the temperature non-uniformity substantially. In Figure 6 and Figure 7, it is easy to find that the value of U keeps almost constant all around the chip. However, after inserting the thermal ridges, there are several values of U on the chip. The largest U is around 4.6 K/cm, but the average U decreases substantially to 1.5 K/cm. The temperature non-uniformity is largely improved at the center and the suburb areas by the values of 0.5 K/cm and 1.5 K/cm, respectively. About 85% of the chip area is covered in the region. This means that around 850 cores have better temperature non-uniformity. Since the tile size is 410 μm × 410 μm, the temperature difference between neighboring cores in the region is less than 0.3 K.

In addition, the insertion of the type-II thermal ridge is performed, as shown in Figure 8(b). The temperature profile is shown in Figure 10. The maximum temperature of 371.8 K is about 1.5 K lower than that shown in Figure 9. It can be further reduced, since the thermal conductivity of the type-I thermal ridge is lower than that of the type-II thermal ridge. The temperature non-uniformity and the temperature profile remain quite similar. Compared with the results from the traditional scheme with mere rotation of the hotspots, the maximum temperature decreases from 408.9 K to 372.8 K, and the temperature non-uniformity decreased from 3.2~4.0 K/cm to 0.5~1.5 K/cm in 80% of the chip area, under the constraint of increasing 20% extra area for the thermal ridges.

3. Chip design and implementation by using metallic thermal skeletons

In this chapter, a realistic thermal dissipation enhancement methodology for NoC system will be introduced. The on-chip virtual 126-core network as the hot-spot dissipates the generated heat through the metallic thermal skeletons. To evaluate the feasibility of the thermal enhancement, 9 arrays of metallic thermal skeletons are designed in the test chip. Essentially, by improving the lateral thermal dissipation path by increasing the thermal metallic skeleton in the back end of line (BEOL) metals, the heat consumed by the virtual core can be conducted into the on-chip heat sink such as the TSVs. The temperature of the hotspot can be lowered substantially if the metallic thermal skeletons arranged properly. In addition, we design thermal sensor-network on chip to facilitate the measurement and evaluation for the capability of heat transfer. Last, some important thermal characteristics of metallic thermal skeleton are listed in this chapter. In order to design a better thermal dissipation path, metallic thermal skeletons can provide alternatives for just increasing the number of thermal TSVs.

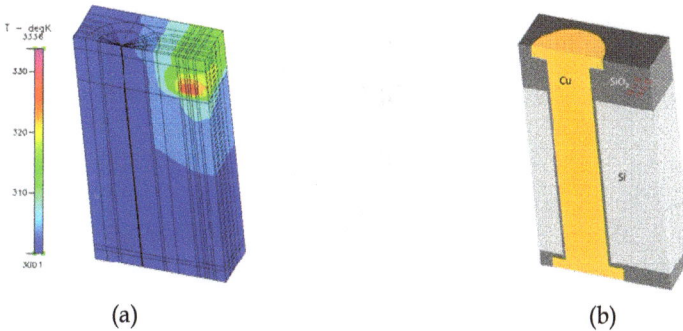

(a)   (b)

Fig. 11. FEM simulation model and result. (a) Temperature profile. (b) Simulation model.

The FEM simulation is performed by using CFD-RC, based on the following assumptions. As shown in Figure 11, a TSV is on the left, and a heat source is on the right. The other half of the structure is mirrored to the cross section. The heat source consists of 12 squares, each with power of 0.5 mW, and area of 1 μm × 1 μm, which run to the top by local interconnects (not shown in the figure for they are buried in the structure), just shy of the front metal layer at the top. It is seen that the neighboring TSV is unconnected electrically and cold. The simulation assumes a TSV with dielectric thickness of 0.5 μm, diameter of 10 μm, and length of 50 μm.

3.1 Design of the proposed test chip

3.1.1 Overall floorplan of the chip

The floorplan of the proposed test chip is depicted in the Figure 12. The metallic thermal skeletons are arranged and enclosed by the core-sensor blocks. The peripheral area is for input/output and power/ground connections which provide external accesses. The test chip is designed without resorting to a complex control scheme. The virtual cores are arranged in three groups, each consisting of three rows and seven columns. The whole chip can be divided into nine regions. Each region consists of two separate areas which are enclosed by core-sensor block named A1-A7, B1-B7 and C1-C7 respectively and represent 3 types of metallic thermal skeletons. to are identical design of the metallic thermal skeleton, so do the to and to . The major differences among these nine regions are the combinations of , and elements, which are shown in Figure 13. In this design as shown in Figure 13(a), elements , and are different in the distribution densities of metal in the BEOL. For better visualization, Figure 13(b) shows the three-dimensional view of the metallic thermal skeletons. The combinations of TSVs with front metals form the on-chip heat sink, and the BEOL metal 1 to metal 4 form the metallic thermal skeletons.

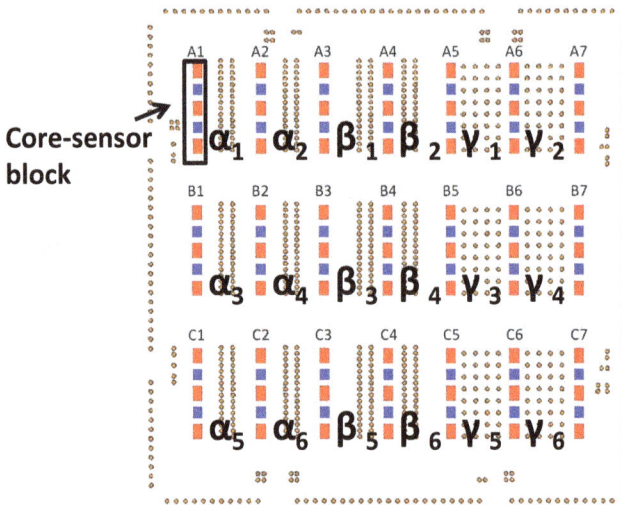

Fig. 12. The floorplan of designed test chip.

In this chapter, the stacking of the identical chips is not included in discussions, only planar die is reported. The future thermal TSV test chip will divide the core area into blocks, each, as shown in Figure 14, consisting of virtual cores, temperature sensors, and a TSV array with metallic thermal skeletons to constructs the on-chip heat sink. The virtual cores and temperature sensors are laid out at the left and right side of the on-chip heat sink. As shown in Figure 14, thermal TSV with front metals will be the on-chip heat sink, and the metallic thermal skeletons play the role as the conduction path for high speed heat transfer. Therefore, the performance of the metallic thermal skeletons are emphasized and compared with each other.

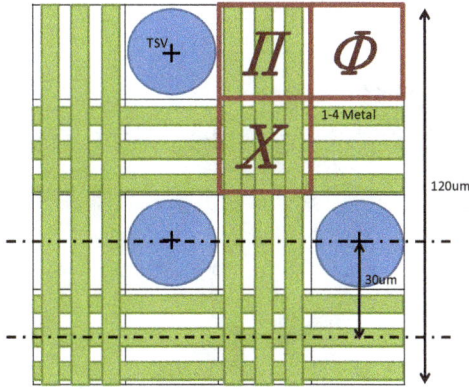

(a)

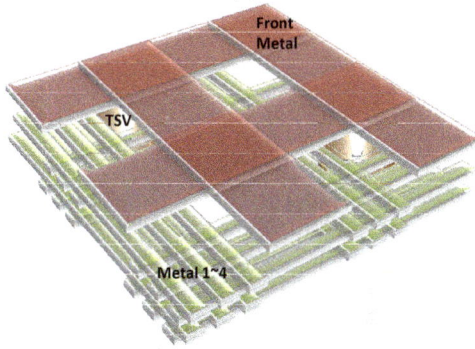

(b)

Fig. 13. The design of TSVs with metallic thermal skeletons. (a) The planar floorplan with , , and TSVs. (b) The three-dimensional view of the metallic thermal skeletons.

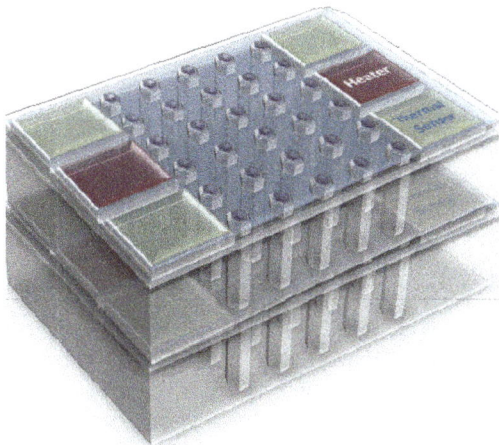

Fig. 14. Concept of virtual block design.

Fig. 15. The layout of the test chip.

In this chapter, to verify the capability of heat conduction, triplet experiments are designed to test the chip. Since A1-A3 is at the corner of the chip, the heat transfers more to the peripherals than to the central area of the chip. Such kind of location factors occur often in the chip measurement of thermal phenomenon. Hence, A1-A3, B1-B3 and C1-C3 are identical combination of the metallic thermal skeletons to avoid the location effects happening. The layout of the designed test chip is shown in Figure 15. The core-sensor blocks, metallic thermal skeletons, peripherals, IOs, and power domains are in one SOC chip as the NoC. The virtual core system composed of on-chip heaters can be operated at the same time. The die size measures 5,040 μm × 5,040 μm, including the seal ring. There are three voltage levels, four power domains, and nine test regions in this chip. Each voltage level can be separately controlled by the programmable logic analysis instrument. All the cores in the chip can be operated independently through the power gating mechanism. In order to precisely observe the temperature distribution of the chip surface, all sensors on the chip are activated simultaneously, and the measured temperature values can be read out as the matrix data.

3.1.2 Design of the core-sensor block

The temperature sensitive ring oscillator (TSRO) thermal sensor in Figure 16 is based on a ring oscillator whose oscillation frequency is sensitive to temperature, albeit not completely linear. In fact, the ring oscillator is also sensitive to supply voltage. Hence, to minimize power droop is important in improving the accuracy. By establishing the relationship between temperature and frequency, and opting for on-die calibration, the thermal sensor can be quite accurate. The frequency is converted by a counter and read out to a register. Figure 16(a) shows the block diagram. The control unit (CU) accepts a reference clock TS_CK and an input TS_EN which enables the sensing operation when transitioning from 0 to 1. As shown in Figure 16(a), four signals a, b, c, and RDY are generated. When the internal signal a changes from 0 to 1, the counter is reset and the count is cleared. When internal signal b changes from 0 to 1, the ring oscillator is activated and the counter starts; when it changes from 1 to 0, the ring oscillator is deactivated and the counter stops. When the internal signal c changes from 0 to 1, the count is loaded into an output register TS_REG to be read out. The handshake signal RDY indicates that the count is ready. The physical view of the thermal sensor used in this test chip is shown in Figure 16(b).

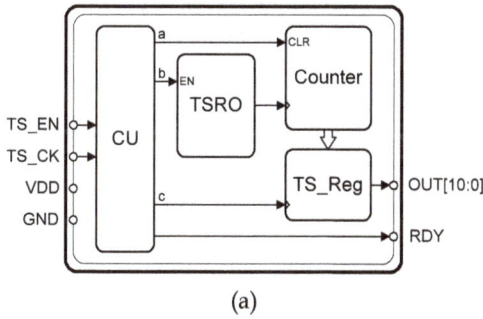

(a)

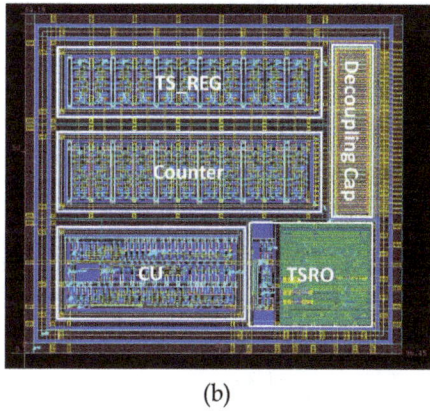

(b)

Fig. 16. Thermal sensor design. (a) The block diagram of the thermal sensor. (b) The layout of the thermal sensor, including a regulator, counter and a control unit.

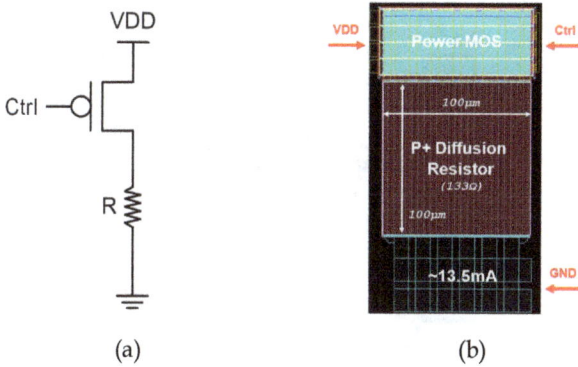

(a)   (b)

Fig. 17. Power gating design. (a) The schematic diagram of the virtual core circuits. (b) The layout view of the virtual core circuits.

The virtual core circuit is composed of a PMOS switch and a p-type diffusion resistor, as shown in Figure 17. The diffusion resistor is non-silicided and placed in an n-well. Consequently, the n-well becomes hot at first, if the heater in the virtual core is turned on, which is slightly different from a conventional CMOS circuit in that the substrate is more likely to be the heat source. The maximum current flowing into the resistor is regulated below 13.5 mA.

3.2 Thermal property analysis of the metallic thermal skeletons

The metallic thermal skeletons are intended to be placed in the regions enclosed by the core-sensor blocks. In this section, we derive analytical expressions for some key parameters.

3.2.1 Analytical model of the metallic thermal skeleton

It is clear that the heat removing rate of the metallic thermal skeletons is assumed to be q. Let us consider a pair of core-sensor blocks as the heat sources. The temperature distribution on the metallic thermal skeletons between any couple of core-sensor blocks can be expressed by (4), and then can be expressed as the following equation.

$$T_k = T_a + \frac{T_b - T_a}{w}x - \frac{q}{2k_{sk}}[w-x]x \tag{8}$$

As shown in Figure 18, where T_a and T_b are the temperatures of CS1 and CS2, respectively, q is the heat conducted to the ambient environment by the metallic thermal skeletons, k_{sk} is the equivalent thermal conductivity of the metallic thermal skeletons, and w is the width of the metallic thermal skeletons. Since T_k denotes the temperature at the location x, examining the mid-point $T_{1/2}$ by substituting x with $w/2$ into (9), we have

$$w = \left(\frac{8k_{sk}}{q}\right)^{1/2}\left[\frac{T_a + T_b}{2} - T_{1/2}\right]^{1/2} \tag{9}$$

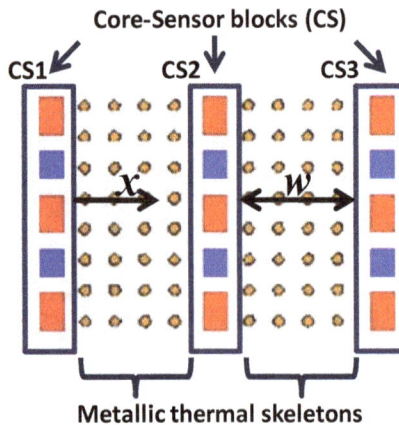

Fig. 18. The theoretical model of the core-sensor blocks with the metallic thermal skeletons.

3.2.2 Effective thermal conductivity of the metallic thermal skeletons

For the die with 9 μm of BEOL and 450 μm of the silicon substrate, we can clearly figure out that k_{sxx} is around 12~68 W/mK and k_{szz} is around 116~147 W/mK, by substituting the thermal conductivities into (6). The variation in the equivalent thermal conductivity depends on the percentage distribution of the metal in BEOL. Thus, the heat flows through the silicon substrate almost dissipates by the metallic thermal skeletons instead of transferring by silicon dioxide in the BEOL. By substituting the equivalent k_{sk} and the temperature values of T_a, T_b and $T_{1/2}$ into (9) we obtained that the widths of the metallic thermal skeleton should be 420 μm. FEM simulations have been performed to see the effectiveness of the proposed metallic thermal skeletons, as shown in Figure 19. For the reason of compatibility, we have combined the simulation results both from CFD-RD and ANSYS, so as to link the design platform for our circuit designers. Hence, to design the metallic thermal skeleton shown in Figure 12, we assumed the type α, β and γ with different distribution densities of metal in the BEOL as following equation.

$$\begin{bmatrix} \alpha \\ \beta \\ \gamma \end{bmatrix} = D \begin{bmatrix} X \\ \Pi \\ \Phi \end{bmatrix} \tag{10}$$

where

$$D = \begin{bmatrix} 0.28 & 0.44 & 0.28 \\ 0.20 & 0.52 & 0.28 \\ 0.36 & 0.36 & 0.28 \end{bmatrix} \tag{11}$$

The matrix D represents the weighting coefficients of the metallic thermal skeletons. The percent contribution of the element Φ is limited by the metal density constraint in the design rule released from the foundry.

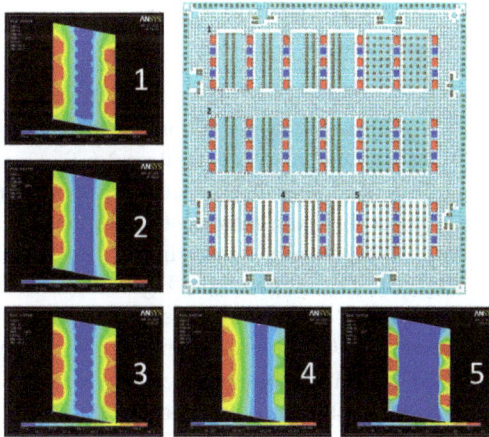

Fig. 19. The simulated results of the selected regions of the proposed architecture are shown. The enable signal H_EN is broadcast to all virtual cores.

3.3 Experimental setup

The die photo of the proposed test chip in this chapter is shown in Figure 20. This chip is fabricated by TSMC in 0.18 μm 1P4M mixed-mode process technology. The package uses 256-pin IST Universal PGA. The front of the chip is covered by the package glue. In order to observe the thermal behavior of the test chip, the back of the chip is exposed to air with a transparent PYREX® glass of 120 μm. There is a 6 cm x 6cm open window in the central area of the evaluation board to facilitate the observation on the temperature measurement.

Fig. 20. The die photo of the proposed test chip in this chapter, the dimension of the chip is 5,040 μm x 5,040 μm, including the seal ring.

The principle measurement environment setup includes DC power supplier (MOTECH PPS 3210), current meter (FLUKE 189), function generator (HP 8166A), temperature-humidity chamber (HOLINK EZ040-72001), logic analyzer (Agilent N6705A), infrared camera (FLIR SC5700), and thermal management total analysis platform. As shown in Figure 21(a), the FLIR SC5700 with a microscope of three μm resolution is responsible for infrared radiation (IR) inspection. The temperature responses are measured by the thermal management total analysis platform designed by ICL, ITRI as shown in Figure 21(b). It is clear in Figure 21(c), the test environment is controlled at a constant ambient temperature, in which the temperature error varies within ± 0.5 ºC. The programmable temperature-humidity chamber HOLINK EZ040-72001 is used to control the operation temperature from 0 ºC to 100 ºC. MOTECH PPS 3210 is the power supply which provides the three voltage levels. The control signals (TS_EN and CLK) are generated from HP 8166A. The current meter FLUKE 189 is utilized for measuring the current consumption. Last, the output signals are collected and analyzed by Agilent N6705A.

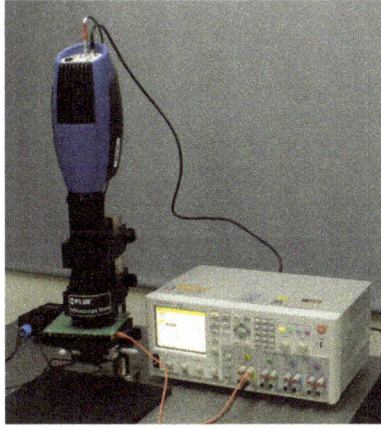

(a)

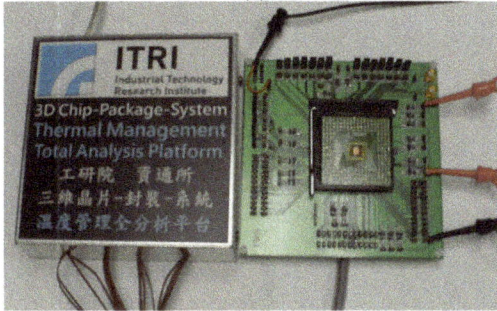

(b)

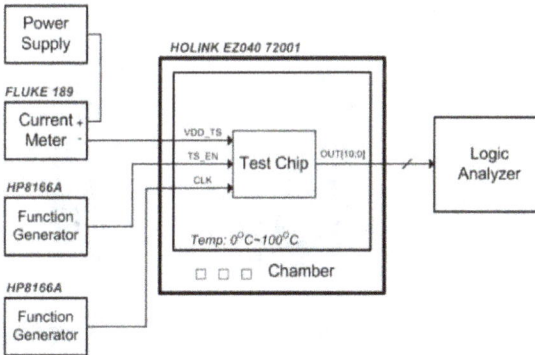

(c)

Fig. 21. The testing environment and setup. (a) The test chip is under the measurement environment with the infrared radiation inspection. (b) The naked die with the evaluation board and thermal management total analysis platform. (c) The test chip is placed in the chamber at a nearly constant ambient temperature.

3.4 Results and discussions

The experimental results are shown in Table 1. When the power density of 7.38 W/cm² is applied to the virtual core, each core is operated at the power of 20 mW. To evaluate the thermal conduction capability of the metallic thermal skeleton, the average temperature of the metallic thermal skeleton is an important index. Since the metallic thermal skeletons are employed to conduct the heat flux generated by the virtual cores, the temperature at $w/2$ (referred to Figure 18) especially represents the results of the lateral thermal diffusion. To compare with the experimental steady state data shown in Table 1, it is clear that the virtual cores with the metallic thermal skeleton type γ have better thermal conductive performance. Moreover, T_{max}-T_{min} denotes the temperature uniformity in the region. The results show that the metallic thermal skeleton γ has the best performance among these three combinations.

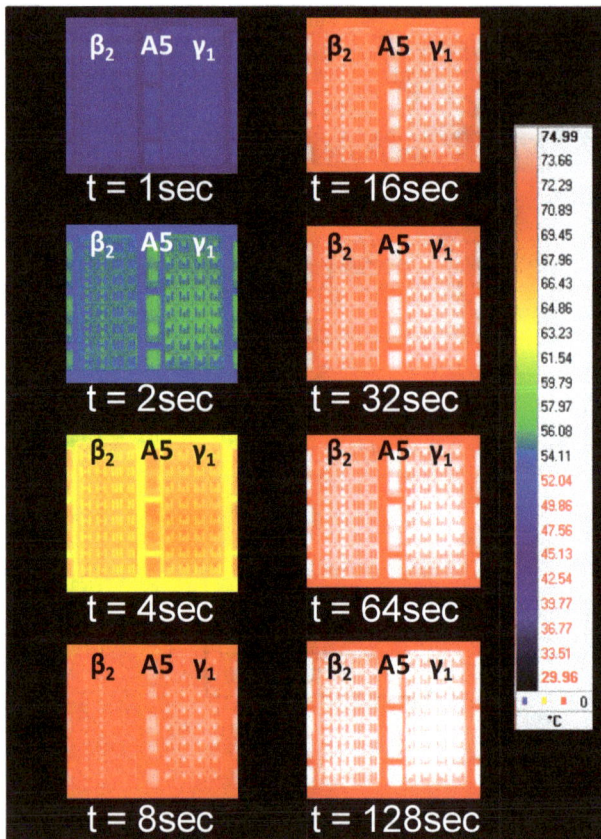

Fig. 22. The transient response of the test chip is taken by the infrared radiation camera when the virtual cores are activated. These are the back views of the test chip.

On the other hand, transient temperature response is recorded by the high speed infrared radiation dynamic photos as shown in Figure 22. Take the region in the photo for example; the β_2 -A5- γ_1 region (referred to Figure 12) includes 2 types of metallic thermal skeletons. It is clear that the temperature of γ_1 is higher than that of β_2. This results show that the thermal conductive capability of γ_1 is better than that of β_2. The area of Φ is limited by the metal density constraint in the design rule released from the foundry, therefore no more metal are allowed to be placed. However, the Φ region may be reserved for the placement of thermal TSVs or front metal stripes during the post CMOS process to be the on-chip heat sink.

Type of metallic thermal skeleton	α	β	γ
T_{max} (at virtual core)	71.00	71.36	70.12
T_{avg} (on the metallic thermal skeleton)	63.13	62.92	63.45
T_{min} (in the region)	57.48	56.38	57.82
$T_{max} - T_{min}$	13.52	14.98	12.30

Temperature in ºC

Table 1. The temperature distribution of the test chip.

4. Conclusion

The cost of thermal ridges and metallic thermal skeletons may be compared with the advanced techniques, such as micro-channel liquid cooling or the thermo-electric cooling (TEC). Since by ITRS, the number of stacked dies is expected to increase in the future, the cooling problem of the inter-layer dies will become more challenging. If the heat should be removed by pumping liquid or external energy into the stacked dies, the cooling cost will grow exponentially. The thermal ridges and metallic thermal skeletons proposed in this chapter will be relatively cost-effective and energy-saving. Moreover, this proposed method locally improves the temperature non-uniformity, and the thermal gradient of the most part of the chip also decreases. Nevertheless, the global temperature non-uniformity which affects the chip operations from the electrical perspective deserves more efforts to pursue. Since the 3D IC with TSV now appears as an emerging technology, the early floorplan for the insertion of thermal ridges and metallic thermal skeletons for thermal management will be discussed more and more widespread. The temperature distributions measured by the infrared radiation and by the thermal sensors are compared in this study. By these results, readers can understand that both of the data could be calibrated with each other if the package of the chip is chosen properly. Meanwhile, authors would like show also that the thermal test chip designed and proposed would be capable to evaluate the thermal properties and thermal characteristics of the packages if desired. In the 3D design of the stacking dies, the thermal measurement and verification are getting much more important. This research may give a direction or inspiration for the engineers to investigate the possibility or feasibility of better thermal designs.

5. References

Beigne, E.; Clermidy, F.; Miermont, S.; and Vivet, P. (2008). Dynamic voltage and frequency scaling architecture for units integration within a GALS NoC, *Proceedings of ACM/IEEE International Symposium on Networks-on-Chip (NoCS)*, ISBN: 0-7695-3098-2, Newcastle upon Tyne, April 2008, pp. 129-138.

Brunschwiler, T.; Michel, B.; Rothuizen, H.; Kloter, U.; Wunderle, B.; Oppermann H. and Reichl, H. (2007). Interlayer cooling potential in vertically integrated packages, *ACM Journal of Microsystem Technologies - Special Issue on MicroNanoReliability*, Vol. 15, No. 1, October 2008, pp. 57-74, ISSN: 0946-7076

Chien, J. H.; Lung, C. L.; Tsai, K. J.; Hsu, C. C.; Chen, T. S.; Chou, Y. F.; Chen, P. H.; Chang, S. C. and Kwai, D. M. (2011). Realization of 3-dimentional virtual 126-core system with thermal sensor-network using metallic thermal skeletons, *Proceedings of International Conference on Electronic Components and Technology Conference (ECTC)*, ISBN: 978-1-61284-497-8, Lake Buena Vista, FL., USA., June 2011, pp. 873-879.

Chien, J. H.; Lung, C. L.; Hsu, C. C.; Chou, Y. F.; and Kwai, D. M. (2010). Floorplanning 1024 cores in a 3D-stacked networkon- chip with thermal-aware redistribution, *Proceedings of 12th IEEE Intersociety Conference on Thermal and Thermomechanical Phenomena in Electronic Systems (ITHERM)*, ISBN: 978-1-4244-5342-9, Las Vegas, NV., USA., June 2010, pp. 1-6.

Cong, J.; Luo, G.; Wei, J.; and Zhang, Y. (2004). A thermal-driven floorplanning algorithm for 3D ICs, *Proceedings of International Conference on Computer Aided Design (ICCAD)*, ISBN: 0-7803-8702-3, San Jose, CA., USA., November 2004, pp. 306-313.

Cong, J.; Luo, G.; Wei, J.; and Zhang, Y. (2007). Thermal-aware 3D IC placement via transformation, *Proceedings of Asia and South Pacific Design Automation Conference (ASPDAC)*, ISBN: 1-4244-0630-7, Yokohama, Japan, June 2007, pp. 780-785.

Hennessy, J. L. and Patterson, D. A. (2007). *Computer Architecture: A Quantitative Approach* (3rd Edition), Morgan Kaufmann, ISBN: 978-1-5586-0596-1, San Francisco, CA., USA.

Kurd, N. A.; Barkatullah, J. S.; Dizon, R. O.; Fletcher, T. D. and Madland, P. D. (2001). A multigigahertz clocking scheme for the Pentium 4 microprocessor, *IEEE Journal of Solid-State Circuits (JSSC)*, Vol. 36, No. 11, November 2001, pp. 1647-1653, ISSN: 0018-9200

Lee, S.; Lemczyk, T. F. and Yovanovich, M. M. (1992). Analysis of thermal vias in high density interconnect technology, *Proceedings of Semiconductor Thermal Measurement and Management Symposium (SEMI-THERM)*, ISBN: 0-7803-0500-0, Austin, TX., USA., February 1992, pp. 55-61.

Lu, J.-Q. (2009). 3-D hyperintergration and packaging technologies for micro-nano systems, *Proceedings of IEEE*, Vol. 97, No. 1, January 2009, pp. 18-30, ISSN: 0018-9219

Lung, C. L.; Ho, Y. L.; Huang, S. H.; Hsu, C. W.; Liao, J. L.; Huang, S. Y. and Chang, S. C. ; (2010). Thermal analysis experiences of a tri-core SoC system, *Proceedings of International Conference on Green Circuits and Systems (ICGCS)*, ISBN: 978-1-4244-6876-8, Shanghai, China, June 2010, pp. 589-594.

Lung, C. L.; Ho, Y. L.; Kwai, D. M. and Chang, S. C. (2011). Thermal-Aware On-Line Task Allocation for 3D Multi-Core Processor Throughput Optimization, *Proceedings of Design, Automation & Test in Europe (DATE)*, Grenoble, France, March 2011,. pp. 1-6

Tiwari, V., Singh, D., Rajgopal, S., Mehta, G., Patel, R., and Baez, F. (1998). Reducing power in high-performance microprocessors, *Proceedings of ACM/IEEE Design Automation Conference (DAC)*, ISBN: 0-89791-964-5, San Fransisco, CA., USA., June 1998, pp. 732-737.

Tran, A. T.; Truong, D. N. and Baas, B. M. (2009). A GALS many-core heterogeneous DSP platform with source-synchronous on-chip interconnection network, *Proceedings of ACM/IEEE International Symposium on Networks-on-Chip (NoCS)*, ISBN: 978-1-4244-4142-6, San Diego, CA., USA., May 2009, pp. 214-223.

Tran, A. T.; Truong, D. N. and Baas, B. M. (2009). A low-cost high-speed source-synchronous interconnection technique for GALS chip multiprocessors, *Proceedings of IEEE International Symposium on Circuits and Systems (ISCAS)*, ISBN: 978-1-4244-3827-3, Taipei, Taiwan, May 2009, pp. 996-999.

Truong, D.; Cheng, W.; Mohsenin, T.; Yu, Z.; Jacobson, T.; Landge, G.; Meeuwsen, M.; Watnik, C.; Mejia, P.; Tran, A.; Webb, J.; Work, E.; Xiao, Z. and Baas, B. (2008). A 167-processor 65 nm computational platform with per-processor dynamic supply voltage and dynamic clock frequency scaling, *IEEE Symposium on VLSI Circuits (VLSIC)*, ISBN: 978-1-4244-1804-6, Honolulu, HI., USA., June 2008, pp. 22-23.

Tsai, C. H. and Kang, S. M. (2000). Cell-level placement for improving substrate thermal distribution, *IEEE Transactions on Computer-Aided Design of Integrated Circuits and Systems (TCAD)*, Vol. 19, No. 2, February 2000, pp. 253-266, ISSN: 0278-0070.

Tsui, Y. K.; Lee, S. W. R.; Wu, J. S.; Kim, J. K. and Yuen, M. M. F. (2003). Three-dimensional packaging for multi-chip modulewiththrough-the-silicon via hole, *Proceedings of Electronics Packaging Technology Conference (EPTC)*, ISBN: 0-7803-8205-6, Singapore, Marcg 2003, pp 1-7.

Yu, Z. and Baas, B. M. (2006). Implementing tile-based chip multiprocessors with GALS clocking styles, *IEEE International Conference on Computer Design (ICCD)*, ISBN: 978-0-7803-9707-1, San Jose, CA., USA., October 2006, pp. 174-179.

Xu, G.; Guenin, B. and Vogel, M. (2004). Extension of air cooling for high power processors, *Proceedings of 9th IEEE Intersociety Conference on Thermal and Thermomechanical Phenomena in Electronic Systems (ITHERM)*, ISBN: 0-7803-8357-5, Las Vegas, NV., USA., August 2004, pp. 186-193.

Xu, G. (2006). Thermal nodeling of multi-core processors, *Proceedings of 10th IEEE Intersociety Conference on Thermal and Thermomechanical Phenomena in Electronic Systems (ITHERM)*, ISBN: 0-7803-9524-7, San Diego, CA., USA., May 2006, pp. 96-100.

5

VLSI Design of Sorting Networks in CMOS Technology

Víctor M. Jiménez-Fernández et al.*
Universidad Veracruzana/Facultad de Instrumentación Electrónica,
México

1. Introduction

Although sorting networks have extensively been reported in literature (Batcher, 1962), there are a few references that cover a detailed explanation about their VLSI (Very Large Scale of Integration) realization in CMOS (Complementary Metal-Oxide-Semiconductor) technology (Turan et al., 2003). From an algorithmic point of view, a sorting network is defined as a sequence of compare and interchange operations depending only on the number of elements to be sorted. From a hardware perspective, sorting networks can be visualized as combinatorial circuits where a set of denoted compare-swap (CS) circuits can be connected in accordance to a specific network topology (Knuth, 1997). In this chapter, the design of sorting networks in CMOS technology with applicability to VLSI design is approached at block, transistor, and layout levels. Special attention has been placed to show the hierarchical structure observed in sorting schemes where the so called CS circuit constitutes the fundamental standard cell. The CS circuit is characterized through SPICE simulation making a particular emphasis in the silicon area and delay time parameters. In order to illustrate the inclusion of sorting networks into specific applications, like signal processing and nonlinear function evaluation, two already reported examples of integrated circuit designs are provided (Agustin et al., 2011; Jimenez et al., 2011).

2. Compare-swap block design in CMOS technology

In an algorithmic context, the CS element is conceived as an ideal operator which is free of the inherent delay time presented when a signal propagates through it. It can be seen as a trivial two-input/two-output component with a general two number sorting capability. Also, it is considered that the CS element works taking in two numbers and, simultaneously, placing the minimum of them at the bottom output, and the maximum at the top output by performing a swap, if necessary (Pursley, 2008). Figure 1 shows the typical Knuth diagram for a CS operator. In this pictorial representation, at the input, the horizontal lines describe

* Ana D. Martínez[1], Joel Ramírez[1], Jesús S. Orea[1], Omar Alba[2], Pedro Julián[3], Juan A. Rodríguez[3],
Osvaldo Agamennoni[3] and Omar D. Lifschitz[3]
[1]Universidad Veracruzana/Facultad de Instrumentación Electrónica, México
[2]Instituto Tecnológico Superior de Xalapa/ Departamento de Electrónica, México
[3]Universidad Nacional del Sur/ Departamento de Ingeniería Eléctrica y de Computadoras, Argentina

the two numbers to be sorted (A and B) and, at the output, max(A,B) and min(A,B) denote the maximum and minimum numbers, respectively. In turn, the vertical connector line represents the element dedicated to compare and interchange (swap) data.

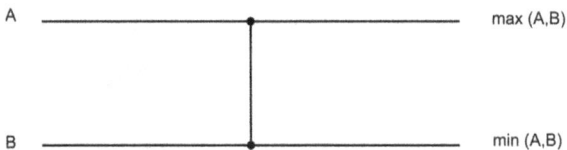

Fig. 1. Knuth diagram for a compare-swap element

However, this is only a theoretical viewpoint, because when the CS element is carried out to a level of silicon realization it is affected by parasitic elements, presenting a different time delay for each output. Due to the fact that in a sorting network the main structural element is the CS circuit, a special attention is given to describe in detail its internal design. In this section, the CS circuit design is covered at schematic transistor and at layout levels; furthermore, the area and delay time are estimated by considering a given 0.5 microns process technology.

2.1 Design at transistor level for a compare-swap standard cell

The CS element is a combinatorial circuit that accepts as input two binary signals (numbers), compares their magnitude, and outputs the maximum in the max(A,B) bus line, whereas the minimum is output in the min(A,B) bus line. This block is integrated by one full-adder and two multiplexers, as shown in Fig. 2.

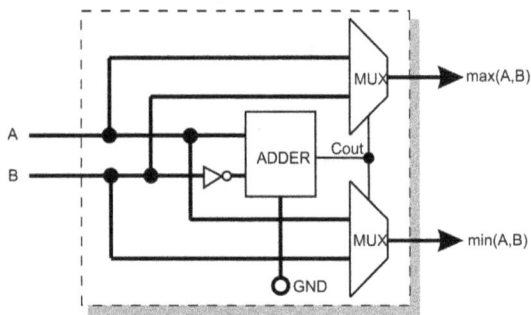

Fig. 2. Block level diagram for the CS circuit

Notice that due to one input is complemented, the full-adder is in fact configured as a subtractor. The most significant bit resulting from the subtraction, carry out (C_{out}), is used to make the selection in the multiplexer. If a greater number is subtracted from a lesser one, then the result is a negative number what, in binary terms, can be identified because the generated C_{out} will be in high ("1" logic). When the C_{out} signal is in high state, a swap data will be performed; otherwise, the input data will not be interchanged.

For translating the diagram of the CS block, in Fig. 2, to a transistor level circuit description, the two well known standard cells for the full-adder and for the multiplexer (Kang &

Leblebici, 2003), are used. Figure 3, shows the transistor level schematic for the one-bit full-adder and for the one-bit multiplexer. The full-adder is composed by 24 MOS transistors, topologically connected in a CMOS configuration, where 12 PMOS transistors (M1...M12) belong to the pull up network and 12 NMOS transistors (M13...M24) are associated to the pull down network. The multiplexor is integrated by two transmission gates (composed by transistors M25-M28) and one NOT-gate with two inputs (In0, In1), one selector (Sw), and one output (Out). The multiplexer output depends only of the C_{out} of the full-adder, since C_{out} is assigned to the selector that will activate one pair of transistors. If the selector is low ("0" logic), the transmission gate at the top (integrated by M25 and M26) switches ON, so In0 becomes the output; otherwise the transmission gate at the bottom (composed by M27 and M28) is ON, hence In1 is the output.

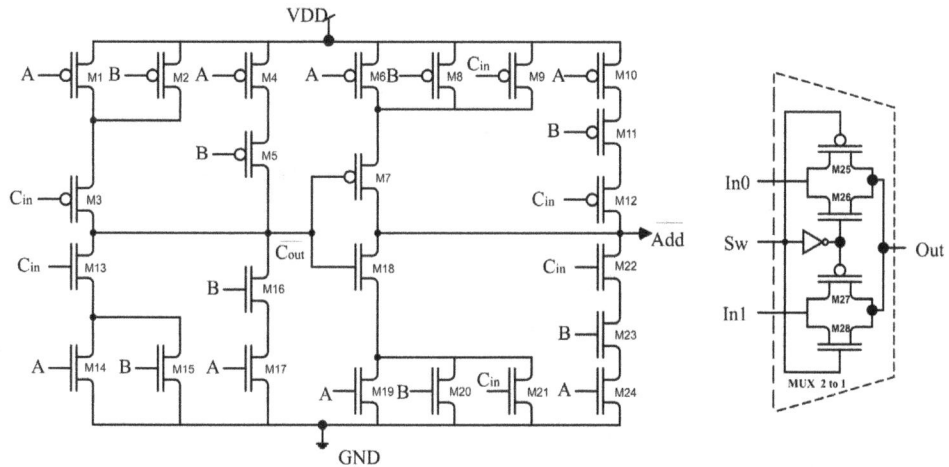

Fig. 3. Transistor level schematic of the one-bit full-adder (on left) and one-bit multiplexer (on right)

2.2 Design at layout level for a compare-swap standard cell

Masks of the CMOS full-adder and multiplexer circuits using minimum size transistor are depicted in Fig. 4 and Fig. 5. It is important to point out that the (W/L) ratio for all the NMOS and PMOS transistors in this layout were computed to optimize the transient performance of the circuit, specifically a balance between the high-to-low and low-to-high propagation times. For PMOS transistors a $(10\lambda/2\lambda)$ ratio was considered while a $(6\lambda/2\lambda)$ ratio was used to NMOS transistors. Since a 0.5 microns process technology is included, the physical dimension of lambda for this design technology is $\lambda=0.35$ microns. The well-known layout style based on a "line of diffusion" rule that is commonly used for standard cells in automated layout systems (Weste & Eshraghian, 1993) is employed in this layout. In this style, four horizontal strips can be identified: a metal ground at the bottom of the cell (GND or VSS), n-diffusion for all the NMOS transistors, a n-well with a corresponding p-diffusion for all the PMOS transistors, and a metal power at the top (VDD). A set of vertical lines of poly-silicon are also used to connect the transistor gates while within the cell metal layers connect the transistors in accordance with a schematic diagram.

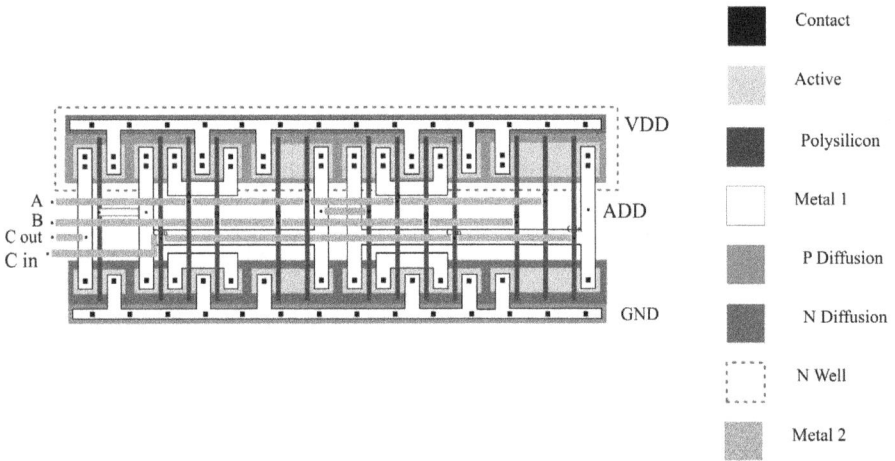

Fig. 4. Mask layout of the CMOS one-bit full-adder circuit

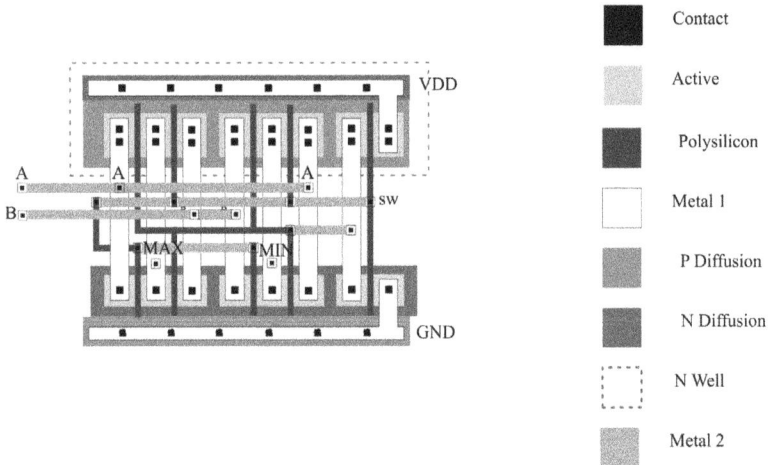

Fig. 5. Mask layout of the one-bit multiplexer circuit

2.3 Delay time and area estimations for the compare-swap cell

In order to have a reference of the switching speed for the one-bit CS circuit, an empirical delay time estimation supported by SPICE simulations is performed. Due to the speed in a CMOS gate is limited by the time taken to charge load capacitances toward VDD and discharge toward GND (Rabaey, 2003), the parasitic capacitances induced by the layout structure are considered. In this sense, a parasitic extractor software (e.g., L-Edit extractor of Tanner EDA) can be used to obtain a circuit netlist file in which all these elements be incorporated. By using SPICE simulation and including the proper test-data fabrication model parameters (AMIS 0.5 microns), an accurate transient response is achieved. The resulting transient responses are analyzed to estimate the switching speed through the delay time (difference between input transition at 50% and the 50% output level). The simulated output voltage obtained for the one-bit CS circuit is shown in Fig. 6. In this simulation, the voltage supply of 5V (VDD) and the overall frequency of 5MHz are considered. Also, the simplest representation 0 or 1 will be hereafter used instead of the "1" logic or the "0" logic notations. After running the SPICE simulation, it can be observed the outputs MAX(A,B)={0,1,1,1} and MIN(A,B)={0,0,0,1} when the inputs A and B are given by A={0,0,1,1} and B={0,1,0,1}. It is important to notice that the signal CARRY_OUT (C_{out}) is only in high when A=0 and B=1 (the unique case where a swap is needed).

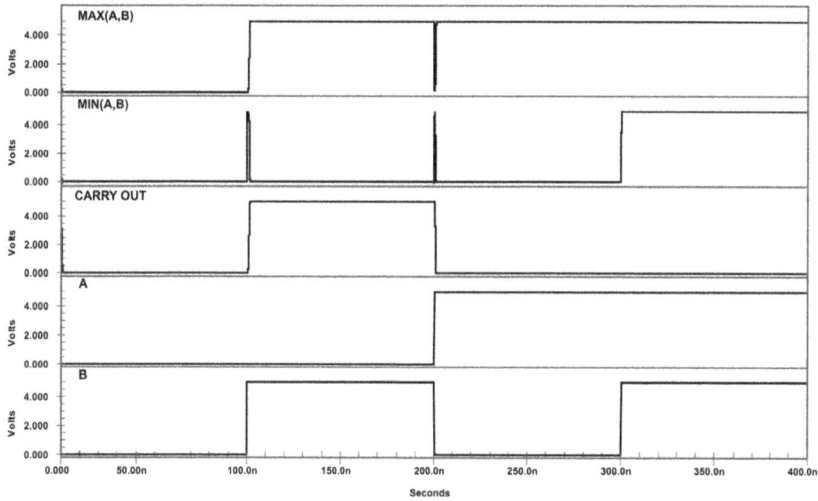

Fig. 6. Simulated output voltage obtained for the one-bit CS circuit

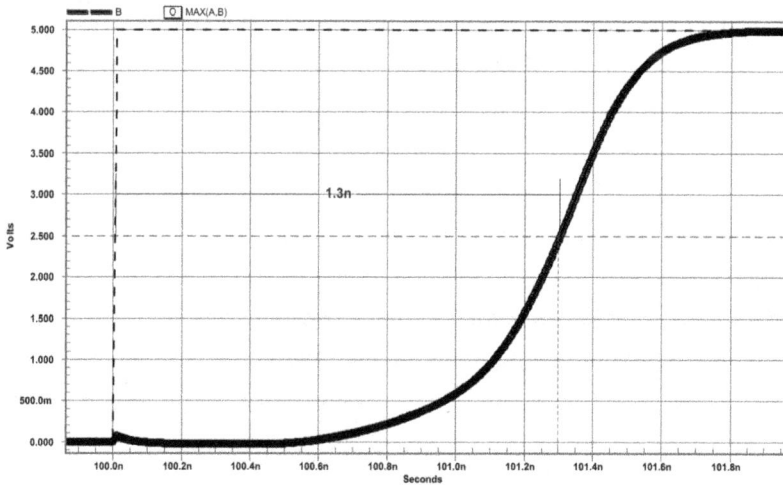

Fig. 7. Worst-case delay time for the one-bit CS circuit

As it was expected, the worst-case of delay time is presented in the swapping case. However, not only the delay time depends on the C_{out} propagation, but also it is related to the delay time added by the transmission gate. In accordance with simulation, a delay of 1.3 ns is exhibited. In Fig. 7 this delay time is showed, the dashed line indicates the input B=1 when A=0 and the solid line represents the propagated B datum after the swap operation.

An accurate silicon area estimation of the CS design can be computed directly from the layout editor by using a ruler tool (usually provided in this software). Figure 8 shows the CS cell layout design that highlights the length and width dimensions expressed in terms of lambda. From this figure the area estimation is given by 30830 λ^2 = 0.0037767 mm^2.

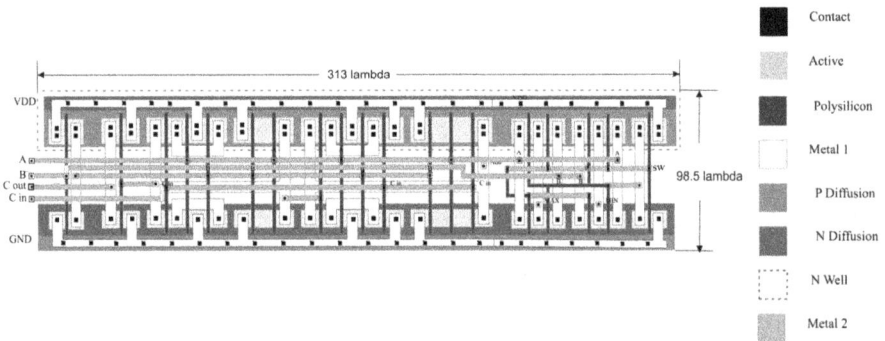

Fig. 8. Silicon area estimation for the one-bit CS layout

2.4 The n-bits compare-swap cell

The one-bit CS circuit in Fig.2 can be easily expanded into an n-bits structure. In order to illustrate how this expansion can be performed, the schematic diagram for a 4-bit CS circuit is shown in Fig. 9. Because of the overall speed of the CS circuit is limited by the delay propagation of the C_{out} bits through the n-bits chain, therefore an estimation of this time becomes essential for determining the speed performance. However, besides to the delay produced due to the critical path of C_{out}, the delay time added by the multiplexer block is also taken into account.

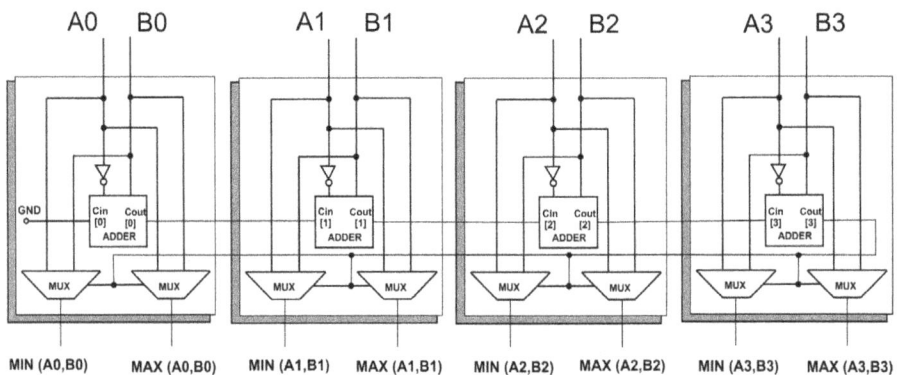

Fig. 9. Block diagram for the 4-bit CS circuit

In Fig. 10, the simulated output voltage obtained for the 4-bit CS circuit is shown when inputs A and B are given by: [A3:A0]={ 0101 (5_{10}), 1001 (9_{10}), 0011 (3_{10}), 0100 (4_{10}) } and [B3:B0]={ 1010 (10_{10}), 0110 (6_{10}), 1100 (12_{10}), 0100 (4_{10}) }.

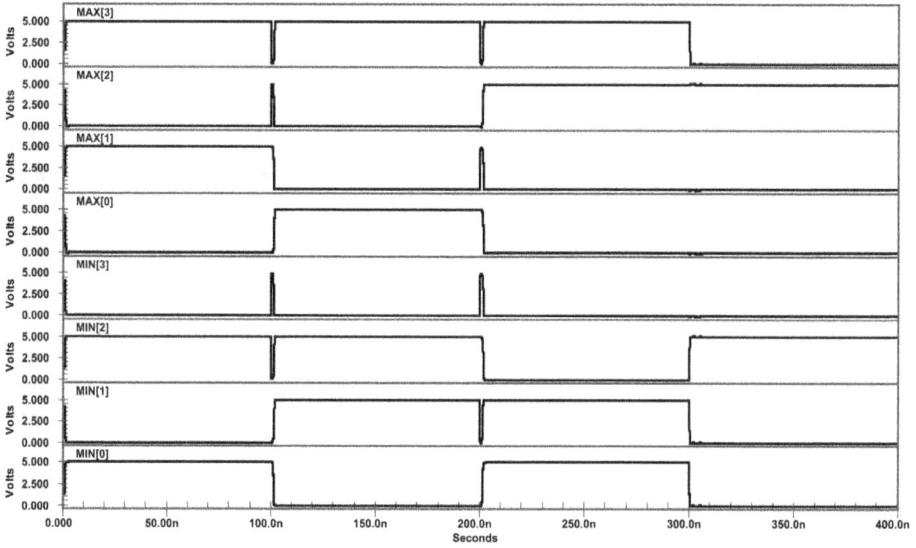

Fig. 10. Simulated output waveforms of the 4-bit CS circuit

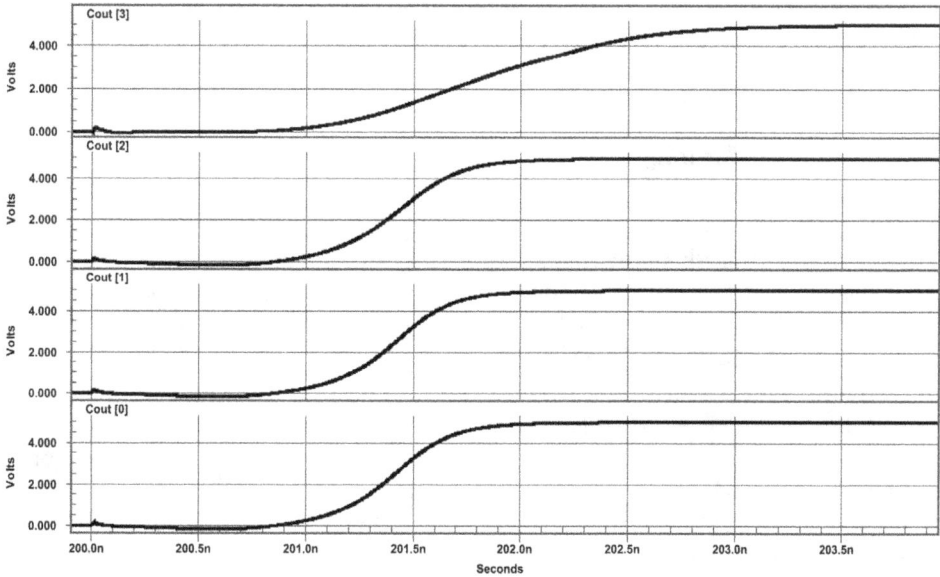

Fig. 11. Carry out (C_{out}) propagation through the 4-bit CS circuit for the C_{out}=[1,1,1,1] case

Figure 11 depicts the C_{out} propagation while Fig.12 indicates the delay time between a signal and its corresponding output after that a swap operation is performed. In these simulations, the delay time was also examined at the overall frequency of 5MHz, VDD=5V, and by considering the worst-case of C_{out} propagation. This case occurs when [A3:A0]=[0000] and [B3:B0]=[1111] what ensures transferring C_{out}= 1 at every one-bit CS basic cell.

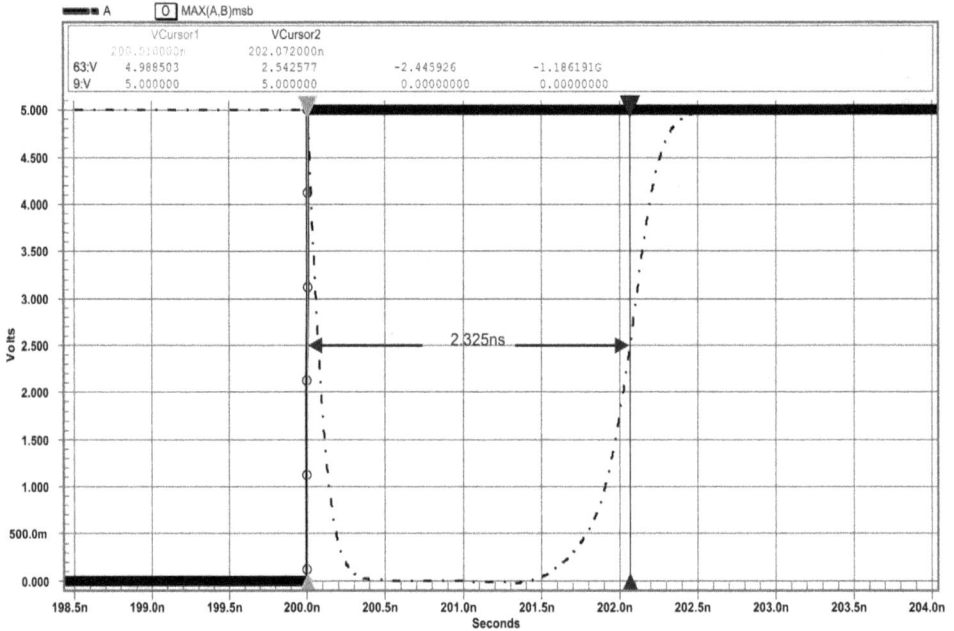

Fig. 12. Delay time between B[0] and MAX[A3,B3]

3. Median filtering for image denoising using sorting network

In order to illustrate the application of the CS circuit to the CMOS design, a digital architecture which is dedicated to median filtering for image denoising, is taken as a reference. This kind of filtering technique is used to reduce impulsive noise in acquired images (Faundez, 2001). Its main advantage consists in diminishing the lossless of information due to the computed pixel values have correspondence to one of the already presented in the image and its main characteristic is the requirement of a sorting operation (Vega et al., 2002).

Before of describing this design, it is important to present a briefly explanation about the algorithm which serves as basis for its digital architecture. The following notational conventions will be used: if $I(x,y)$ is a grayscale image divided in ($m \times n$) pixels (squares) and also $I(x,y)$ is affected by impulsive noise, then by applying a median filter algorithm, a denoised image $IF(x,y)$ can be obtained. In order to achieve $IF(x,y)$, the value of each output pixel must be computed by using iteratively a (3×3) square array (mask) of 9 pixels with center in $I(x,y)$. The position of this mask is shifted along to $I(x,y)$ until the median filtering process is completed. It is worth to mention that because of the mask operates over the neighbour pixels, then it is needed to add elements (for example zeros) around $I(x,y)$, increasing its dimension as $(m+2)\times(n+2)$. At each one of these pixels, a sorting procedure is performed by following three basic steps into the (3×3) mask: firstly, the pixels of the mask are sorted in a column by column sequence, then row by row, and finally along to the diagonal elements. After the sorting task is achieved, the central element (median) of the mask is picked out of $I(x,y)$ and stored in the $IF(x,y)$ to construct the filtered image. An illustrative description for this median algorithm is depicted in Fig. 13. A more formal description of this algorithm can be found in reference (Jimenez et al., 2011).

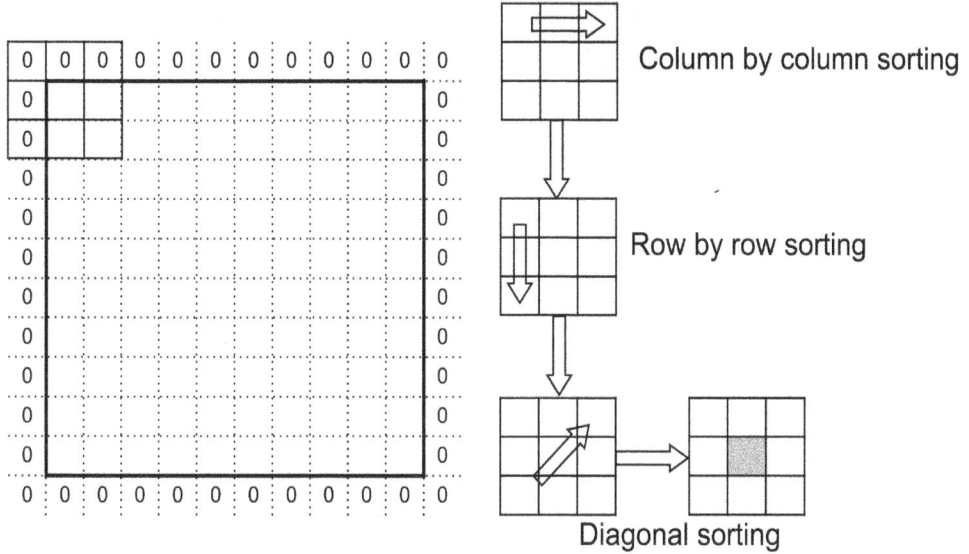

Fig. 13. Graphical description for the median filter algorithm

3.1 The sorting network block in the median filter algorithm

A Knuth diagram for the sorting network procedure which is described in the median filter algorithm is shown in Fig. 14.

Notice that the above sorting network exhibits a very regular structure that is hierarchically partitioned in seven blocks of three-data for median computing. The first stage of three blocks is dedicated to the column by column sorting, the second stage of three blocks is devoted for the row by row sorting, and finally the last block performs the diagonal sorting. It can be also observed that after all data have been propagated through the entire network, the median datum will be appearing in the bus line D4. If the (3×3) mask is defined as follows:

$$\text{MASK}_{I(x,y)} = \begin{vmatrix} D0 & D3 & D6 \\ D1 & D4 & D7 \\ D2 & D5 & D8 \end{vmatrix} \tag{1}$$

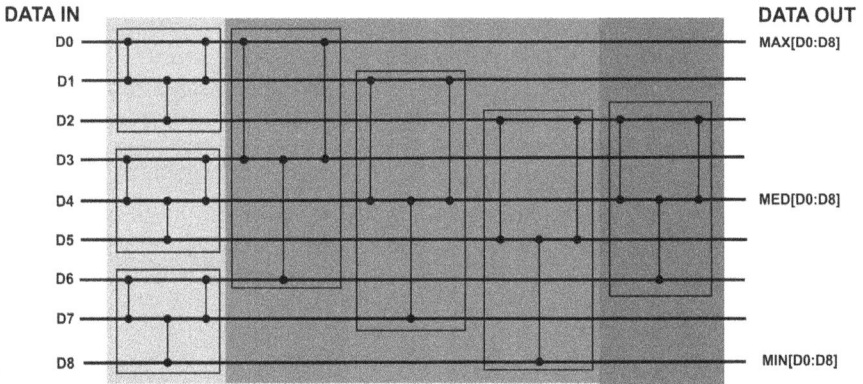

Fig. 14. Knuth diagram for the sorting network included in the median filter algorithm

Then the median datum (collected in D4) is computed trough the next steps:

1. Column by column sorting:

$$SORT\ (D0, D1, D2)$$

$$SORT\ (D3, D4, D5)$$

$$SORT\ (D6, D7, D8)$$

2. Row by Row sorting:

$$SORT\ (D0, D3, D6)$$

$$SORT\ (D1, D4, D7)$$

$$SORT\ (D2, D5, D8)$$

3. Diagonal sorting:

$$SORT\ (D2, D4, D6)$$

3.2 Digital architecture for the image filtering based on sorting network

In reference (Jimenez et al., 2011) a FPGA (Field Programmable Gate Array) implementation for median filtering image based on a sorting algorithm is reported. In such architecture two blocks can be distinguished: a nine-data accumulator and a nine-data sorting network module. The accumulator is a memory register in which the data is received from the (3x3) mask and temporarily stored. The sorting network, which is in fact the kernel of the median filter architecture, is also a nine-inputs/one-output combinational module. It is constituted by an array of seven blocks of three-data comparator modules as corresponds to Fig. 14. This interconnection topology is directly related to the median algorithm because it operates by following the already described three steps: column sorting, row sorting and diagonal sorting. It can be seen that although this block is able to output the nine data in a sorted sequence, only the datum in D4 is collected since it represents the median.

In order to illustrate the correct performance of this architecture, results obtained from the FPGA implementation and from the coded algorithm in Matlab are compared. Figure 15 shows a group of images that have been intentionally corrupted by impulsive noise and then filtered directly by Matlab software and by FPGA hardware.

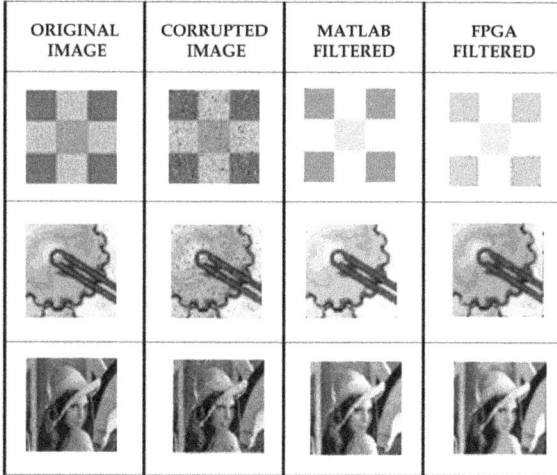

ORIGINAL IMAGE	CORRUPTED IMAGE	MATLAB FILTERED	FPGA FILTERED

Fig. 15. Collection of images filtered by software (Matlab) and by the FPGA device

3.3 Floor planning and design at layout level

The main structural component in the sorting network which is exposed in section 3.1, is a three-data comparator. As shown in Fig. 14, this element can be constituted by a set of interconnected one-bit CS cells. Three 8-bit word-length inputs described as: A, B and C can be identified. Also, three 8-bit CS blocks make possible to collect the median datum in the middle bus denoted by MED(A,B,C), and the corresponding minimum and maximum data into the external buses described as MIN(A,B,C) and MAX(A,B,C). In order to minimize the layout area, the CS modules have been rotated and placed in the position as illustrates the floorplanning and layout of Fig. 16.

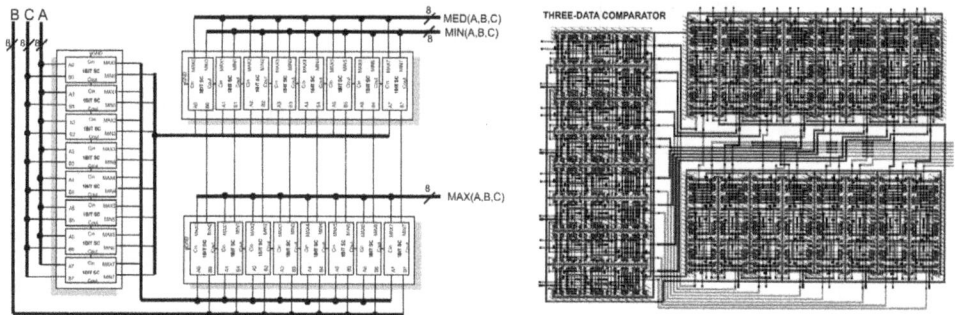

Fig. 16. Floorplanning (on left) and layout (on right) for the 8-bit three-data comparator included in the median filter

A graphical description, about the size and placement of the three-data modules that constitutes the nine-data sorting network is presented in Fig. 17. This floorplanning shows the connectivity between every module without showing internal layout details. It can be observed that some modules should be flipped to improve the routing and also achieving an area minimization. In this layout, two blocks can be recognized: the nine-data serial-in/parallel-out register and the nine-data sorting network. In accordance to the proposed median filter algorithm the unique signal of interest from the sorting network output is the median datum (D4) while the other data (D1,D2,D3,D5,D6,D7, and D8) are discarded.

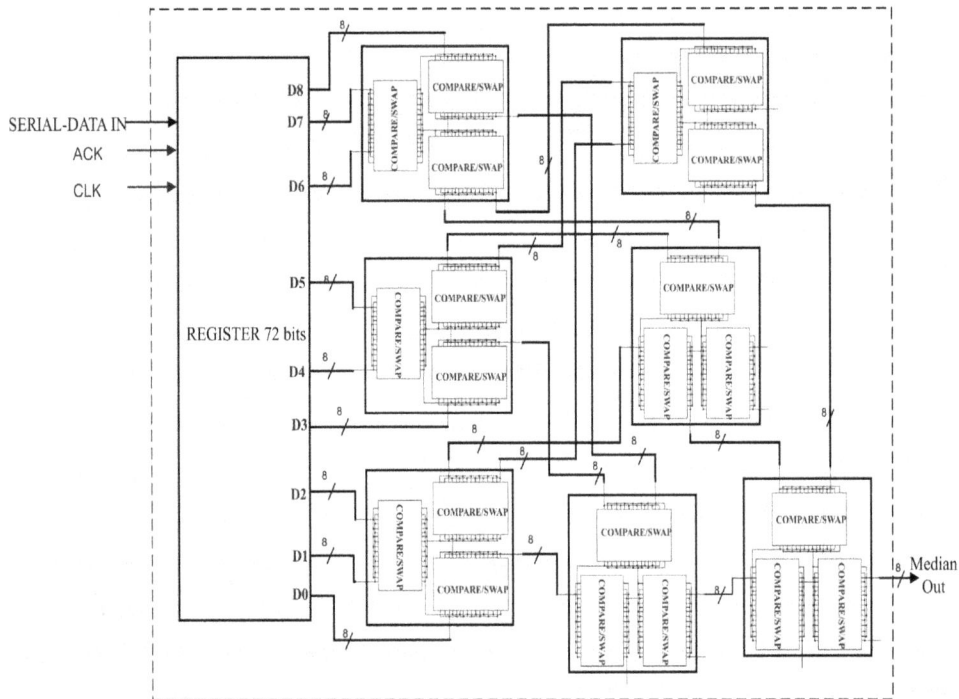

Fig. 17. Floorplanning for the nine-data sorting network in the median filter design

Figure 18 shows the translation to a layout level of the floorplanning design of fig. 17.

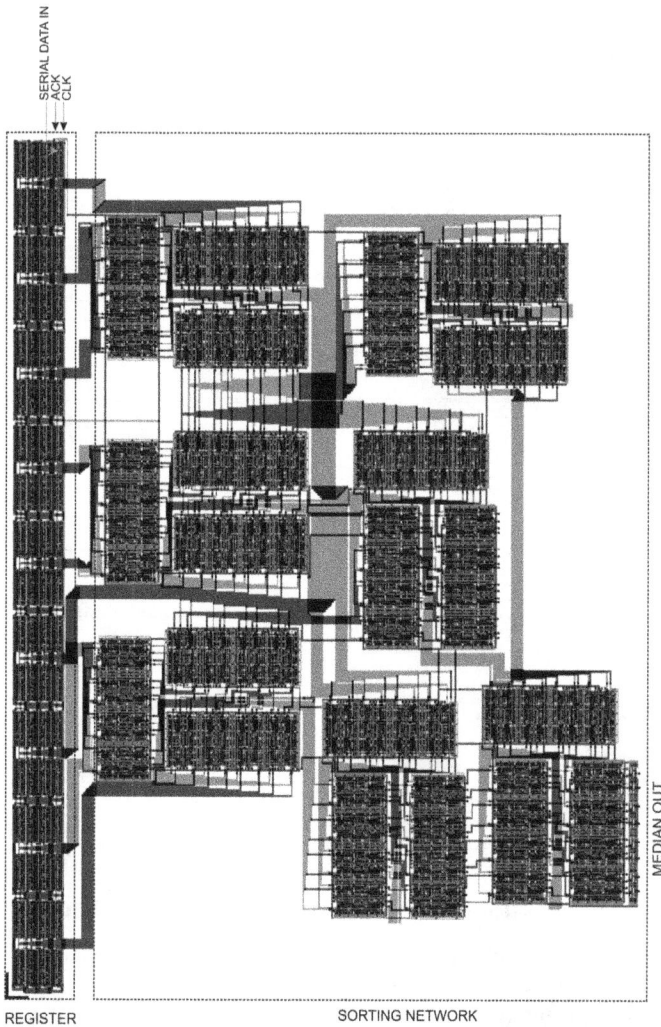

Fig. 18. Layout for the nine-data sorting network in the median filter design

3.4 Spice simulations for delay time estimation and electrical verification

The netlist including parasitic capacitances from the layout of Fig.18 is simulated to verify the circuit operation. The voltage waveforms for the median datum are shown in Fig. 19. The signals S0, S1, ..., and S7, correspond to the 8-bits median datum of the set of inputs {D0, D1, ..., D9} given by: D1=00101011 (212_{10}), D2=01000011(194_{10}), D3=00100010 (69_{10}), D4=00001101 (176_{10}), D5=00011100 (56_{10}), D6=00011110 (120_{10}), D7=00010111 (232_{10}), D8=00001111(240_{10}), and D9=00010101 (168_{10}).

In this simulation, the median for the input described in base-10 { 212_{10}, 194_{10}, 69_{10}, 176_{10}, 56_{10}, 120_{10}, 232_{10}, 240_{10}, 168_{10} } is the datum 176_{10} ,which in a binary base is represented as

D4=0000110110. It is important to clarify that the output median datum will be enabled until the acknowledge signal (ACK) is high and the CLK low.

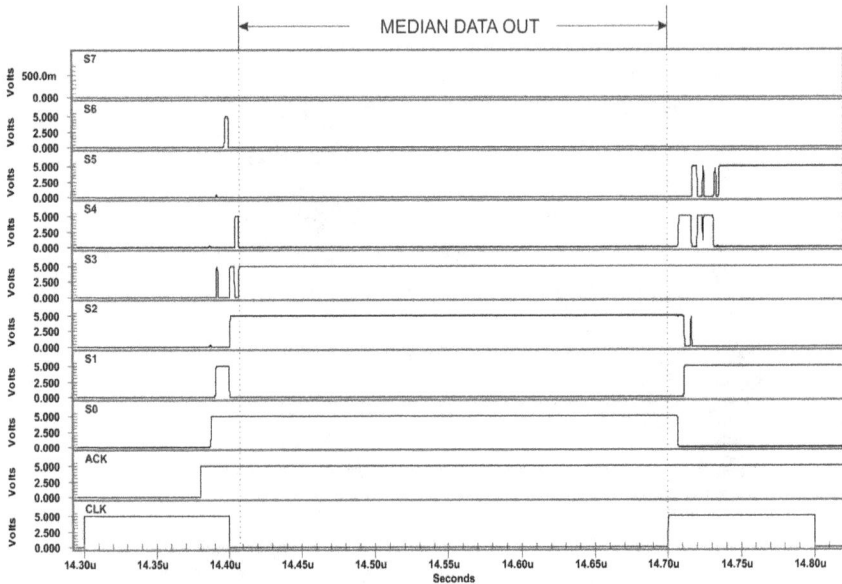

Fig. 19. Simulated output waveforms of the nine data sorting network

4. Piecewise linear function computation using sorting network

As a second example of the application of sorting networks in dedicated VLSI systems, an Application Specific Instruction Processor (ASIP) for piecewise linear function evaluation is described. Piecewise linear (PWL) functions allow the aproximation of multidimensional nonlinear functions or models in a convenient way to be evaluated with computing systems (Julian et al., 1999). The simplicity of the representation and evaluation methods, combined with the scalability in terms of the number of dimensions, impulsed the adoption of PWL functions as the modelling abstraction in a broad spectrum of systems. The first and most traditional area has been the computationally efficient resolution of nonlinear circuits (Chua & Ying, 1983) that require high performace function evaluation in terms of speed, and more recently in communication systems (Kaddoum et al., 2007) and power electronics (Pejovic & Maksimovic, 1995) for nonlinear operations involved in predistortion (Hammi et al., 2005). Motivated by these applications, the ASIP for piecewise linear function evaluation, hereafter denoted as PWLR6-µp, was designed and implemented in CMOS technology to provide a flexible environment for computation of 6-dimentional PWL functions and, due to the fact that PWL evaluation algorithm requires a sorting procedure, sorting networks have been embedded in this design as it will be exposed in this section.

4.1 The sorting networks for nonlinear function evaluation

Given an n-dimensional domain divided into a simplicial partition with a regular grid, a n-dimensional PWL function can be expressed as the weighted sum:

$$F(X) = \sum_{i=0}^{n} \mu_i c_i \tag{2}$$

where $X = \{x_1 ... x_n\}$ is the point in the n-dimensional domain where the function is evaluated, μ_i are scaling parameters depending on X, and c_i are the values of the function at simplex vertices, $i=0,...,n$. In order to compute the PWL equation c_i and μ_i are required. Usually the c_i parameters are stored in a RAM while μ_i need to be computed. For this operation, the algorihmic procedure defined in (Agustin et al., 2011) is followed what involves the descomposition of the x_i componentens into integer and fractional parts, sorting of the fractional parts and performing a sucesive subtraction of the sorted fractional parts.

4.2 Micro-architecture for the PWL evaluation through a sorting scheme

Micro-architecture of an ASIP strongly depends on its target application. A successful ASIP provides the required hardware to solve the target set of computational problems in an optimized way in terms of execution time, power, or chip resources, while maintaining the flexibility and programmability characteristics of a general purpose microprocessor. A trade-off exists among optimization levels and flexibility levels; thus, an ASIP can be considered as an intermediate point between a general purpose microprocessor and an Application Specific Integrated Circuit (ASIC). Three main architectural blocks of the PWLR6-µP, namely, Data Path, I/O, and Control, were designed taking into account the special operations required to perform the PWL calculation. The result was a nearly basic microprocessor with special features that accelerate the PWL computation. In this seccion, the sorting step of the algorithm and its relationship with the resources provieded by the PWLR6-up is addressed. For further details about the rest of the architecture and its special resources for PWL function evaluation, the reader is referred to (Agustin et al., 2011). Sorting constitutes the second part of the PWL function evaluation algorithm (as it was mentioned, it is required to evaluate the μ_i parameters). In this implementation, the 6-fractional parts are sorted following the comparison sequence of the so called Bose–Nelson sorting network (Knuth, 1997). However, in order to maintain the microprocessor structure and to avoid the area overhead of 12 CS blocks, only one comparator (the one provided by the ALU) was used. Consequently, the Bose-Nelson sorting network is embedded in the PWLR6-µp by combining an apropiate harware-software design.

The hardware resouces provided by the PWLR6-µp architecture are showed in Fig. 20 and they are used during the sorting step as follows: firstly the RF (register file), which is composed by six registers, stores the fractional parts to be sorted. Then, the two bidirectional ports, Port A, which is connected to Register A, and port B connected to Register B, transfer data between them and RF. After that, compare operation is performed from these registers and depending on the result, Register A and Register B values may be written back into the RF by switching sources and destinations.

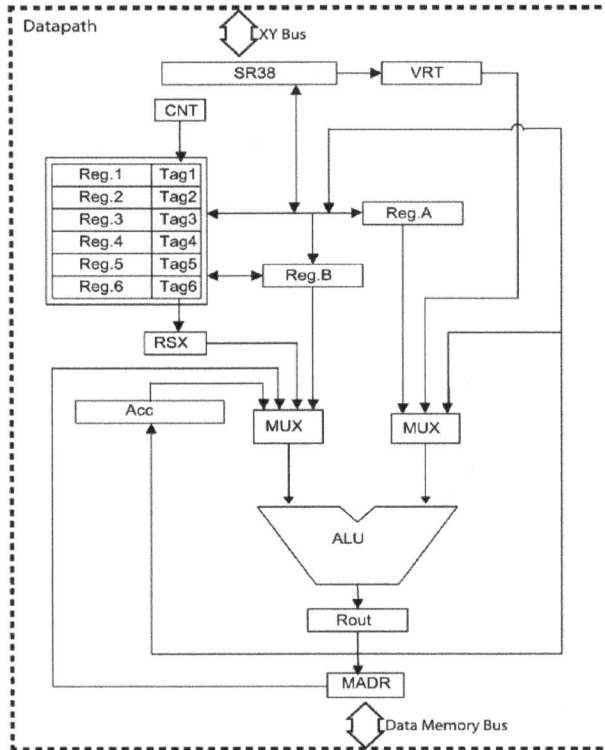

Fig. 20. Architecture for the six-data sorting network embedded in the PWL microprocessor

5. Conclusion

In this chapter, sorting networks have been addressed since a physical CMOS realization perspective with applicability to VLSI design. The CS circuit, analyzed at the beginning of this chapter, was introduced as the fundamental cell from which more complex sorting topologies could emerge. It must be pointed that because the speed in the CS design is limited by the delay of the n-bits carry out critical path, and by the transmission gates delay, a future research proposal for this work must be aimed to achieve higher overall frequencies. The two provided examples: the median filter architecture and the PWL evaluation scheme, allow to show the inclusion of sorting networks, into these specific applications. In these sense, about these examples the following particular conclusions must be observed: firstly, in the sorting network inmerse in the median filter, the main advantage consists in its regular structure beacuse although it is not optimal in the number of comparisons (21), the execution of several CS elements is done in parallel, and finally, the choice of an embeded sorting strategy in the PWL ASIP was due to the simplicity that allows the PWLR6-μP architecture (compared to other sorting algorithms like bubble sort or quick sort, designed to sort bigger datasets) and because of the small size of the input, this strategy is efficient in terms of hardware resources and code length.

6. Acknowledgment

This work has been partially supported by Universidad Veracruzana and by the CB-SEP-CONACyT Project No.102669 of Instituto Tecnológico Superior de Xalapa, México. Some partial research results from the PROMEP/103.5/09/4482 project of Universidad Veracruzana, Mexico, and PICT 2003 No. 13468 of Universidad Nacional del Sur, Argentina, have been referred in this chapter.

7. References

Agustin Rodriguez J., Lifschitz Omar D., Jimenez-Fernandez Victor M., Julian Pedro. (2011). "Application Specific Processor for Piecewise Linear Functions Computation". *IEEE Transactions on circuits and systems*, Vol. 58, No. 5, May 2011, pp. 971-981, 1459-8328

Batcher K. E.. (1962). "Sorting networks and their applications", *Proceedings of AFIPS Spring Joint Computer Conference*, ISBN: n.d., April 1962

Chua L. & R. Ying. (1983). "Canonical piecewise-linear analysis," *IEEE Transactions on Circuits Systems*, Vol. 30, No. 3, Mar 1983, pp. 125–140, 0098-4094

Faundez Zanuy M. (2001). *"Digital voice and image processing with multimedia application"*, First Edition, Alfaomega-Marcombo, 97-88-42671244-8, Barcelona, Spain

Hammi O., S. Boumaiza, M. Jaidane-Saidane, and F. Ghannouchi. (2005). "Digital subband filtering predistorter architecture for wireless transmitters," *IEEE Transactions Microwave Theory Tech.*, Vol. 53, No. 5, May 2005 , pp. 1643–1652, 0018-9480

Jimenez F. Victor, Martinez-Navarrete Denisse, Ventura-Arizmendi Carlos, Hernandez-Paxtian Zulma, Ramirez-Rodriguez Joel. (2011) *International Journal of Circuits, Systems and Signal Processing*, Vol. 5, No. 3, Apr. 2011, pp. 297-304, 1998-4464

Julian P., Desages A., and Agamennoni O.,(1999), "High-level canonical piecewise linear representation using a simplicial partition," *IEEE Transactions on Circuits and Systems-I, Fundam. Theory Appl.*, Vol. 46, No. 4, Apr. 1999 , pp. 463–480, 1057-7122

Kaddoum, G.; Roviras, D.; Charge, P.; Fournier-Prunaret, D. (2007). "Analytical calculation of BER in communication systems using a piecewise linear chaotic map". *Circuit Theory and Design, 2007. ECCTD 2007. 18th European Conference on* . 978-1-4244-1341-6 ,Toulouse, France, Aug. 2007

Knuth E. Donald. (1997).*"The Art of Computer Programming,"* Third Edition. Addison-Wesley, 1997. 0-201-89685-0, Massachussets, USA

Kang S. M. and Yusuf Leblebici, (2003). *"CMOS Digital Integrated Circuits,"* Third Edition, Mc Graw Hill, 0-07-246053-9, New York, N.Y.

Pejovic, P. Maksimovic, D. (2002). A new algorithm for simulation of power electronic systems using piecewise-linear device models, *IEEE Transactions on Power Electronics*, Vol.10 , No. 3, May 1995, pp. 340-348, 0885-8993

Pursley Bryan, (2008), Sorting Networks, *n.d*, Aug 09,2011,
URL: http:// brianpursley.com/Files/CSC204_FinalProject_BrianPursley.pdf

Rabaey J. M., Chandrakasan A. P., and Nikolic B. (2003). *"Digital Integrated Circuits: A Design Perspective"*, Second Edition., Prentice Hall Electronics and VLSI Series, Upper Saddle River, NJ: Prentice Hall/Pearson Education, 0-13-0909960-3,n.d.

Turan Demirci, Ilhan Hatirnaz, Yusuf Leblebici. (2003)."Full-custom CMOS realization of a high-performance binary sorting engine with linear area-time complexity". *"In Proceedings of ISCAS (5)'2003"*. ISBN: 0-7803-7761-3, Bangkok, Thailand, May 2003

Vega M., Sanchez J., Gomez J. (2002). "An FPGA-based implementation for median filter meeting the real-time requierements of automated visual inspection systems", *Proceedings of the 10th Mediterranean Conference on Control and Automation-MED2002*, 972-9027-03-X, Lisbon, Portugal, July 2002.

Weste N. H.E., and Eshraghian K., (1993),*"Principles of CMOS VLSI desing"*, Second Edition, Addison-Wesley, 0-201-53376-6

Part 2

Modeling, Simulation and Optimization

6

Algorithms for CAD Tools VLSI Design

K.A. Sumithra Devi

R.V. College of Engineering, Bangalore
Vishweshvaraya Technological University, Karnataka
India

1. Introduction

Due to advent of Very Large Scale Integration (VLSI), mainly due to rapid advances in integration technologies the electronics industry has achieved a phenomenal growth over the last two decades. Various applications of VLSI circuits in high-performance computing, telecommunications, and consumer electronics has been expanding progressively, and at a very hasty pace. Steady advances in semi-conductor technology and in the integration level of Integrated circuits (ICs) have enhanced many features, increased the performance, improved reliability of electronic equipment, and at the same time reduce the cost, power consumption and the system size. With the increase in the size and the complexity of the digital system, Computer Aided Design (CAD) tools are introduced into the hardware design process. The early paper and pencil design methods have given way to sophisticated design entry, verification and automatic hardware generation tools. The use of interactive and automatic design tools significantly increased the designer productivity with an efficient management of the design project and by automatically performing a huge amount of time extensive tasks. The designer heavily relies on software tools for every aspect of development cycle starting from circuit specification and design entry to the performance analysis, layout generation and verification. Partitioning is a method which is widely used for solving large complex problems. The partitioning methodology proved to be very useful in solving the VLSI design automation problems occurring in every stage of the IC design process. But the size and the complexity of the VLSI design has increased over time, hence some of the problems can be solved using partitioning techniques. Graphs and hyper-graphs are the natural representation of the circuits, so many problems in VLSI design can be solved effectively either by graph or hyper-graph partitioning. VLSI circuit partitioning is a vital part of the physical design stage. The essence of the circuit partitioning is to divide a circuit into number of sub-circuits with minimum interconnection between them. Which can be accomplished recursively partitioning the circuits into two parts until the desired level of complexity is reached. Partitioning is a critical area of VLSI CAD. In order to build complex digital logic circuits it is often essential to sub-divide multi –million transistor design into manageable pieces. The presence of hierarchy gives rise to natural clusters of cells. Most of the widely used algorithms tend to ignore this clustering and divide the net list in a balanced partitioning and frequently the resulting partitions are not optimal.

The demand for high-speed field-programmable gate array (FPGA) compilation tools has escalated in the deep-sub micron era. Tree partitioning problem is a special case of graph partitioning. A general graph partitioning though fast, is inefficient while partitioning a tree structure. An algorithm for tree partitioning that can handle large trees with less memory/run time requirement will be a modification of Luke's algorithm. Dynamic program mining based tree partition, which works well for small trees, but because of its high memory and run time complexity, it cannot be used for large trees. In order to optimize above mentioned issues this chapter concentrates on different methodologies starting with Memetic Approach in comparison with genetic concept, Neuro-Memetic approach in comparison with Memetic approach, then deviated the chapter to Neuro EM model with clustering concept. After that the topic concentration is on Fuzzy ARTMAP DBSCAN technique and finally there is a section on Data mining concept using two novel Clustering algorithms achieving the optimality of the partition algorithm in minimizing the number of inter-connections between the cells, which is the required criteria of the partitioning technique in VLSI circuit design. Memetic algorithm (MA) is population based heuristic search approach for combinatorial optimization problems based on cultural evolution. They are designed to search in the space of locally optimal solutions instead of searching in the space of all candidate solutions. This is achieved by applying local search after each of the genetic operators. Crossover and mutation operators are applied to randomly chosen individuals for a predefined number of times. To maintain local optimality, the local search procedure is applied to the newly created individuals.

Neuro-memetic model makes it possible to predict the sub-circuit from circuit with minimum interconnections between them. The system consists of three parts, each dealing with data extraction, learning stage and result stage. In data extraction, a circuit is bipartite and chromosomes are represented for each sub circuit. Extracted sequences are fed to Neuro-memetic model that would recognize sub-circuits with lowest amount of interconnections between them.

Next method focuses on the use of clustering k-means (J. B. MacQueen, 1967) and Expectation-Maximization (EM) methodology (Kaban & Girolami, 2000), which divides the circuit into a number of sub-circuits with minimum interconnections between them, and partition it into 10 clusters, by using k-means and EM methodology. In recognition stage the parameters, centroid and probability are fed into generalized delta rule algorithm separately.

Further, a new model for partitioning a circuit is explored using DBSCAN and fuzzy ARTMAP neural network. The first step is concerned with feature extraction, where it uses DBSCAN algorithm. The second step is classification and is composed of a fuzzy ARTMAP neural network.

Finally, two clustering algorithms Nearest Neighbor (NNA) and Partitioning Around Medoids (PAM) clustering algorithms are considered for dividing the circuits into sub circuits. Clustering is alternatively referred to as unsupervised learning segmentation. The clusters are formed by finding the similarities between data according to characteristics found in the actual data. NNA is a serial algorithm in which the items are iteratively merged into the existing clusters that are closest. PAM represents a cluster by a medoid.

2. Circuit partitioning concept

VLSI circuit partitioning is a vital part of physical design stage. The essence of circuit partitioning is to divide the circuit into a number of sub-circuits with minimum interconnections between them. This can be accomplished by recursively partitioning a circuit into two parts until we reach desired level of complexity. Thus two way partitioning is basic problem in circuit partitioning, which can be described as (Dutt& Deng, 1996).

Logic netlist can be represented as a hypergraph H (V,Eh) where

- Each node $v \in V$ in hypergraph represents a logic cell of the netlist, and
- Each hyperedge $e \in$ Eh represents a net connecting various logic cells
- Various weights representing different attributes are attached to all nodes and edges of the hypergraph H, these are: On nodes: area estimates, On edges: length (after global placement)

The problem is:

- To partition the set V of all nodes $v \in V$ into a set of disjoint subsets, of V, such that each node v is present in exactly one of these subsets. These subsets are referred to as blocks of the partition.
- The partition on V induces a cut of the set of all hyper edges, that is, Eh. A cut is subset of Eh, such that for every hyper edge h present in the cut there are at least two nodes adjacent to h, which belong to separate blocks of the partition.
- The objective function of partitioning approach has to address the following issues:
- It should be able to handle multi-million node graphs in a reasonable amount of computation time
- It should attempt to balance the area attribute of all the blocks of the partition with the additional constraint that there is an area penalty associated with every hyperedge that get cut.
- It should try to minimize interconnections between different clusters so as to satisfy the technological limit on the maximum number of interconnects allowed.

3. Memetic approach in VLSI circuit partitioning

A new approach Memetic Algorithm is described in this section to solve problem of circuit partitioning pertaining to VLSI.

3.1 A model to solve circuit partitioning

The circuit partitioning problem can be formally represented in graph theoretic notation as a weighted graph, with the components represented as nodes, and the wires connecting them as edges, the weights of the node represent the sizes of the corresponding components, and the weights of the edges represent the number of wires connecting the components. In its general form, the partitioning problem consists of dividing the nodes of the graph into two or more disjoint subsets such that the sum of weights of the nodes in each subset does not exceed a given capacity, and the sum of weights of edges connecting nodes in different subsets is minimized. But generally the circuits are represented as bipartite graphs consisting of two sets of nodes, the cells and the nets/ Edges connect each cell to several nets, and each net to several cells as shown in Fig1.Let G= (M, N, E), mi $\in$ is a cell, ni$\in$N is a net, and eij=(mi,nj) $\in$E is an edge which represents that mi and nj are connected electrically. For any nj for all I for which eij exists, we say that the cells mi are connected by net nj.

Conversely, for any mi for all j for which eij exists, we say that the nets nj are connected of cell mi. Each cell mi has an area ai, and each net nj has a cost cj. The edges of the bipartite graph are un weighted. In this case, the partitioning problem is to divide the set of cells into disjoint subsets, M1, M2,.......Mk, such that the sum of cell areas in each subset Mi is less than a given capacity Ai, and the sum of costs of nets connected to cells in different subsets is minimized. That is,

$$\bigcup_{n=1}^{k} M \quad n = M, \tag{1}$$

∀Mn, Σ∀miεMnai ≤ An

and ∀ nj, if nj is connected to cells in p different partitions, then,

$$C = \Sigma(p-1)cj \text{ is minimized} \tag{2}$$

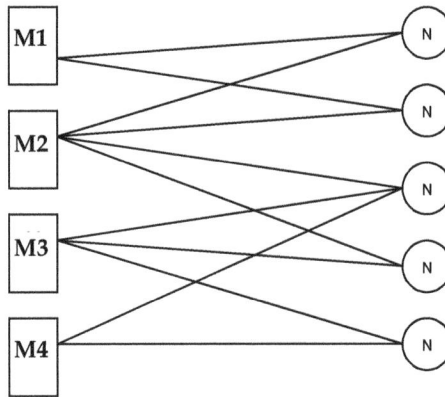

Fig. 1. Bipartite graph model for partitioning

3.2 Memetic algorithms applied to circuit partitioning

i.   Chromosome Representation

1 bit in the chromosome represents each cell, the value of which determines the partition in which the cell is assigned (Krasnogor & Smith, 2008). The chromosome is sorted as an array of 32 bit packed binary words. The net list is traversed in a breadth-first search order, and the cells are assigned to the chromosome in this order. Thus, if two cells are directly connected to each other, there is a high probability that their partition bits are close to each other in the chromosome. An example is the breadth-first search sequence and the corresponding chromosome as shown in Fig. 2.

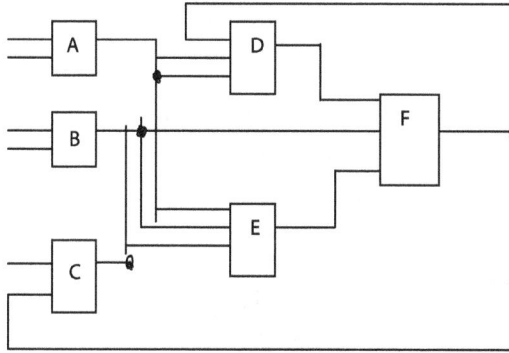

Cell	Connected To	Bipartition
A	D,E	1
B	E,F	1
C	D,E,F	1
D	A,C,F	0
E	A,B,C,F	0
F	B,C,D,E	0

BFS Sequence   A D E C F B

Bipartitioning Chromosome | 1 0 0 1 0 1 |

Fig. 2. Breadth-first search sequence and the corresponding chromosome

ii.   Fitness Scaling

Fitness scaling is used to scale the raw fitness values of the chromosomes so that the GA sees a reasonable amount of difference in the scaled fitness values of the best versus the worst individuals.

The following fitness algorithm applies to evaluation functions that determine the cost, rather than the fitness, of each individual (Univesity of New Mexico, 1995). From this cost, the fitness of each individual is determined by scaling as follows.

A referenced worst cost is determined by

$$Cw = C + S\sigma \tag{3}$$

Where C is the average cost of the population, S is the user defined sigma-scaling factor, and σ is the standard deviation of the cost of the population. In case Cw is less than the real worst cost in the population, they only the individuals with cost lower than Cw are allowed to participate in the crossover.

Then, the fitness of each individual is determined by

$$\begin{cases} \text{Cw-C} & \text{if Cw>0} \\ \text{F=0} & \text{otherwise} \end{cases} \qquad (4)$$

This scales the fitness such that, if the cost is ±k standard deviations from the population average, the fitness is

$$F = (S \pm K)\sigma \qquad (5)$$

This means that may individuals worse than S standard deviation from the population mean (k=s) are not selected at all. If S is small, the ratio of the lowest to the highest fitness in the population increases, and then the algorithm becomes more selective in choosing parents. On the other hand, if S is large, then Cw is large, and the fitness values of the members of the population are relatively close to each other. This causes the difference in selection probabilities to decrease and the algorithm to be less selective in choosing parents.

iii. Evaluation

The cut cost is calculated as the number of nets cuts. If the net is present in both partitions, or if the net is present in the partition opposite to its I/O pad, then it is said to have a cut (Merz & Freisleben, 2000).

Counting number of 1's in the chromosome does partition imbalance evaluation.   A quadratic penalty has been used for imbalance, so that large imbalance is penalized more than a small imbalance. The user specifies the relative weights for cut and imbalance Wc and Wb.

Thus the total cost is:

$$C = Wc(\text{cut}) + Wb(\text{imbalance})2 \qquad (6)$$

iv. Incorporation and Duplicate Check

The two new offspring formed in each generation are incorporated into the population only if they are better than the worst individuals of the existing population. Before entering a new offspring into the population, it is checked against all other members of the population having the same cost, in order to see whether it is duplicate. Duplicates can result due to the same crossover operation (T. Jones, 1995).

Duplicates have two disadvantages:

- First they occupy storage space that could otherwise be used to store a population with more diverse feature.
- Second whenever crossover occurs between two duplicates, the offspring is identical to the parents, regardless of the cut point, and this tends to fill the population with even more duplicates.

v. Mutation

After crossover and incorporation, mutation is performed on each bit of the population with a very small probability Pm. We go through the entire population once (Krasnogor et al., 1998a). For each mutation the location in bits is determined from previous location and a random number as follows,

$$Next = Prev + \left[\frac{Log(rand[0,1])}{Log(1.0\text{-}PM)} \right] \tag{7}$$

Where PM is the mutation probability.

Each mutation is evaluated and accepted separately, and this process is continued until end of population is reached. The mutated version replaces the unmutated version of the same individual in the population. The acceptance of mutation operation has some probabilistic characteristics similar to simulated annealing. If the change in the cost C is negative, signifying that the fitness has increased, the mutation is always accepted, as in simulated annealing. If change in the cost is positive, then mutations are accepted probabilistically.

4. Evolutionary time series model for partitioning using Neuro-Memetic approach

An evolutionary time-series model for partitioning a circuit is discussed using Neuro Memetic algorithm owing to its local search capability.

Sample Data Set

A sample example and the corresponding chromosome representation is shown in Fig 3 and Fig 4.

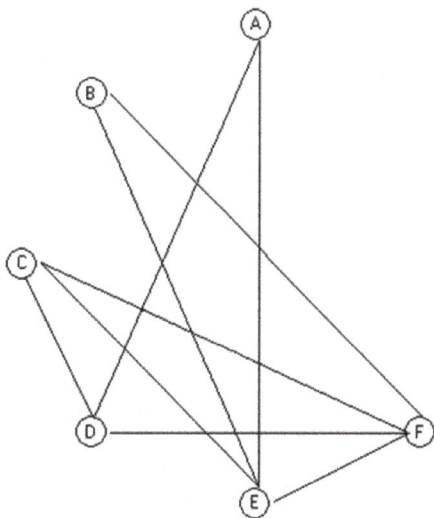

Fig. 3. Sample Circuit

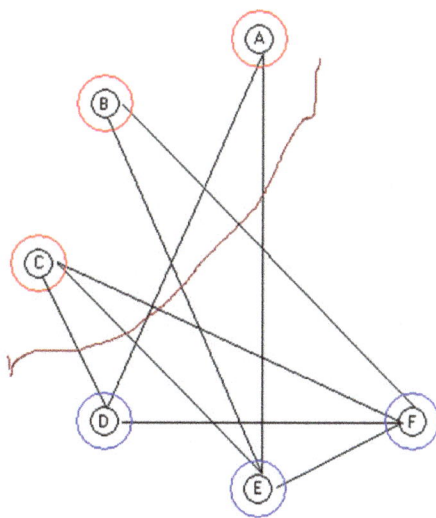

Fig. 4. Bipartition Circuit

Sub circuit 1   A, B, C   total edges = 7

Sub circuit 2   D, E, F   total edges =10

Cell	No of edges	Bipartition
A	2	1
B	2	1
C	3	1
D	3	0
E	3	0
F	4	0

Chromosomes representation: Sub circuit1 (A, B, C) 0010 0010 0011, Sub circuit2 (D, E, F) 0011 0011 0100

- Neuro-Memetic Model: Neuro-memetic model makes it possible to predict the sub circuit from circuit with minimum interconnections between them.
- Training Procedure: The purpose of the training process is to adjust the input and output parameters of the NN (Neural Network) model, so that the MAPE (Mean Absolute Percentage Error) measure is minimized. Training of the feed-forward neural network models is usually performed using back propagation learning algorithms. Most often, the error surface becomes trapped to local minima, usually not meeting the desired convergence criterion. The termination at a local minimum is a serious problem while the neural network is learning. In other words, such a neural network is not completely trained (Oxford Univ Press, 1995). Another issue where care must be taken is "the receptiveness to over-fitting". But, memetic algorithms offer competent search method for intricate (that is, possessing many local optima) spaces to find nearly local optima. Thus, its ability to find a better suboptimal solution or have a higher probability to obtain the local optimal solution makes it one of the preferred candidates to solve the learning problem.
- Training with MA: The parameters of the neural network are tuned by a memetic algorithm (Krasnogor et al., 1998b) with arithmetic crossover and non uniform mutation. A population (P) with 200 genotypes is considered. They are randomly initialized, with maximum number of iterations fixed at 200 and MA is run for 100 generations with the same population size. The best model was found after 63 generations. In this method, the probability of crossover is 0.6 and the probability of mutation is 0.2. These probabilities are chosen by trial and error through experiments for good performance. The new population thus generated replaces the current population. The above procedures are repeated until a certain termination condition is satisfied. The number of the iterations required to train the MA-based neural network is 2000. The range of the fitness function of neural network is (0, 1).
- Evaluate individuals using the fitness function: The objective of the fitness function is to minimize the prediction error. In order to prevent over-fitting and to give more exploration to the system, the fitness evaluation framework is changed and use the weight imbalance to calculate the fitness of a chromosome. The fitness of a chromosome for the normal class is evaluated as shown   in the example   below.

Take the testing samples

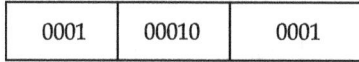

0001	00010	0001

Now take the sub circuit 1 with data set (d1)

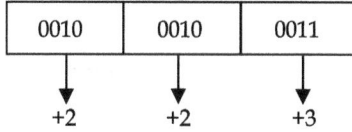

0010	0010	0011

+2   +2   +3

For sub circuit 2 data set d2

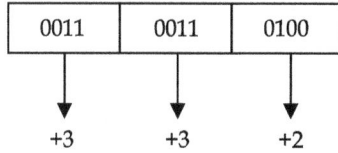

0011	0011	0100

+3   +3   +2

Calculate the sum of (+) credit & (-) debit for each sample data d1 & d2

For d1=+2+2+3=7

d2=+3+3+2=8 so it is found that **sample fitness of data d1** is best sample.

4.1 Design of the system to recognize sub circuit with minimum interconnection

The present task involves the development of Neural Network, which can train to recognize sub circuit with minimum interconnection between them, from a large circuit given. Following are the steps involved in design of the system

1. Create a input data file which consists of training pairs.
2. In data extraction, a circuit is bipartite and chromosomes are represented for each sub circuit.
3. Design the neural network based upon the requirement and availability.
4. Simulate the software for network.
5. Initialize count=0, fitness=0, number of cycles.
6. Generation of Initial Population. The chromosome of an individual is formulated as a sequence of consecutive genes, each one coding an input parameter.
7. Initialize the weight for network. Each weight should be set to a random value between –0.1 to 1.
8. Calculates activation of hidden nodes.

$$xJh = \frac{1}{1 + e - \left(\sum wjkh\right) * xpn} \tag{8}$$

9. Calculate the output from output layers

$$xio = \frac{1}{1 + e - \left(\sum wijo\right) * xjh} \tag{9}$$

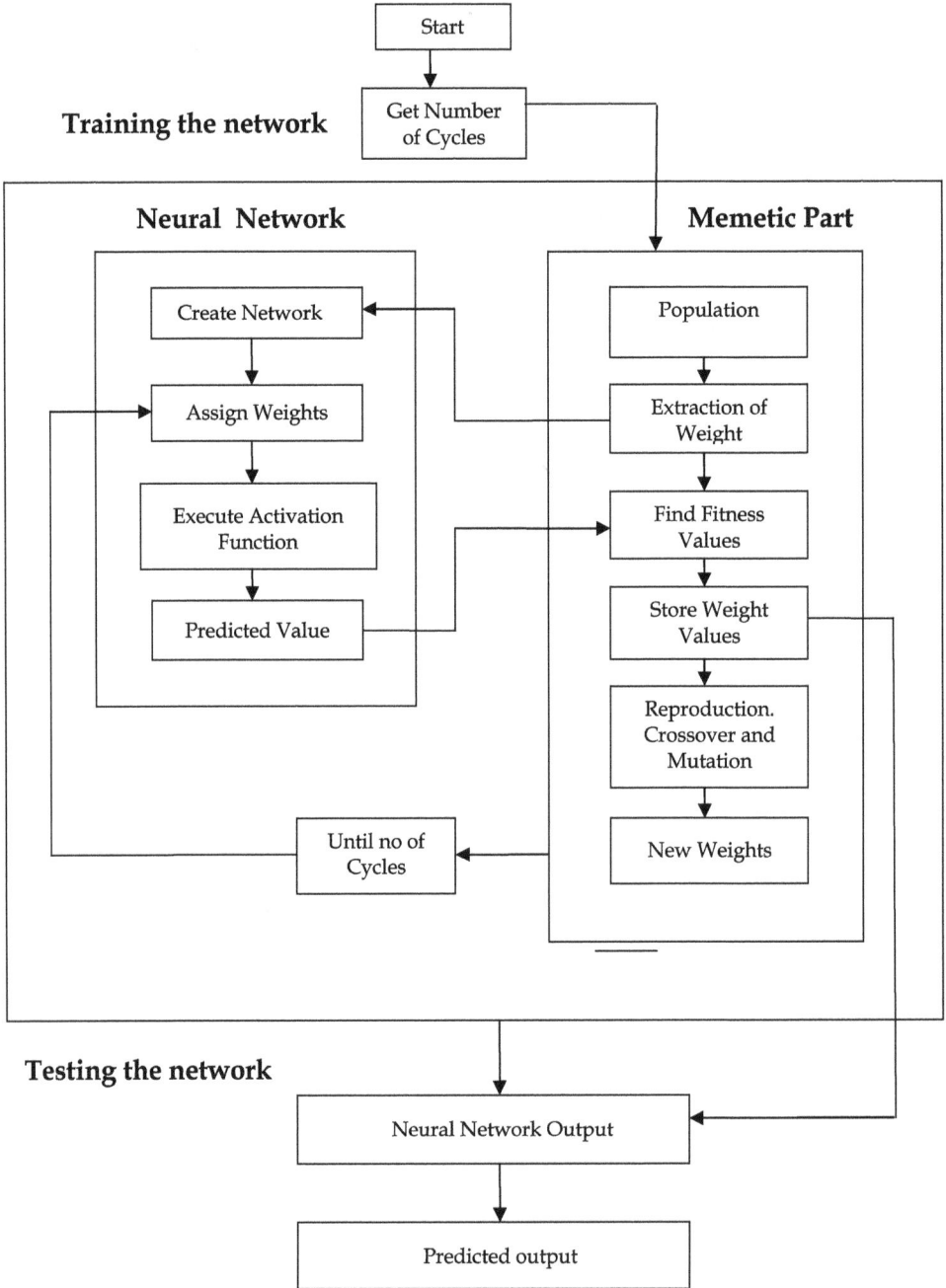

Fig. 5. Recognize Sub Circuit with Minimum Interconnection

10. Compares the actual output with the desired outputs and find a measure of error. The genotypes are evaluated on the basis of the fitness function.
11. If (previous fitness < current fitness value) then store current weights.
12. Count = Count + 1
13. Selection: Two parents are selected by using the Roulette wheel mechanism.
14. Genetic Operations: Crossover, Mutation and Reproduction to generate new weights (Apply new weights to each link).
15. If (number of cycles> count) Go to Step 7
16. training set is reduced to an acceptable value.
17. Verify the capability of neural network in recognition of sub circuit with minimum interconnection between them.
18. End.

- Development of Neural Network: In the context of recognition of sub circuit with minimum interconnection, the 3-layer neural network is employed to learn the input-output relationship using the MA. The layers of input neuron are responsible for inputting. The number of neurons in this output layer is determined by the size of set of desired output, with each possible output being represented by separate neuron. Neural network contains 12 input nodes, 20 neurons in the first hidden layer, 14 neurons in the second hidden layer and the output layer has 2 neurons. It results in a 12-14-2 Back propagation neural network. Sigmoid function is used as the activation function. Memetic Algorithm is employed for learning (Holstein & Moscato, 1999). For the back-propagation with momentum and adaptive learning rate, the learning rate is 0.2, the momentum constant is 0.9. During the training process the performance of 0.00156323 was obtained at 2000 epochs.

5. Neuro–EM and neuro-k mean clustering approach for VLSI design partitioning

This section is focused in use of clustering methods k-means (J. B. MacQueen, 1967) and Expectation-Maximization (EM) methodology (Kaban & Girolami, 2000).

5.1 Neuro-EM model

The system consists of three parts each dealing with data extraction, Learning stage and recognition stage. In data extraction, a circuit is bipartite and partitions it into 10 clusters, a user-defined value, by using K-means (J. B. MacQueen, 1967) and EM methodology (Kaban & Girolami, 2000), respectively. In recognition stage the parameters, that is, centroid and probability are fed into generalized delta rule algorithm separately and train the network to recognize sub-circuits with lowest amount of interconnections between them. Block diagram of model to recognize sub-circuits with lowest amount of interconnections between them using two techniques K-means and EM methodology with neural network are shown in Fig.6 and Fig.7.

In recognition stage the parameters, that is, centroid and probability are fed into generalized delta rule algorithm separately and train the network to recognize sub circuit with minimum interconnection between them. Block diagram of model for Partitioning a Circuit are depicted in Fig. 8.

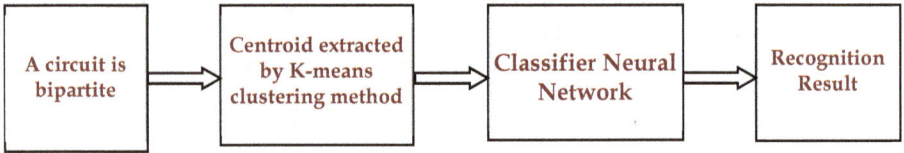

Fig. 6. Block diagram of K-means with neural network

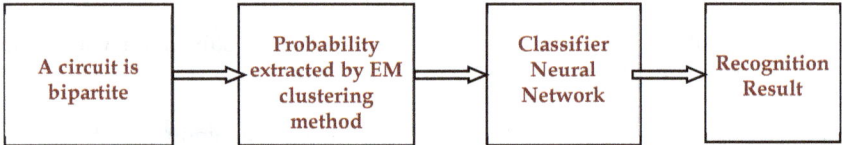

Fig. 7. Block diagram of EM methodology with neural network

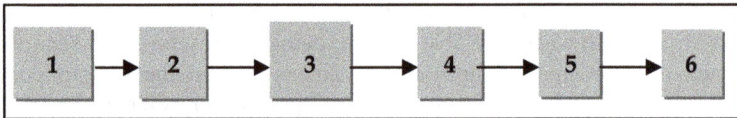

1.   Circuit is bipartite and data represented
2.   Applying K-means
3.   Applying EM methodology
4.   centroid and probability
5.   Neural network
6.   Recognition Result

Fig. 8. Block Diagram of Model for Partitioning a Circuit

5.2 Sample data set

A sample example representation is shown in Fig.9 and Fig 10

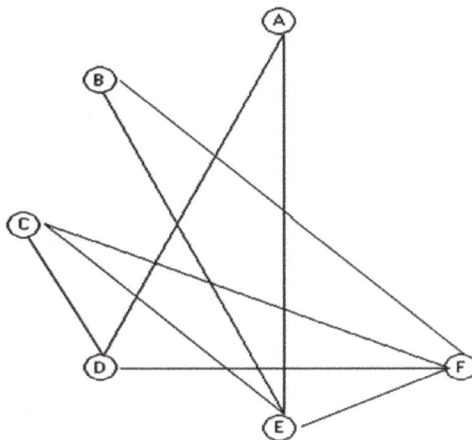

Fig. 9. Sample Circuit

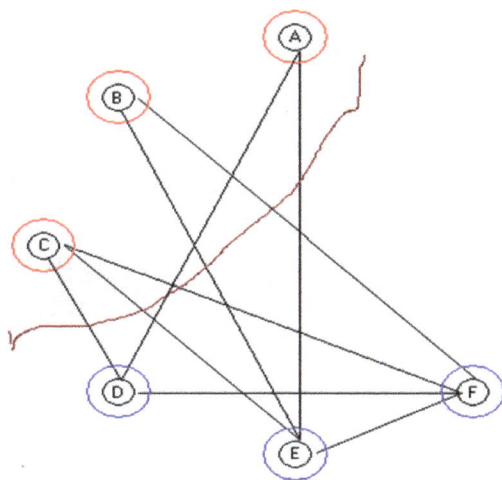

Fig. 10. Bipartition Circuit
Sub circuit 1   A, B, C   total edges = 7,   Sub circuit 2   D, E, F   total edges =10

Cell	No of edges	Bipartition
A	2	1
B	2	1
C	3	1
D	3	0
E	3	0
F	4	0

Table 1. Bipartation Matrix
Data representation: Sub circuit1 (A, B, C) 0010 0010 0011,
Sub circuit2 (D, E, F) 0011 0011 0100

5.3 Expectation Maximization algorithms

The EM algorithm was explained and given its name in a classic 1977 paper by Arthur Dempster, Nan Laird, and Donald Rubin in the Journal of the Royal Statistical Society (Arthur et al.,1997). They pointed out that method had been "proposed many times in special circumstances" by other authors, but the 1977 paper generalized the method and developed the theory behind it.

The EM algorithm for clustering is described in detail in Witten and Frank (2001) (Witten & Frank, 2005). The Expectation-Maximization (EM) algorithm is part of the Weka clustering package. EM is a statistical model that makes use of the finite Gaussian mixtures model. The basic approach and logic of this clustering method is as follows. Suppose   a single continuous variable in a large sample of observations is measured. Further, suppose that the sample consists of two clusters of observations with different means (and perhaps different standard deviations) within each sample, the distribution of values for the continuous variable follows the normal distribution. The resulting distribution of values (in the population) may look as shown in Fig.11.

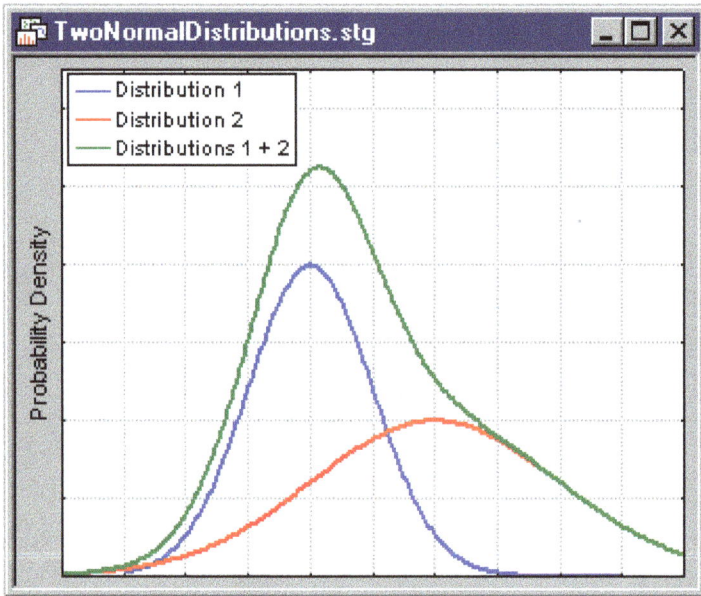

Fig. 11. Two normal distributions of EM Algorithm (Screen Shot)

i. Mixtures of distributions. The illustration in Fig 5.12 shows two normal distributions with different means and different standard deviations and the sum of the two distributions. Only the mixture (sum) of the two normal distributions (with different means and standard deviations) would be observed. The goal of EM clustering is to estimate the means and standard deviations for each cluster so as to maximize the likelihood of the observed data (distribution). Put another way, the EM algorithm attempts to approximate the observed distributions of values based on mixtures of different distributions in different clusters.
    With the implementation of the EM algorithm in some computer programs, one may be able to select (for continuous variables) different distributions such as the normal, log-normal, and Poisson distributions (Karlis, 2003) and can select different distributions for different variables, thus derive clusters for mixtures of different types of distributions.

ii. Categorical variables. The EM algorithm can also accommodate categorical variables. The method will at first randomly assign different probabilities (weights, to be precise) to each class or category, for each cluster. In successive iterations, these probabilities are refined (adjusted) to maximize the likelihood of the data given the specified number of clusters (Kim, 2002).

iii. Classification probabilities instead of classifications. The results of EM clustering are different from those computed by k-means clustering. The latter will assign observations to clusters to maximize the distances between clusters. The EM algorithm does not compute actual assignments of observations to clusters, but classification probabilities. In other words, each observation belongs to each cluster with a certain probability. Of course, as a final result one can usually review an actual assignment of observations to clusters, based on the (largest) classification probability (Gyllenberg et al., 2000).

The algorithm is similar to the K-means procedure in that a set of parameters are recomputed until a desired convergence value is achieved. The parameters are recomputed until a desired convergence value is achieved. The finite mixtures model assumes all attributes to be independent random variables.

A mixture is a set of N probability distributions where each distribution represents a cluster. An individual instance is assigned a probability that it would have a certain set of attribute values given it was a member of a specific cluster. In the simplest case N=2 the probability distributes are assumed to be normal and data instances consist of a single real-valued attribute. Using the scenario, the job of the algorithm is to determine the value of five parameters specifically,

1. The mean and standard deviation for cluster 1
2. The mean and standard deviation for cluster 2
3. The sampling probability P for cluster 1 (the probability for cluster 2 is 1-P)

the general procedure is given below,

1. Guess initial values for the five parameters.
2. Use the probability density function for a normal distribution to compute the cluster probability for each instance. In the case of a single independent variable with mean μ and standard deviation σ, the formula is:

$$f(x) = \frac{1}{(\sqrt{2\pi}\sigma)e^{\frac{-(?-\mu)^2}{2\sigma^2}}} \tag{10}$$

In the two-cluster case, there are two probability distribution formulae each having differing mean and standard deviation values.

1. Use the probability scores to re-estimate the five parameters.
2. Return to Step 2

The algorithm terminates when a formula that measures cluster quality no longer shows significant increases. One measure of cluster quality is the likelihood that the data came from the dataset determined by the clustering. The likelihood computation is simply the multiplication of the sum of the probabilities for each of the instances. With two clusters A and B containing instances x1, x2, ... xn where PA = PB = 0.5 the computation is:

$$[.5P(x_1 \mid A) + .5P(x_1 \mid B)][.5P(x_2 \mid A) + .5P(x_2 \mid B)]...[.5P(x_n \mid A) + .5P(x_n \mid B)] \tag{11}$$

Algorithm is similar to K-mean procedure, in that sets of parameters are re-computed until desired convergence value is achieved. General procedure is

- Initialize parameters.
- Use the probability density function for normal distribution to compute cluster probability for each instance. For example in the case of two-cluster one will have the two probability distribution formulae each having different mean and standard deviation values.
- Use the probability scores to re-estimate the parameter.
- Return to step 2.
- The algorithm terminates when formula that measure cluster quality exists no longer.

The tool shed output of this algorithm would be the probability for each cluster. EM assigns a probability distribution to each instance, which indicates the probability of it belonging to each of the clusters.

In the context of recognizing the sub circuit from circuit with minimum interconnections between them, artificial neurons is structured into three normal types of layers input, hidden and output which can create artificial neural networks. The layers of input neuron are responsible for inputting a feature vectors that is, centroid and probability, which are extracted from K-means and EM algorithms respectively. The number of neurons in this output layer is determined by size of set of desired output, with each possible output being represented by separate neuron. Between these two layers there can be many hidden layers. These internal layers contain many of the neuron in various interconnected structures.

5.4 Design of the system to recognize sub circuit with minimum interconnections

The present task involves the development of neural network, which can train to recognize sub circuit with minimum interconnection between them from large circuit given.

Following are the steps involved in design of the system,

1. Create a input data file which consists of training pairs.
2. In data extraction, a circuit is bipartite and data are represented for each sub circuit.
3. Centroid and probability features are extracted from K-means and EM algorithms
4. Design the neural network based upon the requirement and availability.
5. Simulate the software for network.
6. Train the network using input data files until error falls below the tolerance level.
7. Verify the capability of neural network in recognition of test data

Algorithm:

The learning algorithm of back propagation network is given by "generalized delta rule".

Step 1. The algorithm takes input vector (features) to the back propagation network.

Step 2. let K be number of nodes in the layer determined by length of training vectors that is number of feature N. Let j be number of nodes in hidden layer. Let I be number of nodes in output layer. Denote activation of hidden layer as xjh and in output layer is xio. Weight connecting input layer and hidden layer are wjkh and weight connecting hidden layer and output layer is wijo.

Step 3. Initialize the weight for network. Each weight should be set to a random value between –0.1 to 1.

Step 4. Calculates activation of hidden nodes

$$xJh=g(\sum wjkh * xpn)=\frac{1}{1 + e - (\sum wjkh * xpn)} \qquad (12)$$

Step 5. Calculate the output from output layers

$$xio=g(\sum wjko * xjh)=\frac{1}{1 + e - (\sum wijo * xjh)} \qquad (13)$$

Step 6. Compares the actual output with desired outputs and finds a measure of error.

Step 7. After comparison it finds in which direction (+ or -) to change each weight in order to reduce error.

Step 8. Find the amount by which to change each weight. It applies the corrections to the weight and repeat all above steps with all training vectors until the error for all the vectors in training set is reduced to an acceptable value.

Step 9: End.

6. Evaluation of fuzzy ARTMAP with DBSCAN in VLSI partition application

This section describes a new model for partitioning a circuit using DBSCAN and fuzzy ARTMAP neural network.

6.1 Overview of art map

The basic ART system is an unsupervised learning model. It typically consists of a comparison field and a recognition field composed of neurons, a vigilance parameter, and a reset module. The vigilance parameter has considerable influence on the system, higher vigilance produces highly detailed memories (many, fine-grained categories), while lower vigilance results in more general memories (fewer, more-general categories). The comparison field takes an input vector (a one-dimensional array of values) and transfers it to its best match in the recognition field. Its best match is the single neuron whose set of weights (weight vector) most closely matches the input vector. Each recognition field neuron outputs a negative signal (proportional to that neuron's quality of match to the input vector) to each of the other recognition field neurons and inhibits their output accordingly. In this way the recognition field exhibits lateral inhibition, allowing each neuron in it, to represent a category to which input vectors they are classified. After the input vector is classified, the reset module compares the strength of the recognition match to the vigilance parameter. If the vigilance threshold is met, training commences. Otherwise, if the match level does not meet the vigilance parameter, the firing recognition neuron is inhibited until a new input vector is applied. The training commences only upon completion of a search procedure. In the search procedure, recognition neurons are disabled one by one, by the reset function until the vigilance parameter is satisfied by a recognition match. If no committed recognition neuron's match meets the vigilance threshold, then an uncommitted neuron is committed and adjusted towards matching the input vector.

There are two basic methods of training ART-based neural networks: slow and fast. In the slow learning method, the degree of training of the recognition neuron's weights towards the input vector is calculated to continuous values with differential equations and is thus dependent on the length of time the input vector is presented. The basic structure of the ART based neural network is shown in Fig 12 With fast learning, algebraic equations are used to calculate degree of weight adjustments to be made, and binary values are used. While fast learning is effective and efficient for a variety of tasks, the slow learning method is more biologically plausible and can be used with continuous-time networks (that is, when the input vector can vary continuously). Fig 13 shows the fast learning ART-based neural network.

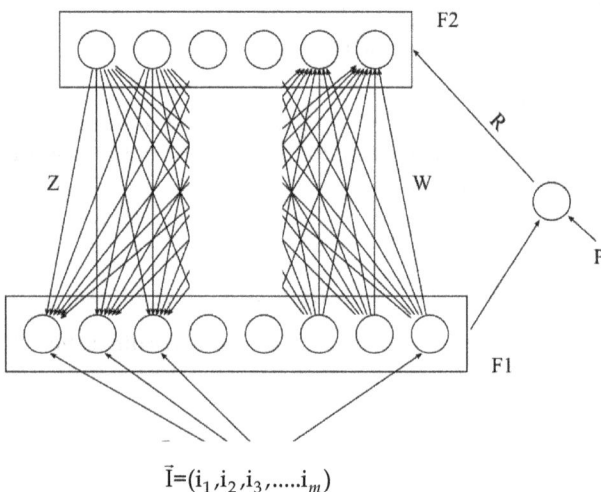

$$\vec{I}=(i_1,i_2,i_3,.....i_m)$$

Fig. 12. Basic ART Structure

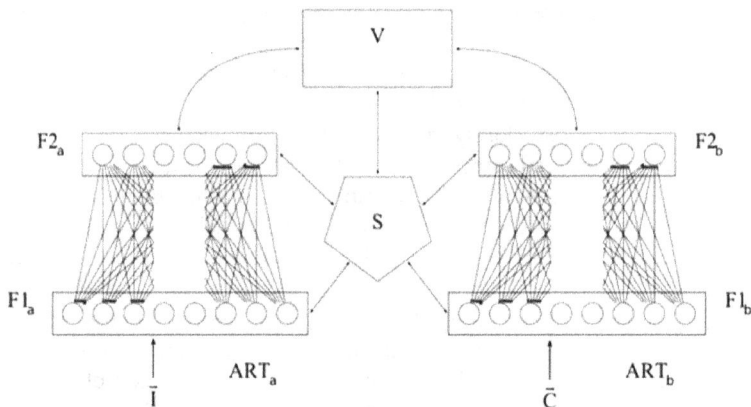

Fig. 13. Fast learning ART-based neural network

The first principle of Adaptive Resonance Theory (ART) was first introduced by Grossberg in 1976 (Carpenter,1997), whose structure resembles those of feed-forward networks. The simplest variety of ART networks is accepting only binary inputs which is called as ART (Grossberg,1987,2003). It was then extended for network capabilities to support continuous inputs called as ART-2 (Carpenter & Grossberg ,1987).ARTMAP (Carpenter et al.,1987),also known as Predictive ART, combines two slightly modified ART-1 or ART-2 units into a supervised learning structure where the first unit takes the input data and the second unit takes the correct output data and then used to make the minimum possible adjustment of the vigilance parameter in the first unit in order to make the correct classification.

6.2 Fuzzy ARTMAP

Fuzzy logic with the combination of Adaptive Resonance Theory gives Fuzzy ARTMAP, is a class of neural network that perform supervised training of recognition pattern and maps in response to input vectors generated. Fuzzy ART (Carpenter et al.,1991) implements fuzzy logic into ART's pattern recognition, thus enhancing generalizability. An optional (and very useful) feature of fuzzy ART is complement coding, a means of incorporating the absence of features into pattern classifications, which goes a long way towards preventing inefficient and unnecessary category proliferation. The performance of fuzzy ARTMAP depends on a set of user-defined hyper-parameters, and these parameters should normally be fine-tuned to each specific problem (Carpenter et al.,1992). The influence of hyper-parameter values is rarely addressed in ARTMAP literature. Moreover, the few techniques that are found in the literature for automated hyper-parameter optimization,example(Canuto et al., 2000; Dubrawski, 1997; Gamba & DellAcqua, 2003; C. Lim,1999), focus mostly on the vigilance parameter, even though there are four inter-dependent parameters (vigilance, learning, choice, and match tracking). A popular choice consists in setting hyperparameter values such that network resources (the number of internal category neurons, the number of training epochs, etc.) are minimized (Carpenter,1997). This choice of parameters may however lead to overtraining and significantly degrade the network. An effective supervised learning strategy could involve co-jointly optimizing both network (weights and architecture) and all its hyper-parameter values for a given problem, based on a consistent performance objective. Fuzzy ARTMAP neural networks are known to suffer from overtraining or over fitting, which is directly connected to a category proliferation problem. Overtraining generally occurs when a neural network has learned not only the basic mapping associated training subset patterns, but also the subtle nuances and even the errors specific to the training subset. If too much learning occurs, the network tends to memorize the training subset and loses its ability to generalize on unknown patterns. The impact of overtraining on fuzzy ARTMAP performance is two fold that is, an increase in the generalization error and in the resources requirements.

6.3 DBSCAN (Density-Based Spatial Clustering Of Applications with Noise)

DBSCAN is a data clustering algorithm proposed by Martin Ester, Hans-Peter Kriegel, Jörg Sander and Xiaovei Xui in 1996(Ester, 1996). It is a density based clustering algorithm because it finds a number of clusters starting from the estimated density distribution of corresponding nodes. DBSCAN is one of the most common clustering algorithms and also most cited in scientific literature. The basic DBSCAN algorithm has been used as a base for many other developments.

The overall structure of model is illustrated in Fig14 and Fig 15, Fig 16 show a sample circuit bipartite with related data set used . The feature extractor obtains feature vector for subcircuit, and is sent to training or inference module. The SFAM (simplified fuzzy ARTMAP) (Carpenter,1997) has two modules, that is, training and inference module. The feature vector of training subcircuits and the categories to which they belongs are specified to SFAM's training module. Once the training phase is complete, the vector represents the subcircuit with minimum interconnection. The test subcircuit pattern which is to be recognized with minimum interconnection is fed to inference module. Classifications of sub circuits are done by associating the feature vector with the top-down weight vectors (Carpenter et l., 1992;Caudell et al., 1994) in SFAM. The system can handle both symmetric and asymmetric circuit. In symmetric pattern, only distinct portion of circuit is trained whereas in asymmetric (1/2n)th portion of circuit is considered.

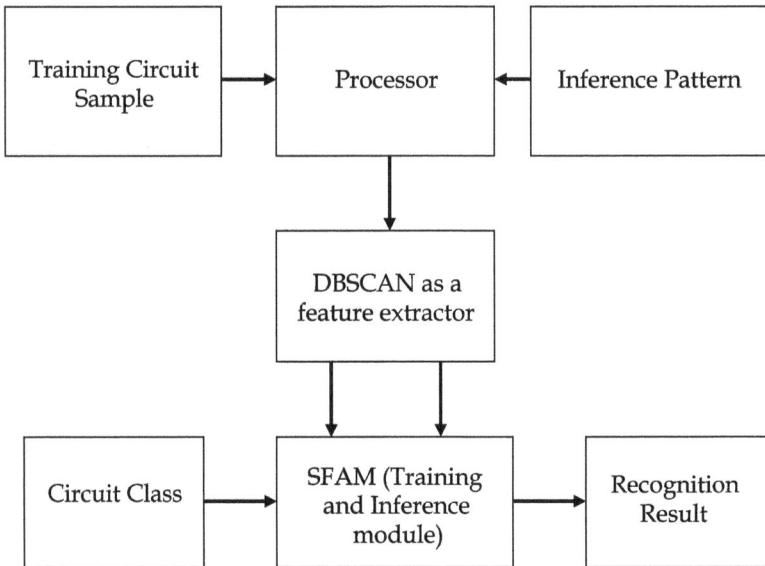

Fig. 14. Block diagram of recognition module for partitioning in VLSI Design

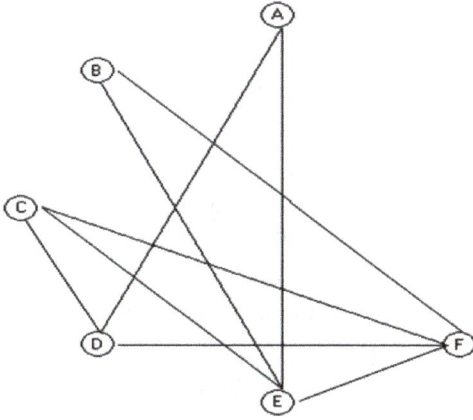

Fig. 15. Sample Circuit

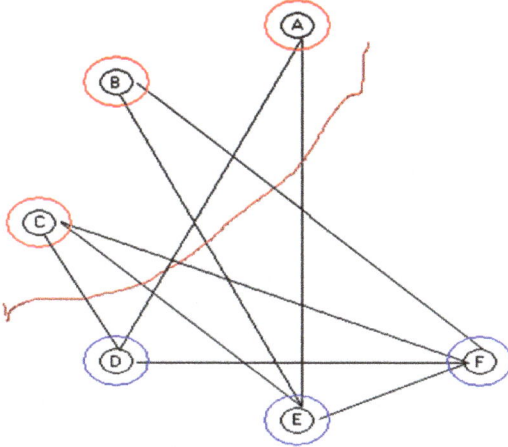

Fig. 16. Sample Bi-parted Circuit with data
Sample circuit bi parted Sub circuit 1 A, B, C   total edges = 7   Sub circuit 2 D, E, F
total edges =10

Cell	No of edges	Bipartition
A	2	1
B	2	1
C	3	1
D	3	0
E	3	0
F	4	0

Table 2. Bipartition Matrix
Data representation: Sub circuit1 (A, B, C) 0010 0010 0011,
Sub circuit2 (D, E, F) 0011 0011 0100

6.4 Overview of DBSCAN algorithm as a feature exactor

DBSCAN and clustering algorithm is used for feature Exactor which works on the densities (International Workshop on Text- Based information Retrieval (TIR 05),University of Koblenz-Landau, Germany). It separates the set D into subsets of similar densities. In the best case they can find out the cluster number k routinely and categorize the clusters of random shape and size.The runtime of this algorithms is in magnitude of O(n log(n)) for low-dimensional data (Busch,2005). A density-based cluster algorithm is based on two properties given below (TIR 05,University of Koblenz-Landau, Germany).

1. One is to define a region $C \subseteq D$, which forms the basis for density analyses.
2. Another is to propagate density information (the provisional cluster label) of C.

In DBSCAN a region is defined as the set of points that lie in the ϵ-neighborhood of some point p. if $|C|$ exceeds a given Min Points-threshold Cluster label propagates from p to the other points in C. The complete description of DBSCAN algorithm is provided in (Ester et al., 1996;Tan et al., 2004;Dagher. I et al., 1999).

6.5 Simplified fuzzy ARTMAP module

In context of the circuit partitining in VLSI design to recognize the subcircuit with minimum interconnection between them, the size of input layer is 4 and output layer is 10. Hence it outcomes in 2-10 layered Fuzzy ARTMAP model.

Match and choice function for fuzzy ARTMAP in context to circuit partitioning is defined by,

For input vector I and cluster j from DBSCAN algorithm, Choice function given by

$$CFj(I) = \frac{|I^\wedge Wj|}{\alpha + |Wj|} \tag{14}$$

Where α is small constant about 0.0000001,Wj is top-down weight

Winner node is one with highest activation /choice function, that is,

$$Winner = max(CFj) \tag{15}$$

Match function which is very much used to find out whether the network must adjust its learning parameters is given by,

$$Fj(I) = \frac{|I^\wedge Wj|}{|I|} \tag{16}$$

If MF j (I) $\geq$ vigilance parameter (ρ) then Network is in state of resonance, where ρ is in range $0 \leq \rho \leq 1$.

If MF j (I) $\leq$ vigilance parameter (ρ) then Network is in state of mismatch reset.

7. A new clustering approach for VLSI circuit partitioning

The vital problem in VLSI for physical design algorithm is circuit partitioning. In this section concentration is on improving the partitioning technique using data mining approach. This section deals with a range of partitioning methodological aspects which predicts to divide the circuit into sub circuits with minimum interconnections between them. This approach considers two clustering algorithms proposed by (Li & Behjat, 2006) Nearest Neighbor(NN) and Partitioning Around Mediods(PAM) clustering algorithm for dividing the circuits into sub circuits. The experimental results show that PAM clustering algorithm yields better subcircuits than Nearest Neighbour. The experimental results are compared using benchmark data provided by MCNC standard cell placement bench netlists.

7.1 Considerations in choosing the right algorithm

Data mining algorithms have to be adapted to work on very large databases. Data reside on hard disks because they are too large to fit in main memory, therefore, algorithms have to make as few passes as possible over the data, as secondary memory fetch cycle increases the computational time and therefore reduces the run time performance. Quadratic algorithms are too expensive, that is the execution time of the operations in clustering algorithms is quadratic and so it becomes an important constraint in choosing an algorithm for the problem at hand. The aim in the thesis is to reduce the interconnections between the circuits with minimum amount of error,hence prototype based clustering is used. The attributes in the data set were less important, so the proximity matrix was created. Since both PAM and NNA belong to partitional and prototype based clustering and also the intention was to get the partition with the minimum interconnections these two algorithms were used.

7.2 Implementation

The implementation consists of three stages consisting of data extraction, partitioning and result using VHDL (VHSIC (Very High Speed Integrated Circuit) Hardware Description Language) as a tool. In data extraction, a VLSI circuit represented as a bipartite graph is considered. The bipartite graph considered for the approach is shown in Fig 17.

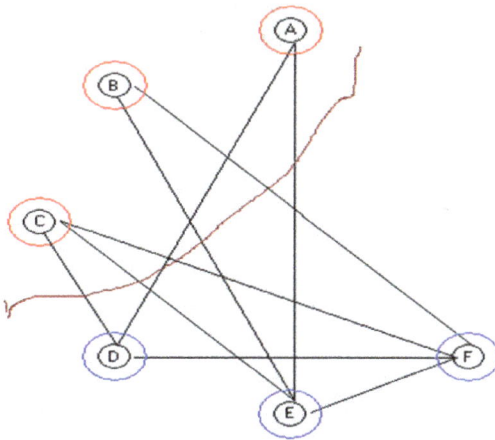

Fig. 17. Bipartition Circuit

| Sub circuit 1 | A, B, C | total edges = 7 |
| Sub circuit 2 | D, E, F | total edges =10 |

Cell	No of edges	Bipartition
A	2	1
B	2	1
C	3	1
D	3	0
E	3	0
F	4	0

Table 3. Bipartition Matrix

The block diagram to recognize sub-circuits with minimum interconnections using two techniques(Nearest Neighbor , PAM).A new clustering algorithm is explored.

7.3 Applying clustering techniques to VLSI circuit partitioning

In adapting the two cluster partitioning algorithms to the area of VLSI circuit partitioning, the following considerations are of utmost importance.

The two algorithms take as input an adjacency matrix, which gives an idea of the similarity measure in the form of distances between the various data that are to be clustered. This approach uses this tool to partition circuits, so the circuit to be partitioned is the effective data to be clustered and the basic unit on which the algorithms will act are the nodes in a circuit.

Similarity between nodes in a circuit

Here, the input is the adjacency matrix, which defines the similarity between different nodes in the circuit. The attributes of nodes that are to be quantified as similarity between different nodes are based on several characteristics of logic gates such as,

1. Interconnections between nodes
2. Common signals as input
3. Functionality
4. Physical distance
5. Presence of the node on the maximum delay path

For example, if two nodes are interconnected, then the similarity between them is increased and the distance between them is reduced compared to two nodes which are not connected together.

Also, if some nodes get a common signal, such as a set of flip-flops sharing a common clock signal, it is desirable to have them partitioned into the same sub-circuit so as to reduce problems due to signal delay of synchronous control inputs. So, the distances between such nodes are also low.

The distance of a node to itself is taken as 0 and a low value of distance means the highest similarity. A high value of distance means maximum dissimilarity, and therefore least similarity, such nodes can be placed in different sub-circuits.

This adjacency or distance matrix is acted upon by the two algorithms, to effectively divide the circuit into sub-circuits, with the objective that is minimum interconnection under check. Adapting and applying data mining tools to VLSI circuit partitioning is a new approach. Improvisations and optimizations to the two algorithms are necessary and is essential to make them workable and viable as CAD tools.

Circuit chosen for implementation and testing

The circuit on which the two data mining algorithms are implemented (NNA and PAM) is as shown below. The circuit is a Binary Coded Decimal (BCD) code to seven segment code converter (Fig18). It has 4 inputs and 7 outputs. In this figure each rectangular block is considered as a node. A node is one which performs a defined function (Fig 19), it may be a simple AND gate or it may contain many interconnected flip-flops. So, a node contains one or more components and performs a logical function, the level of abstraction of a node can be changed to suit the basic unit understandable by a CAD tool.

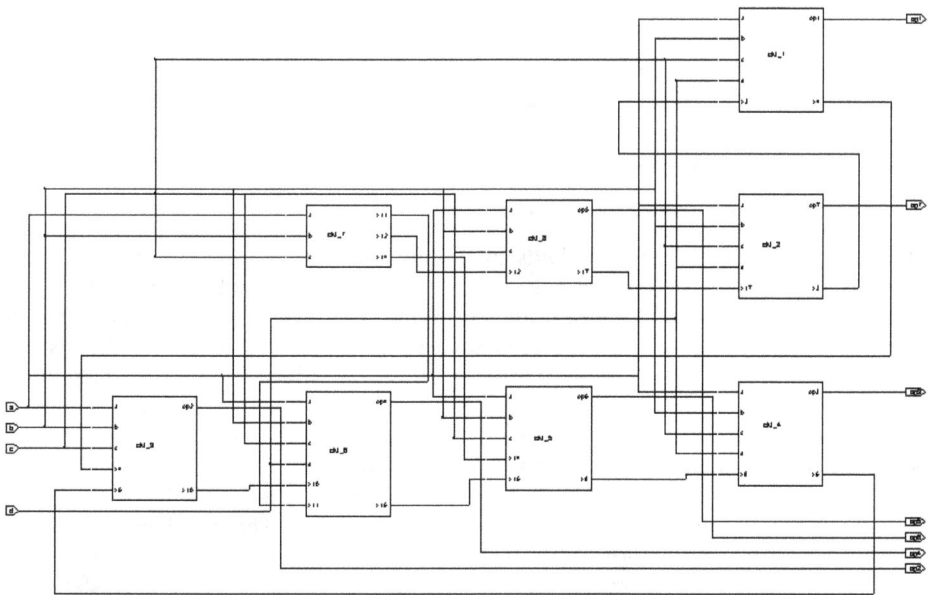

Fig. 18. Circuit of BCD code to Seven Segment code converter

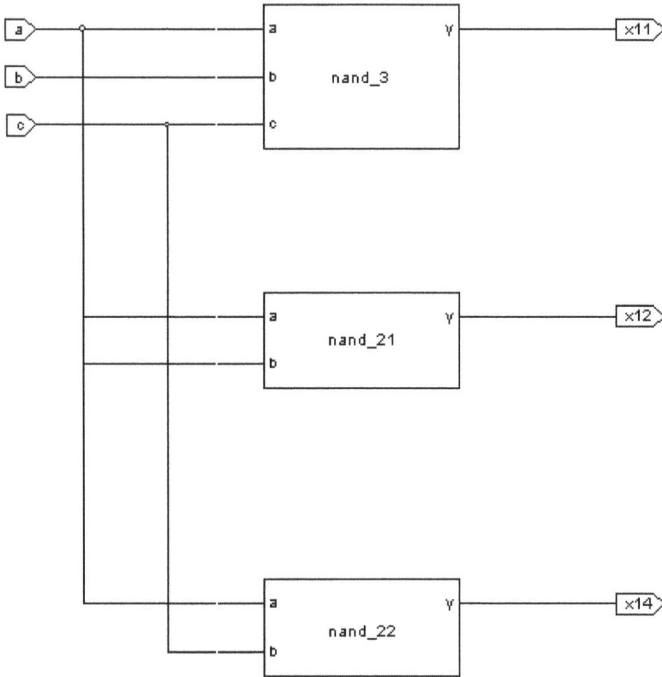

Fig. 19. A node (node 5) enlarged.

This shows that a node which is part of the main circuit consists of gates, such as Nand gate and or gates, or one which performs a logical function.

7.4 How to choose k and threshold value

7.4.1 PAM Algorithm – Choosing initial medoids

PAM starts from an initial set of medoids, by finding representative objects, called medoids, in clusters and iteratively replaces one of the medoids by one of the non-medoids if it improves the total distance of the resulting clustering. The PAM algorithm is based on the search for k medoids which are representative of the sequences based on the distance matrix. These k values should represent the structure of the sequences. After defining the set of k medoids, they would be used to construct the k clusters and partition the nodes by assigning each observation to the nearest medoid. In doing this, the target would be to identify the medoids that minimize the sum of the dissimilarities in the observations. As it can be seen, the choice of the initial medoids is very important. Medoid is the most centrally located point in a cluster, as a representative point of the cluster. The initial medoids chosen decides the quality of the formed clusters and the computational speed. If the initial medoids chosen are close to the final optimal medoids, yielding the final clusters with reduced cost, the computational cost will be reduced. Otherwise the number of iterations to find the final medoids will increase, this in turn increasing the time taken to obtain results and computational cost.

The representation by k-medoids has two advantages. First, it presents no limitations on attributes types and second, the choice of medoids is dictated by the location of a predominant fraction of points inside a cluster, therefore it is less sensitive to the presence of outliers. Therefore, PAM is iterative optimization that combines relocation of points between perspective clusters with re nominating the points as potential medoids.Earlier the task is done to find out the optimum value of threshold "t", which decides the cluster density and quality, shows that the value of threshold from 2 to 5 gives optimal minimization of interconnections between sub-circuits. Therefore, for the two algorithms NNA and PAM, the threshold value of 2 and 3 are respectively chosen based on this task.

7.4.2 Details of the partitioned Circuits - Results on a Circuit with 8 Nodes is discussed

Fig.20 is an example of a Testing Circuit 1 with 8 nodes before applying the partitioning and the circuits after partitioning using the NN algorithm and applying the PAM algorithms are shown in Fig. 21. and Fig. 22. respectively.

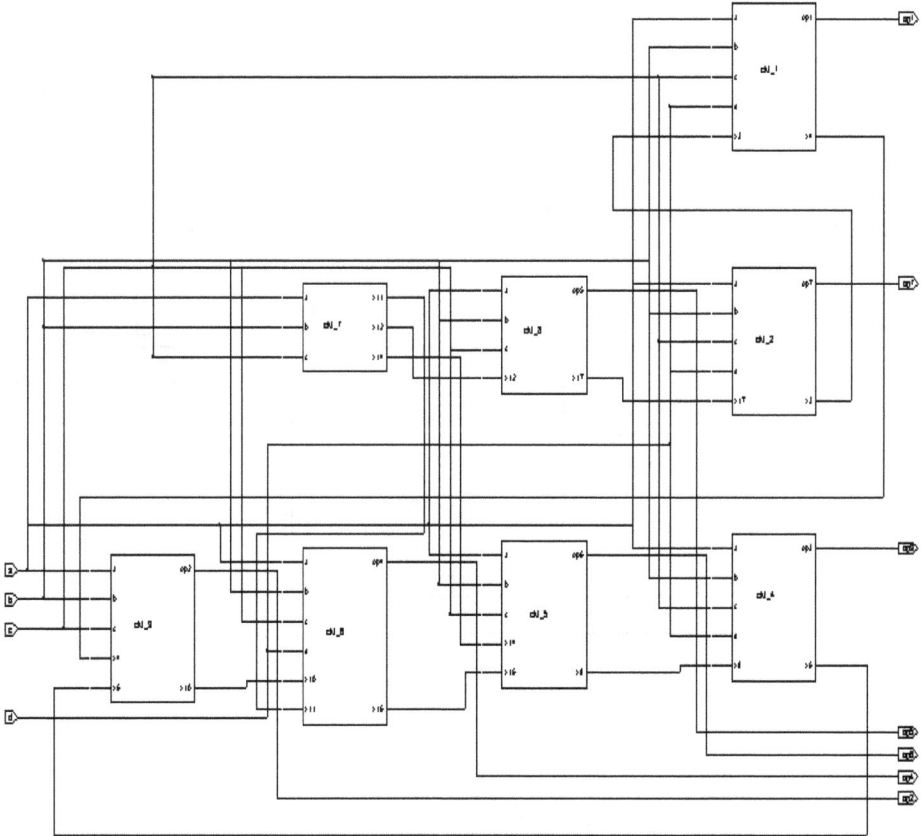

Fig. 20. Circuit before applying partitioning techniques

The circuit shown in Fig7.10 is a BCD to seven-segment code converter before applying the partitioning algorithms and it has 8 nodes as shown in Fig 7.10. This circuit is tested in hardware and the functionality is concluded to be correct.

Partitioned circuit obtained after applying Nearest Neighbor Algorithm

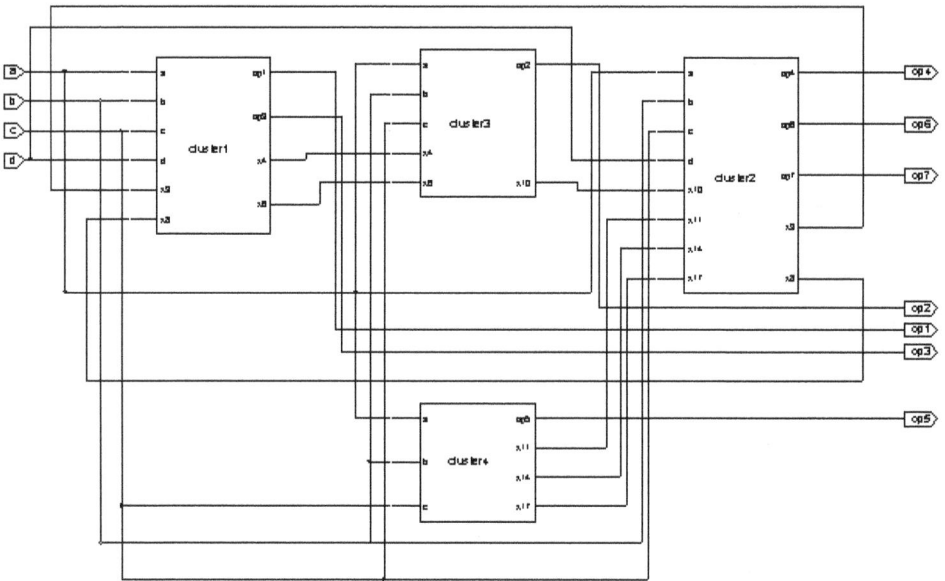

Fig. 21. NNA Partitioned circuit showing 4 sub-circuits

Partitioned circuit obtained after applying Partitioning Around Medoids algorithm

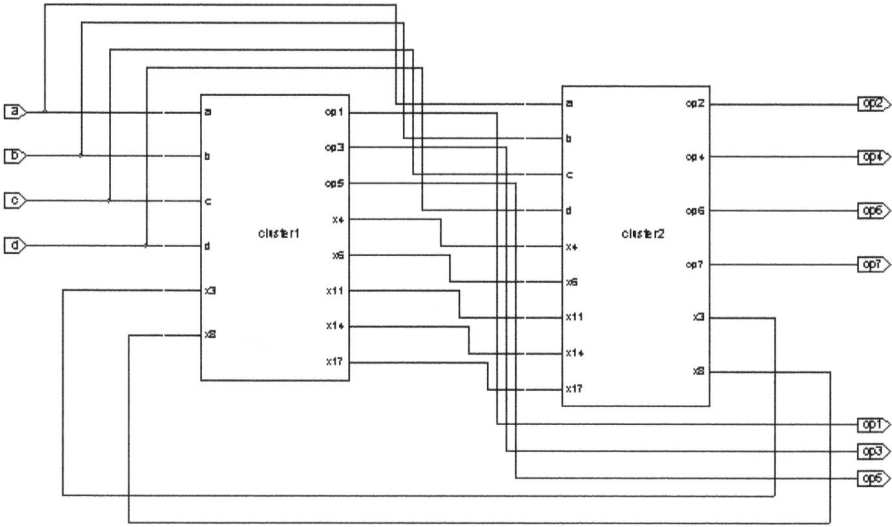

Fig. 22. PAM Partitioned circuit showing 2 sub-circuits

Results on a Circuit with 15 Nodes

Example Testing Circuit 2 with 15 nodes:

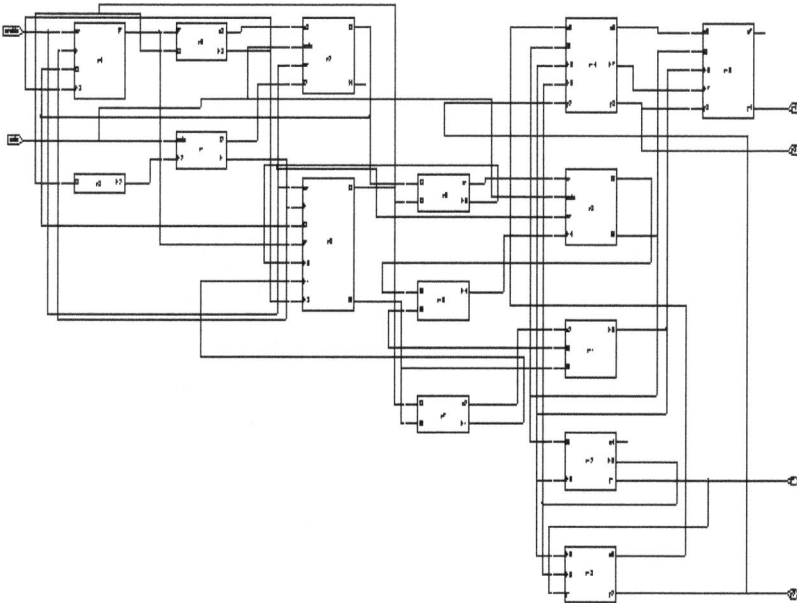

Fig. 23. Circuit before applying partitioning techniques (Rubin, Willy Publications)

Partitioned circuit obtained after applying Nearest Neighbor Algorithm

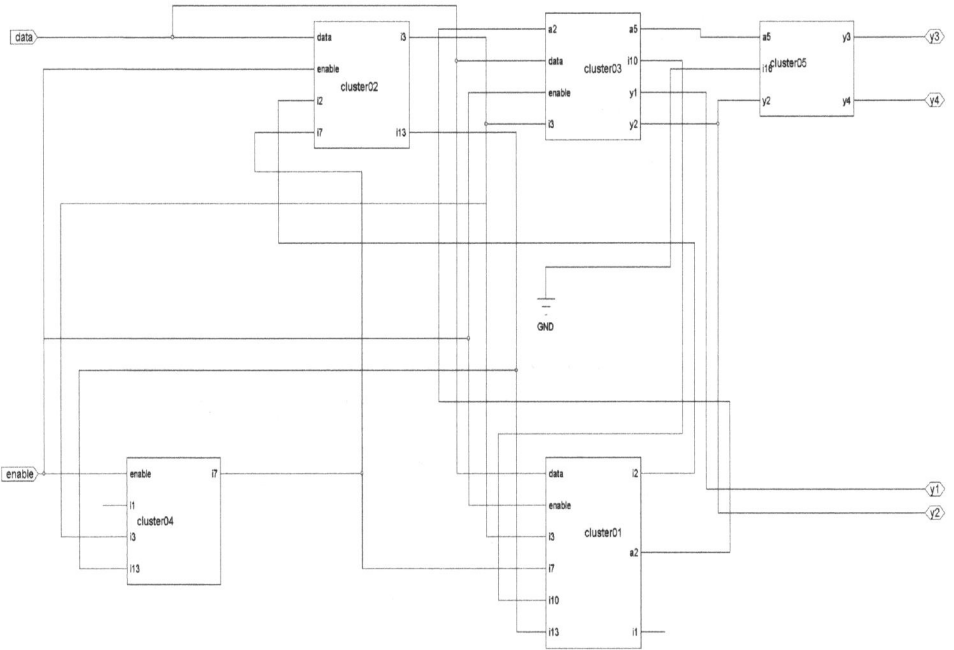

Fig. 24. NNA partitioned circuit showing 5 sub-circuits

Partitioned circuit obtained using Partitioning Around Medoids algorithm

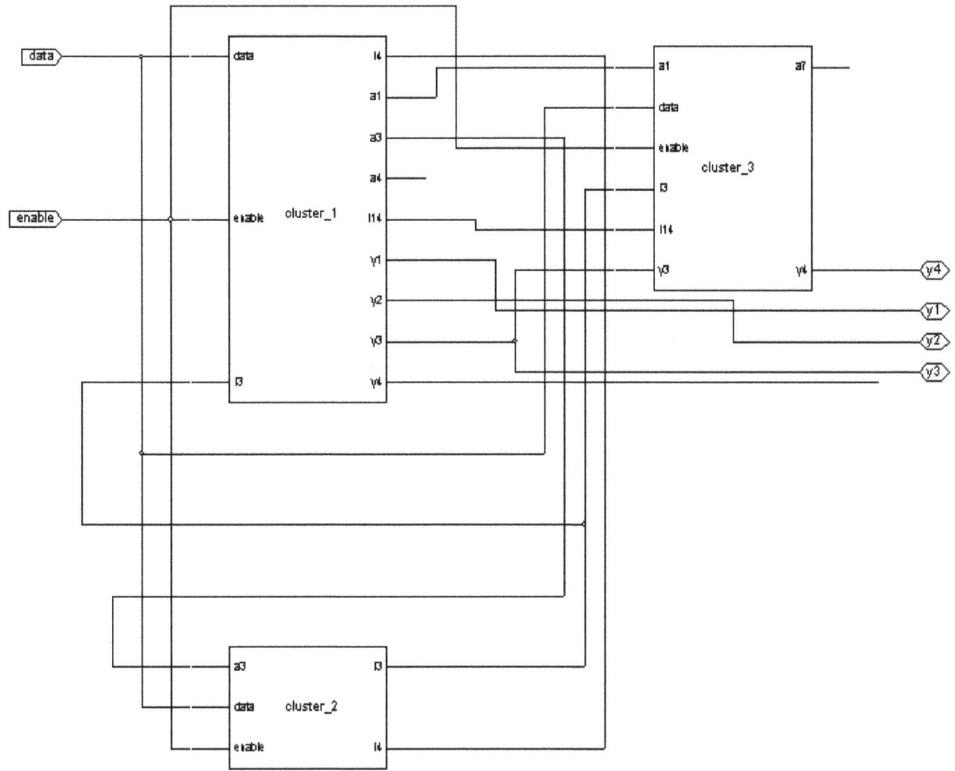

Fig. 25. PAM Partitioned circuit showing 3 sub-circuits

8. Conclusion

This section provides observations about the various techniques explained in this chapter with a detailed results based explaination of the Nearest Neighbor and Partitioning Around Medoids Clustering Algorithms.

8.1 Memetic approach to circuit partitioning

Memetic algorithm (MA) are population based heuristic search approaches for combinatorial optimization problems based on cultural evolution. They are designed to search in the space of locally optimal solutions instead of searching in the space of all candidate solutions. This is achieved by applying local search after each of the genetic operators. Crossover and mutation operators are applied to randomly chosen individuals for a predefined number of times. To maintain local optimality, the local search procedure is applied to the newly created individuals.

Neuro-Memetic Approach to Circuit Partitioning makes it possible to predict the sub-circuit from circuit with minimum interconnections between them. The system consists of three parts, each dealing with data extraction, learning stage & result stage. In data extraction, a circuit is bipartite and chromosomes are represented for each sub circuit. Extracted sequences are fed to Neuro-memetic model that would recognize sub-circuits with lowest amount of interconnections between them.

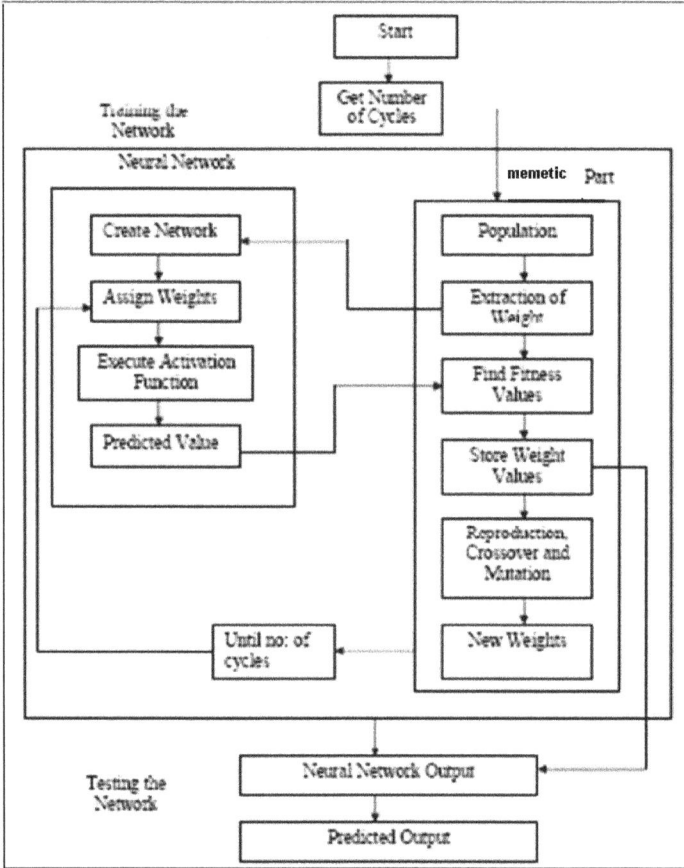

Fig. 26. Working procedure of Neuro-Memetic approach

8.2 Neuro-EM model

The system consists of three parts each dealing with data extraction, Learning stage and recognition　stage. In data extraction, a circuit is bipartite and partitions it into 10 clusters, a user-defined value, by using K-means (J. B. MacQueen, 1967) and EM methodology (Kaban & Girolami, 2000), respectively. In recognition stage the parameters, centroid and probability are fed into generalized delta rule algorithm separately and train the network to recognize sub-circuits with lowest amount of interconnections between them

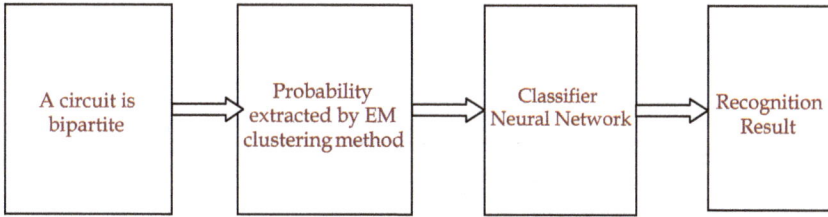

Fig. 27. Block diagram of K-means with neural network

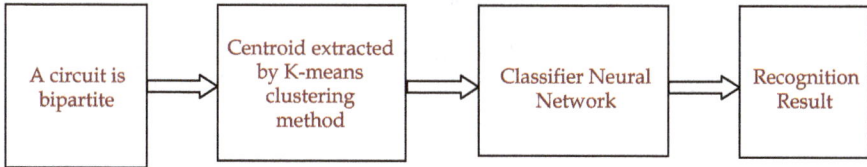

Fig. 28. Block diagram of EM methodology with neural network

In recognition stage the parameters, that is, centroid and probability are fed into generalized delta rule algorithm separately and train the network to recognize sub circuit with minimum interconnection between them

8.3 Fuzzy ARTMAP with DBSCAN

A new model for partitioning a circuit is proposed using DBSCAN and fuzzy ARTMAP neural network. The first step is concerned with feature extraction, where it uses DBSCAN algorithm. The second step is classification and is composed of a fuzzy ARTMAP neural network.

8.4 Nearest Neighbor and Partitioning Around Medoids clustering Algorithms

Two clustering algorithms Nearest Neighbor (NNA) and Partitoning Around Medoids (PAM) clustering algorithms are considered for dividing the circuits into sub circuits. Clustering is alternatively referred to as unsupervised learning segmentation. The clusters are formed by finding the similarities between data according to characteristics found in the actual data. NNA is a serial algorithm in which the items are iteratively merged into the existing clusters that are closest. PAM represents a cluster by a medoid.

- Criteria Used: Clustering/Unsupervised learning segmentation
- Testing: The algorithms are tested using VHDL,Xilinx xc9500 CPLD/FPGA tool and MATLAB simulator using a test netlist matrix .
- Results and Observations:

As the number of clusters increases, the time taken for PAM increases but is less than Nearest Neighbor algorithm. PAM performs better than Nearest Neighbor algorithm. PAM has been very competent, especially in the case of a large number of cells when compared with Nearest Neighbor. The proposed model based algorithm has achieved sub-circuits with minimum interconnections, for the Circuit Partitioning problem.

No. of Clusters	Time taken(Secs)	
	PAM	Nearest Neighbor
2	0.045	0.5
4	0.067	0.7
6	0.075	0.8
8	0.087	0.9

Table 4. Comparison of time taken to partition using PAM and Nearest Neighbor

Completion time: Graphs depict completion time increases proportionately with respect of number of clusters. Nearest neighbor algorithm takes more completion time as number of iterations increase when compared to PAM.

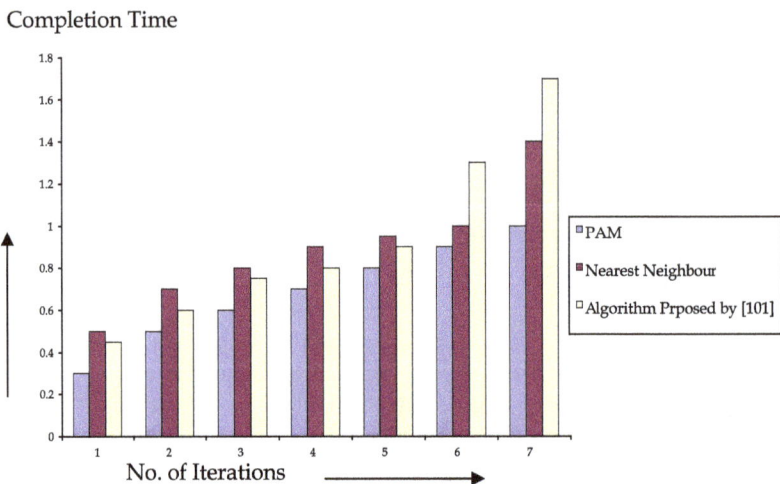

Fig. 29. Graph depicting the Completion time of NNA and PAM compared with algorithm proposed in (Gerez, 1999) over number of clusters

CPU Utilization: A graph depicts CPU utilization increases proportionately with respect of number of iterations. PAM takes less CPU utilization as number of iteration increase compared to NNA algorithm.

Completion Time

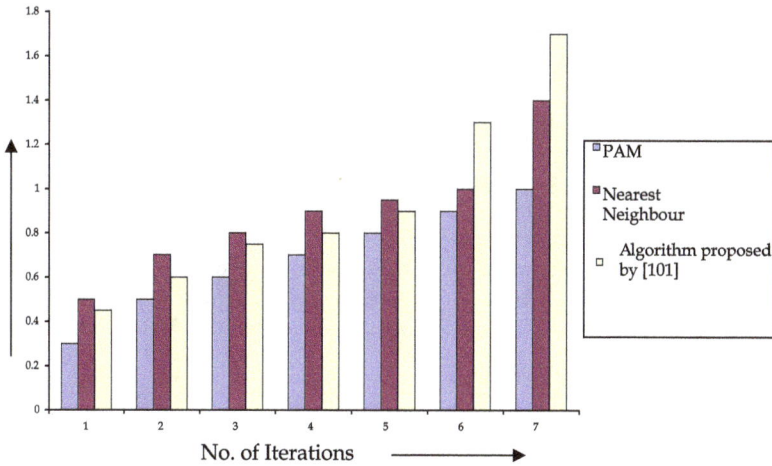

Fig. 30. Graph depicting CPU utilization of NNA, PAM, algorithm proposed in (Gerez, 1999) over number of iterations

From the implementation of the two algorithms, Nearest Neighbor and Partitioning Around Medoids, some fundamental observations are made. There is a reduction of 1 interconnection when a circuit with 8 nodes is partitioned and when a circuit with 15 nodes is partitioned, there is a reduction of 5 interconnections between the sub-circuits obtained using NNA and PAM. Therefore, it is concluded that the number of nodes in a circuit and the number of interconnections are inversely proportional. That is, as the number of nodes in a circuit increases, the number of interconnections between sub-circuits decreases for both partitioning methods. This reduction is not consistent since the complexity of any circuit will not be known a priori. One of the future enhancements would be to analyze the percentage ratio of the number of nodes in a circuit to the number of interconnections that get reduced after the circuit is partitioned.

9. Future enhancements

Future enhancements envisaged are using of distance based classification data mining concepts and other data mining concepts, Artificial/ Neural modeled algorithm in getting better optimized partitions.

10. Acknowledgement

I would like to acknowledge my profound gratitude to the Management, Rashtreeya Sikshana Samithi Trust, Bangalore.I am indebted to Dr. S.C. Sharma, Vice Chancellor, Tumkur University, Karnataka for his unending support.I would like to thank all my official colleagues at R V College of Engineering and specifically the MCA department staffs for their remarkable co-ordination. Last but not the least, I would like to thank my family members.

11. References

J. B. MacQueen, "Some Methods for classification and Analysis of Multivariate Observations", *Proceedings of 5-th Berkeley Symposium on Mathematical Statistics and Probability*, Berkeley, University of California Press, 1967, pp. 281-297.

Ata Kaban, Mark Girolami, "Initialized and Guided EM-Clustering of Sparse Binary Data with Application to Text Based Documents," ICPR, *15th International Conference on Pattern Recognition (ICPR'00)*, Vol. 2, 2000, pp. 2744.

Shantanu Dutt , Wenyong Deng, "A probability-based approach to VLSI circuit partitioning", *Proceedings of the 33rd annual conference on Design automation*, Las Vegas, Nevada, United States, June 03-07, 1996 , pp.100-105.

N. Krasnogor, J. Smith, "Memetic algorithms: The polynomial local search complexity theory perspective," *Journal of Mathematical Modelling and Algorithms*, Springer Netherlands, ISSN: 1570-1166 (Print) 1572-9214 (online), Vol. 7, Number 1 / March, 2008, pp. 3 - 24.

J. T., "Evolutionary Algorithms, Fitness Landscapes and Search". *Ph.D. thesis, Univesity of New Mexico*, Albuquerque, NM, 1995.

P. Merz and B. Freisleben, "Fitness landscapes, memetic algorithms, and greedy operators for graph bipartitioning," *Journal of Evolutionary Computation*, vol. 8, no. 1, 2000, pp. 61–91.

T. Jones, "Crossover, macromutation, and population based search," *in Proceedings of the Sixth International Conference on Genetic Algotihms*, M. Kauffman, Ed., 1995, pp. 73–80.

N. Krasnogor, P. M. L´opez, E. de la Canal, D. Pelta, "Simple models of protein folding and a memetic crossover," *in Exposed at INFORMS CSTS, Computer Science and Operations Research: Recent Advances in the Interface meeting*, 1998.

Christopher M. Bishop, "Neural Networks for Pattern Recognition", *Oxford Univ Press*, ISBN: 0198538642,1995.

N. Krasnogor, P. Mocciola, D. Pelta, G. Ruiz, W. Russo, "Arunnable functional memetic algorithm framework," *in Proceedings of the Argentinian Congress on Computer Sciences*. Universidad Nacionaldel Comahue, 1998, pp. 525–536.

D. Holstein, P. Moscato, "Memetic algorithms using guided local search: A case study," *in New Ideas in Optimization*, D. Corne, F. Glover, M. Dorigo, Eds. McGraw-Hill, 1999.

Arthur Dempster, N Laird, D B Rubin, "Maximum likelihood estimation for incomplete data via the EM algorithm", Journal of the *Royal Statistical Society* (B) 39, 1977, pp.1-38.

Witten I, Frank E, "Data Mining: Practical Machine Learning Tools and Techniques", *Second Edition, Morgan Kaufmann*, San Francisco, 2005.

Dimitris Karlis, "An EM algorithm for multivariate Poisson distribution and related models", *Journal of Applied Statistics*, Vol. 30, No. 1, 2003 , pp. 63-77(15)

Carpenter, G.A, "Distributed learning, recognition, and prediction by ART and ARTMAP neural networks", *Neural Networks*, 10:, 1997, pp. 1473- 494.

Sung Ho Kim, "Calibrated initials for an EM applied to recursive models of categorical variables", *Computational Statistics & Data Analysis*, Vol. 40, Issue 1, July 2002, pp. 97-110 .

Gyllenberg M, Koski T, Lund T, "Applying the EM-algorithm to classification of bacteria", *Proceedings of the International ICSC Congress on Intelligent Systems and Applications*, 2000, pp. 65-71.

Carpenter, G.A., Grossberg, S., "Adaptive Resonance Theory, In Michael A. Arbib" (Ed.), *The Handbook of Brain Theory and Neural Networks*, Second Edition,Cambridge, MA: MIT Press, 2003, pp. 87 – 90.

Grossberg, S., "Competitive learning: From interactive activation to adaptive resonance", Cognitive Science

Carpenter, G.A, Grossberg, S., "ART 2: Self-organization of stable category recognition codes for analog input patterns", *Applied Optics, 26(23)*, 1987, pp. 4919-4930.

Carpenter, G.A., Grossberg, S., Reynolds J.H., "ARTMAP: Supervised real-time learning and classification of nonstationary data by a self-organizing neural network", *Neural Networks (Publication)*, 4, 1991, pp. 565-588.

Carpenter, G.A., Grossberg, S., Rosen, D.B., "Fuzzy ART: Fast stable learning and categorization of analog patterns by an adaptive resonance system", *Neural Networks (Publication)*, 4, 1991, pp. 759-771.

Carpenter, G.A., Grossberg, S., Markuzon, N., Reynolds, J.H., Rosen D.B., "Fuzzy ARTMAP: A neural network architecture for incremental supervised learning of analog multidimensional maps", *IEEE Transactions on Neural Networks*, 3, 1992, pp. 698-713.

A. Canuto, G. Howells, M. Fairhurst, "An investigation of the effects of variable vigilance within the RePART neuro-fuzzy network," Journal of Intelligent and Robotic Systems, 29:4, 2000, pp. 317-334.

A. Dubrawski, "Stochastic validation for automated tuning of neural network's hyper-parameters," *Robotics and Automated Systems*, 21, 1997, pp. 83-93.

P. Gamba and F. DellAcqua, "Increased accuracy multiband urban classification using a neuro-fuzzy classifier," *International Journal of Remote Sensing*, 24:4, 2003, pp. 827-834.

C. P. Lim, H. H. Toh, T. S. Lee, "An evaluation of the fuzzy ARTMAP neural network using offline and on-line strategies," *Neural Network World*, 4, 1999, pp. 327-339.

Carpenter, G.A., Grossberg, S., Markuzon, N., Reynolds, J.H., and   Rosen, D.B., "Fuzzy ARTMAP: neural network architecture for Incremental supervised   learning of analog multidimensional maps", *IEEE   Transactions on Neural Networks*, 1992, 3:, pp. 698-713.

Caudell, T.P., Smith, S.D.G., Escobedo, R., Anderson, M, "NIRS: Large scale ART–1 neural architectures for engineering design Retrieval", *Neural Networks, 7:*, 1994, pp. 1339-1350.

Benno Stein, Michael Busch, "Density based cluster Algorithms in Low-dimensional and High-dimensional Applications", *Second International Workshop on Text- Based information Retrieval (TIR 05)*, Stein, Meyer zu Eißen (Eds.) Fachbericht,e, Informatik, University of Koblenz-Landau, Germany, ISSN 1860-4471c, pp. 45-56.

Busch. Analyse dichtebasierter Clusteralgorithmen am Beispiel von DBSCAN und MajorClust. Study work, Paderborn University, Institute for Computer Science, March 2005.

Martin Ester, Hans-Peter Kriegel, Jörg Sander, Xiaowei Xu, "A density-based algorithm for discovering clusters in large spatial databases with noise". *Proceedings of the Second International Conference on Knowledge Discovery and Data Mining (KDD-96)*: AAAI Press, 1996, pp. 226-231.

Tan, S.C., Rao, M.V.C., Lim C.P, "An adaptive fuzzy min-max conflict-resolving classifier,"
    *in Proceedings of the 9th Online World Conference on Soft Computing in Industrial
    Applications, WSC9*, 20 September – 8 October 2004.

Dagher. I, Georgiopoulos. M, Heileman. G.L., Bebis, G, "An ordering algorithm for pattern
    presentation in fuzzy ARTMAP that tends   to improve generalization
    performance," *IEEE Trans Neural Networks*, vol. 10, 1999, pp. 768-778.

Jianhua Li, Laleh Behjat, "A Connectivity Based Clustering Algorithm with Applications to
    VLSI Circuit Partitioning", *IEEE Transactions on Circuits and Systems-II.Express
    Briefs*, Vol.53, No.5, May 2006.

Steven M Rubin, "Computer Aids for VLSI Design", Second Edition, Willy Publications.

Sabin H Gerez, "Algorithms for VLSI design automation", University of Tacut, department
    of electrical engineering, Netherlands, John Wiley & Sons, 1999, pp. 1-15.

7

Parallel Symbolic Analysis of Large Analog Circuits on GPU Platforms *

Sheldon X.-D. Tan[1], Xue-Xin Liu[1], Eric Mlinar[1] and Esteban Tlelo-Cuautle[2]
[1]*Department of Electrical Engineering, University of California, Riverside, CA 92521*
[2]*Department of Electronics, INAOE*
[1]*USA*
[2]*Mexico*

1. Introduction

Graph-based symbolic technique is a viable tool for calculating the behavior or the characterization of an analog circuit. Traditional symbolic analysis tools typically are used to calculate the behavior or the characteristic of a circuit in terms of symbolic parameters (Gielen et al., 1994). The introduction of determinant decision diagrams based symbolic analysis technique allows exact symbolic analysis of much larger analog circuits than any other existing approaches (Shi & Tan, 2000; 2001). Furthermore, with hierarchical symbolic representations (Tan et al., 2005; Tan & Shi, 2000), exact symbolic analysis via DDD graphs essentially allows the analysis of arbitrarily large analog circuits. Some recent advancement in DDD ordering technique and variants of DDD allow even larger analog circuits to be analyzed (Shi, 2010a;b). Once the circuit's small-signal characteristics are presented by DDDs, the evaluation of DDDs, whose CPU time is proportional to the sizes of DDDs, will give exact numerical values. However, with large networks, the DDD size can be huge and the resulting evaluation can be very time consuming.

Modern computer architecture has shifted towards designs that employ multiple processor cores on a chip, so called multi-core processors or chip-multiprocessors (CMP) (AMD Inc., 2006; Intel Corporation, 2006). The graphic processing unit (GPU) is one of the most powerful many-core computing systems in mass-market use (AMD Inc., 2011a; NVIDIA Corporation, 2011a). For instance, NVIDIA Telsa T10 chip has a peak performance of over 1 TFLOPS versus about 80–100 GFLOPS of Intel i5 series Quad-core CPUs (Kirk & Hwu, 2010). In addition to the primary use of GPUs in accelerating graphics rendering operations, there has been considerable interest in exploiting GPUs for general purpose computation (Göddeke, 2011). The introduction of new parallel programming interfaces for general purpose computations, such as Computer Unified Device Architecture (CUDA) (NVIDIA Corporation, 2011b), Stream SDK (AMD Inc., 2011b) and OpenCL (Khronos Group, 2011), has made GPUs a powerful and attractive choice for developing high-performance numerical, scientific computation and solving practical engineering problems.

*This work is funded in part by NSF grants NSF OISE-0929699, OISE-1130402, CCF-1017090 and part by CN-11-575 UC MEXUS-CONACYT Collaborative Research Grants.

In this chapter, we present an efficient parallel DDD evaluation technique based on general purpose GPU (GPGPU) computing platform to explore the parallelism of DDD structures. We present a new data structures to present the DDD graphs in the GPUs for massively threaded parallel computing of the numerical values of DDD graphs. The new method explores data parallelism in the DDD numerical evaluation process where DDD graphs are traversed in a depth-first fashion. Numerical results show that the new evaluation algorithm can achieve about one to two orders of magnitude speedup over the serial CPU based evaluations of some analog circuits. The presented parallel techniques can be used for the parallelization of many decision diagrams based applications such as logic synthesis, optimization, and formal verification, all of which are based on binary decision diagrams (BDDs) and its variants (Bryant, 1995; Minato, 1996).

This chapter is organized as follows. Section 2 reviews DDD-based symbolic analysis techniques. Section 3 briefly review the GPU architectures and CUDA computing. Section 4 introduces the new parallel algorithm, and then the results are demonstrated in Section 5. Lastly, Section 6 summarizes this chapter.

2. DDDs and DDD-based symbolic analysis

Before we introduce our GPU-base parallel analysis method, we first provide a brief overview of determinant decision diagram (DDD) Shi & Tan (2000) in this section.

Determinant decision diagrams (DDDs) was introduced to represent determinants symbolically Shi & Tan (2000). A DDD is essentially a zero-suppressed Binary Decision Diagrams (ZBDDs) — introduced originally for representing sparse subset systems Minato (1993). A ZBDD is a variant of a Binary Decision Diagram (BDD) introduced by Akers Akers (1976) and popularized by Bryant Bryant (1986). BDDs have brought great success to formal verification and testing for combinational and sequential digital circuits Bryant (1986); Hachtel & Somenzi (1996). DDD representation has several advantages over both the expanded and arbitrarily nested forms of a symbolic expression.

- First, similar to the nested form, DDD representation is compact for a large class of analog circuits. A ladder-structured network can be represented by a diagram where the number of vertices in the diagram (called its *size*) is equal to the number of symbolic parameters. As indicated by Shi & Tan (2000), the typical size of DDD is dramatically smaller than that of product terms. For instance, 5.71×10^{20} terms can be represented by a diagram with 398 vertices.

- Second, similar to the expanded form, DDD representation is canonical, i.e., every determinant has a *unique* representation, and is amenable to symbolic manipulations. This property of canonical representation for matrix determinants is similar to BDD for representing *binary functions* and ZBDD for representing *subset systems*.

A key observation is that the circuit matrix is sparse and that for many times, a symbolic expression may share many sub-expressions. For example, consider the following determinant

$$\det(\mathbf{M}) = \begin{vmatrix} a & b & 0 & 0 \\ c & d & e & 0 \\ 0 & f & g & h \\ 0 & 0 & i & j \end{vmatrix} = adgj - adhi - aefj - bcgj + cbih. \tag{1}$$

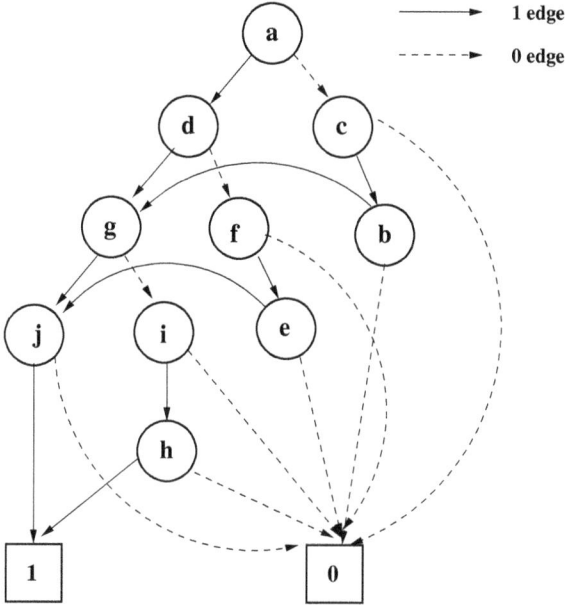

Fig. 1. A ZBDD representing $\{adgi, adhi, afej, cbgj, cbih\}$ under ordering
$a > c > b > d > f > e > g > i > h > j$

Note that sub-terms ad, gj, and hi appear in several product terms, and each product
term involves a subset (four) out of ten symbolic parameters. Therefore, we view each
symbolic product term as a subset, and use a ZBDD to represent the subset system
composed of all the subsets each corresponding to a product term. Fig. 1 illustrates
the corresponding ZBDD representing all the subsets involved in $\det(\mathbf{M})$ under ordering
$a > c > b > d > f > e > g > i > h > j$. It can be seen that sub-terms ad, gj, and ih have been
shared in the ZBDD representation.

Following directly from the properties of ZBDDs, we have the following observations. First,
given a fixed order of symbolic parameters, all the subsets in a symbolic determinant can be
represented uniquely by a ZBDD. Second, every 1-path in the ZBDD corresponds to a product
term, and the number of 1-edges in any 1-path is n. The total number of 1-paths is equal to
the number of product terms in a symbolic determinant.

We can view the resulting ZBDD as a graphical representation of the recursive application
of the determinant expansion with the expansion order $a, c, b, d, f, e, g, i, h, j$. Each vertex
is labeled with the matrix entry with respect to which the determinant is expanded, and
it represents all the subsets contained in the corresponding sub-matrix determinant. The
1-edge points to the vertex representing all the subsets contained in the cofactor of the current
expansion, and 0-edge points to the vertex representing all the subsets contained in the
remainder.

To embed the signs of the product terms of a symbolic determinant into its corresponding
ZBDD, we associate each vertex v with a sign, $s(v)$, defined as follows:

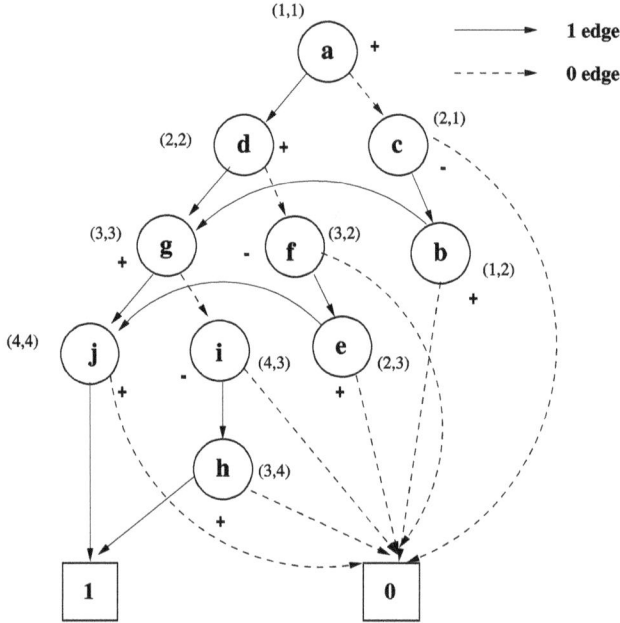

Fig. 2. A signed ZBDD for representing symbolic terms from matrix **M**

- Let $P(v)$ be the set of ZBDD vertices that originate the 1-edges in any 1-path rooted at v.
  Then

$$s(v) = \prod_{x \in P(v)} \text{sign}(r(x) - r(v)) \, \text{sign}(c(x) - c(v)), \qquad (2)$$

where $r(x)$ and $c(x)$ refer to the absolute row and column indices of vertex x in the original matrix, and u is an integer so that

$$\text{sign}(u) = \begin{cases} +1, & \text{if } u > 0, \\ -1, & \text{if } u < 0. \end{cases}$$

- If v has an edge pointing to the 1-terminal vertex, then $s(v) = +1$.

This is called the *sign rule*. For example, in Fig. 2, shown beside each vertex are the row and column indices of that vertex in the original matrix, as well as the sign of that vertex obtained by using the sign rule above. For the sign rule, we have following result:

Theorem 1. *The sign of a DDD vertex v, $s(v)$, is uniquely determined by (2), and the product of all the signs in a path is exactly the sign of the corresponding product term.*

For example, consider the 1-path $acbgih$ in Fig. 2. The vertices that originate all the 1-edges are c, b, i, h, their corresponding signs are $-, +, -$ and $+$, respectively. Their product is $+$. This is the sign of the symbolic product term $cbih$.

With ZBDD and the sign rule as two foundations, we are now ready to introduce formally our representation of a symbolic determinant. Let **A** be an $n \times n$ sparse matrix with a set of distinct m symbolic parameters $\{a_1, ..., a_m\}$, where $1 \le m \le n^2$. Each symbolic parameter a_i

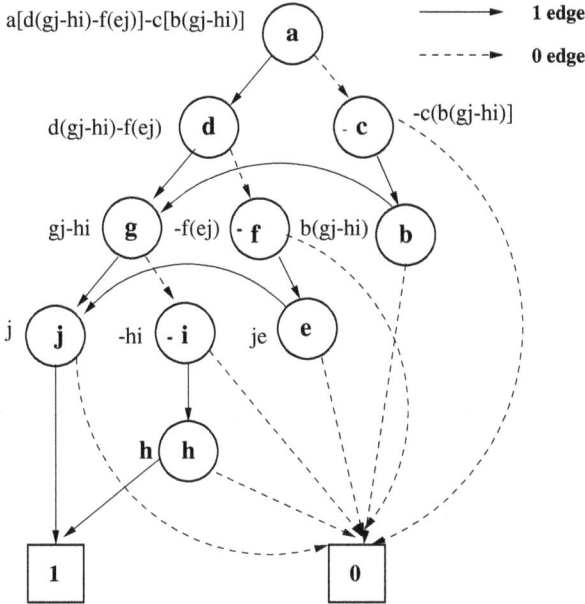

Fig. 3. A determinant decision diagram representation for matrix **M**

is associated with a unique pair $r(a_i)$ and $c(a_i)$, which denote, respectively, the row index and column index of a_i. A *determinant decision diagram* is a signed, rooted, directed acyclic graph with two terminal vertices, namely the 0-terminal vertex and the 1-terminal vertex. Each non-terminal vertex a_i is associated with a sign, $s(a_i)$, determined by the sign rule defined by (2). It has two outgoing edges, called 1-edge and 0-edge, pointing, respectively, to D_{a_i} and $D_{\bar{a}_i}$. A determinant decision graph having root vertex a_i denotes a matrix determinant D defined recursively as

- If a_i is the 1-terminal vertex, then $D = 1$.
- If a_i is the 0-terminal vertex, then $D = 0$.
- If a_i is a non-terminal vertex, then

$$D = a_i s(a_i) D_{a_i} + D_{\bar{a}_i} \tag{3}$$

Here $s(a_i)D_{a_i}$ is the *cofactor* of D with respect to a_i, D_{a_i} is the *minor* of D with respect to a_i, $D_{\bar{a}_i}$ is the *remainder* of D with respect to a_i, and operations are algebraic multiplications and additions. For example, Fig. 3 shows the DDD representation of det(**M**) under ordering $a > c > b > d > f > e > g > i > h > j$.

To enforce the uniqueness and compactness of the DDD representation, the three rules of ZBDDs, namely, zero-suppression, ordered, and shared are adopted. This leads to DDDs having the following properties:

- Every 1-path from the root corresponds to a product term in the fully expanded symbolic expression. It contains exactly n 1-edges. The number of 1-paths is equal to the number of product terms.

- For any determinant D, there is a unique DDD representation under a given vertex ordering.

We use $|DDD|$ to denote the *size of* a DDD, i.e., the number of vertices in the DDD.

Once a DDD has been constructed, its numerical values of the determinant it represents can be computed by performing the depth-first type search of the graph and performing (3) at each node, whose time complexity is linear function of the size of the graphs (its number of nodes). The computing step is call *Evaluate(D)* where D is a DDD root.

For each vertex, there are two values, *vself* and *vtree*. As above mentioned, vself represents the value of the element itself; while vtree represents the value of the whole tree (or subtree). For each vertex, the vtree equals vself multiplying vtree of 1-subtree plus vtree of 0-subtree as shown in (3). In this example, the value of the determinant equals vtree of a; while vtree of a equals vself of a multiplying vtree of b plus vtree of c. In a serial implementation, the tree value of a is computed by recursively computing all vtree of subtrees, which is very time-consuming when the tree becomes large.

One key observation for DDD structure is that the data dependence is very clear. The data dependency is very simple: a node can be evaluated only after its children are evaluated. Such dependency implies the parallelism where all the nodes satisfying this constraint can be evaluated at the same time. Also, in frequency analysis of analog circuits, evaluation of a DDD node at different frequency runs can be performed in parallel. In the following section we show how we can explore such parallelism to speed up the DDD evaluation process.

3. Review of GPU architectures

CUDA (short for Compute Unified Device Architecture) is the parallel computing architecture for NVIDIA many-core GPU processors. The architecture of a typical CUDA-capable GPU is consisted of an array of highly threaded streaming multiprocessors (SM) and comes with more than 4 GBytes DRAM, referred to as global memory. Each SM has eight streaming processor (SP) and two special function units (SFU), and possesses its own shared memory and instruction cache. The structure of a streaming multiprocessor is shown in Fig. 4.

As the programming model of GPU, CUDA extends C into CUDA C, and supports tasks such as threads calling and memory allocation, which makes programmers able to explore most of the capabilities of GPU parallelism. In CUDA programming model, threads are organized into blocks; blocks of threads are organized into grids. CUDA also assumes that both the host (CPU) and the device (GPU) maintain their own separate memory spaces in DRAM, referred to as host memory and device memory, respectively. For every block of threads, a shared memory is accessible to all threads in that same block. And the global memory is accessible to all threads in all blocks. Developers can write programs running millions of threads with thousands of blocks in a parallel approach. This massive parallelism forms the reason that programs with GPU acceleration can be multiple times faster than their CPU counterparts.

One thing to mention is that for some series of CUDA GPU, a multiprocessor has eight single-precision floating point ALUs (one per core) but only one double-precision ALU (shared by the eight cores). Thus, for applications whose execution time is dominated by floating point computations, switching from single-precision to double-precision will decrease performance by a factor of approximately eight. However, this situation is being improved

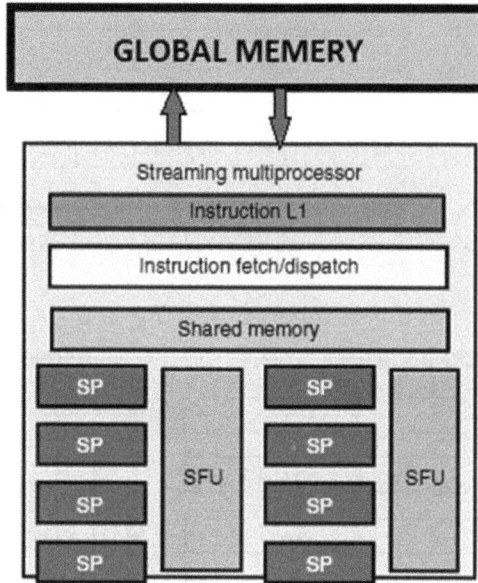

Fig. 4. Structure of streaming multiprocessor.

in NVIDIA T20 series, the Fermi family. These most recent GPU from NVIDIA can already provide much better double-precision performance than before.

4. New GPU-based DDD evaluation

In this section, we present the new GPU-based DDD evaluation algorithm. Before the details of GPU-based DDD evaluation method, we first discuss the new DDD data structure for GPU parallel computing.

One key observation for DDD structure is that the data dependence is very clear. The data dependency is very simple: a node can be evaluated only after its children are evaluated. Such dependency implies the parallelism where all the nodes satisfying this constraint can be evaluated at the same time. Also, in frequency analysis of analog circuits, evaluation of a DDD node at different frequency runs can be performed in parallel. In the following subsections we show how we can explore such parallelism to speed up the DDD evaluation process.

4.1 New data structure

To achieve the best performance on GPU, linear memory structure, i.e., data stored in consequent memory addresses, is preferable. For CPU serial computing, the data structure is based on dynamic links in a linked binary tree. For parallel computing, the data will be stored in linear arrays which can be more efficiently accessed by different threads based on thread ids.

As we discussed above, the DDD representation stores all product terms of the determinant of the MNA matrix in a binary linked tree structure. The vertex in the tree structure is known

DDD node 'A'

sign	MNA idx	RCL	freq	vself	vtree	'left node	'right node

DDD node 'B' ◄

sign	MNA idx	RCL	freq	vself	vtree	'left node	'right node

DDD node 'C' ◄

sign	MNA idx	RCL	freq	vself	vtree	'left node	'right node

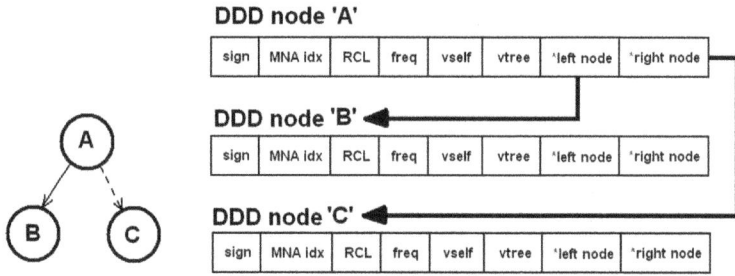

Fig. 5. Illustration of the data structure for serial method

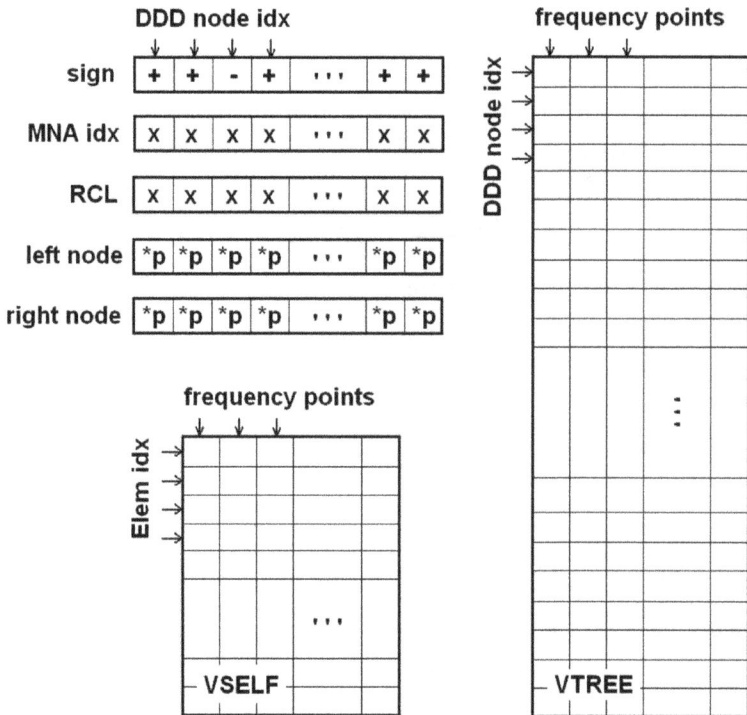

Fig. 6. Illustration of the data structure for parallel method

as DDD node that represents element in MNA matrix which is identified by its index. For each DDD node, the data structure includes the sign value, the MNA index, the RCL values, corresponding frequency value, *vself*, and *vtree*. In the serial approach, these values are stored in a data structure and connected through links, as shown in Fig. 5. On the other side, in the parallel approach, all of these data are stored separately in corresponding linear arrays and each element is identified by the DDD node index (not necessarily the same as the MNA element index). Figure 6 illustrates the new data structure.

Two choices are available for *vself* data structure. One is similar to the data structure of *vtree*. The *vself* value for each DDD node is stored consecutively. This data structure is called the linear version of *vself* data structure. The other method is as shown in Fig. 6. The array is organized per MNA element. Due to the fact that some of the DDD nodes share the same MNA element value, the second data structure is more compact in memory than the linear version. So it is called the compact version of *vself* data structure. The compact version is suitable for small circuits because it reduces the global memory traffic when computing *vself*. However, for large circuits, the calculation of *vtree* dominates the time cost. And we can implement a strategy to reduce the global memory traffic for computing *vtree* using the linear version of *vself* data structure to further improve the GPU performance. Therefore, for larger circuits, the linear version is preferable. The performance comparison is discussed later in the next section.

4.2 Algorithm flow

The parallel evaluating process consists of two stages. First, the *vself* values for all DDD nodes are computed and stored. In this stage, a set of 2D threads are launched on GPU devices. The X-dimension of the 2D threads represents different frequencies; the Y-dimension represents different elements (for compact *vself*) or DDD nodes (for linear *vself*). Therefore, all elements (or DDD nodes) can be computed under all frequencies in massively parallel manners. In the second stage, we simultaneously launch GPU 2D threads to compute all the *vtree* values for DDD nodes based on (3). Notice that a DDD node *vtree* value becomes valid when all its children's *vtree* values are valid. Since we compute all the *vtree* for all the nodes at the same time, the correct *vtree* values will automatically propagate from the bottom of the DDD tree to the top node. The number of such *vtree* iterative computing are decided by the number of layers in DDD tree. A layer represents a set of DDD nodes whose distance from *1-terminal* or *0-terminal* are the same. The number of layers equal to the longest distance between non-terminal nodes and *1-terminal/0-terminal*. Algorithm 1 shows the flow of parallel DDD evaluation using compact *vself* data structure.

Line 3 and Line 4 load frequency index and element index respectively with CUDA built-in variables (Thread.X and Thread.Y are our simplified notations). These built-in variables are the mechanism for identifying data within different threads in CUDA. Then, line 5 and Line 6 compute the *vself* with the RCL value of the element under given frequency. From line 8, loop for computing *vtree* is entered. Line 13 and Line 14 load *vtree* values for left/right branch using function *Then()/Else()*. Line 15 through Line 26 explains themselves. Line 27 computes *vtree* with *vself* and *Left/Right* and ends the flow.

4.3 Coalesced memory access

The GPU performance can be further improved by making proper use of coalesced global memory access to prevent the global memory bandwidth from being performance limitation. Coalesced memory access is one efficient method reducing global memory traffic. When all threads in a warp execute a load instruction, the hardware detects whether the threads access the consecutive global memory address. In such case, the hardware coalesces all of these accesses into a consolidated access to the consecutive global memory. In the implementation of GPU-accelerated DDD evaluation, such favorable data access pattern is fulfilled for the linear version of *vself* data structure to gain performance enhancement. The *vself* data structure is in a linear pattern so that the *vself* values for a given DDD node under a series of frequency

---

Algorithm 1 Parallel DDD evaluation algorithm flow

1: **if** Launch GPU threads for each node **then**
2: {Computing vself:}
3: $FreqIdx \leftarrow Thread.X$
4: $ElemIdx \leftarrow Thread.Y$
5: $(R, C, L) \leftarrow GetRCL(ElemIdx)$
6: $vself \leftarrow (R, C * Freq + L/Freq)$
7: **end if**
8: **for all** lyr such that $0 \leq lyr \leq NumberOfLayers$ **do**
9: {Computing vtree:}
10: **if** Launch GPU threads for each node **then**
11:   $FreqIdx \leftarrow Thread.X$
12:   $DDDIdx \leftarrow Thread.Y$
13:   $Left \leftarrow Then(DDDIdx)$
14:   $Right \leftarrow Else(DDDIdx)$
15:   **if** is $0 - terminal$ **then**
16:    $Left \leftarrow (0, 0)$
17:    $Right \leftarrow (0, 0)$
18:   **else**
19:    **if** is $1 - terminal$ **then**
20:     $Left \leftarrow (1, 0)$
21:     $Right \leftarrow (1, 0)$
22:    **end if**
23:   **end if**
24:   **if** $sign(DDDIdx) < 0$ **then**
25:    $vself \leftarrow -1 * vself$
26:   **end if**
27:   $vtree \leftarrow vself * Left + Right$
28: **end if**
29: **end for**

---

values are stored in coalesced memory. Therefore, threads, in the same block, with consecutive thread index will access consecutive global memory locations, which ensure that the hardware coalesces these accessing process in just one reading operation. In this example, this technique reduces the global memory traffic by a factor of 16. However, for the compact version of *vself* data structure, the *vself* values are stored per elements, which means that for consecutive DDD nodes, their respective *vself* values are not stored in consecutive locations. So, for the compact version of *vself* data structure, the global memory access is not coalesced. The performance comparison for both of versions is discussed in experimental result section.

5. Numerical results

We have implemented both CPU serial version and GPU version of the DDD-based evaluation programs using C++ and CUDA C, respectively.

The serial and parallel versions of programs have been tested under the same hardware and OS configuraions. The computation platform is a Linux server with two Intel Xeon E5620 2.4 GHz Quad-Core CPUs, 36 GBytes memory, equipped with NVIDIA Tesla S1070

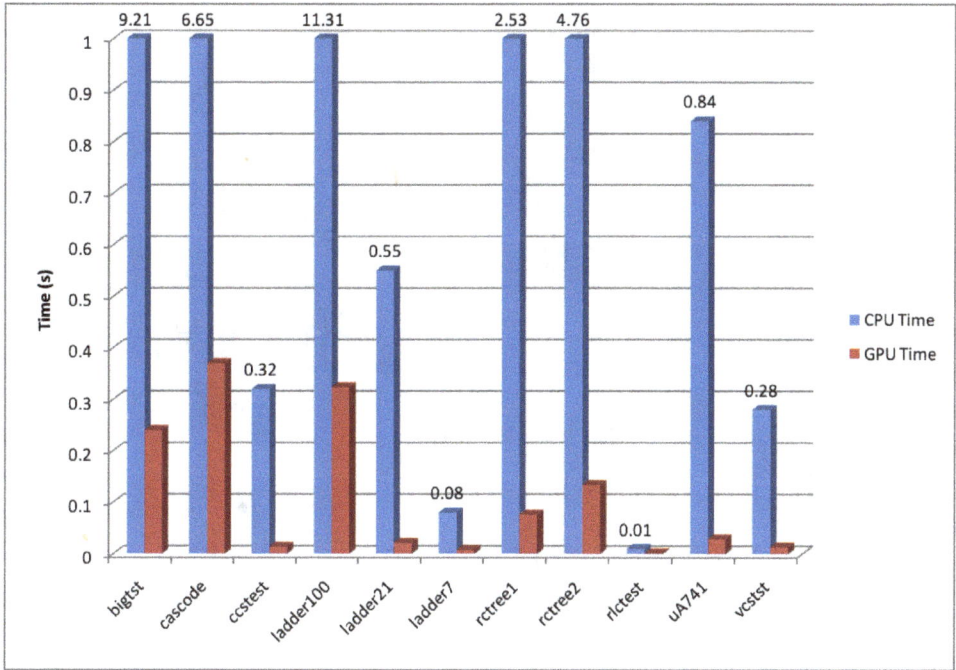

Fig. 7. Performance comparison

1U rack-mounted system (containing four T10 GPUs). The software environment is Red Hat 4.1.2-48 Linux, gcc version 4.1.2 20080704, and CUDA version 3.2.

For the purpose of performance comparison, the programs with CPU-serial and GPU-parallel algorithm are both tested for the same set of circuits. The testing circuits include: μA741 (a bipolar opamp), Cascode (a CMOS Cascode opamp), ladder7, ladder21, ladder100 (7-, 21-, 100-section cascade resistive ladder networks), rctree1, rctree2 (two RC tree networks), rlctest, vcstest, ccstest, bigtst (some RLC filters).

In the two implementations, the same DDD construction algorithm is shared. The numerical evaluation process is done under serial and parallel version separately. The performance comparison for each of the given circuit is listed in Table 1 and illustrated in Fig. 7. In our experimental results, the overhead for data transferring between host and GPU devices are not included as their costs can be amortized over many DDD evaluation processes and can be partially overlapped with GPU computing in more advanced parallelization implementation. The statistics information for DDD representation is also included in the same table. The first column indicates the name of each circuit tested. The second to fourth columns represent the number of nodes in circuit, the number of elements in the MNA matrix and the number of DDD nodes in the generated DDD graph, respectively. The number of determinant product terms is shown in fifth column. CPU time is the time cost for the calculation of DDD evaluation in serial algorithm. The GPU time is the computing time cost for GPU-parallelism (the kernel parts). The final column summerizes the speedup of parallel algorithm over serial algorithm.

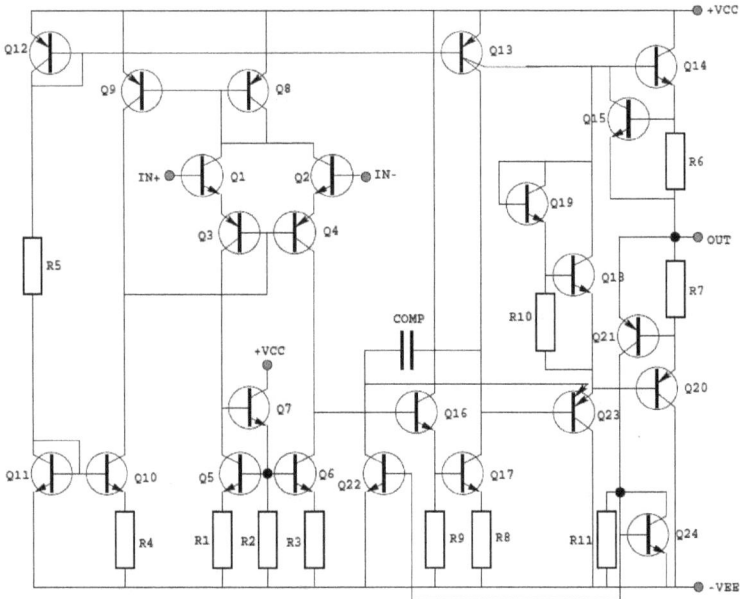

Fig. 8. The circuit schematic of μA741

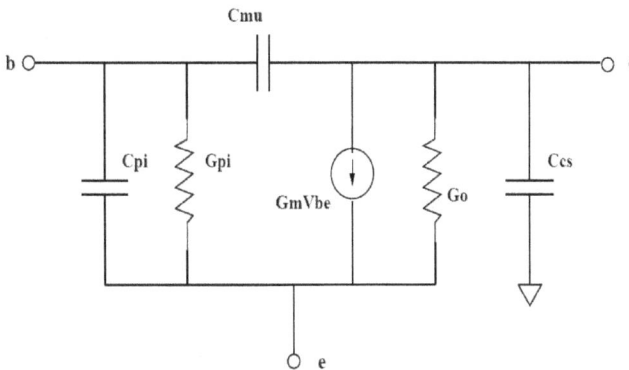

Fig. 9. The small signal model for bipolar transistor

Now let us investigate one typical example in detail. Fig. 8 shows the schematic of a μA741 circuit. This bipolar opamp contains 26 transistors and 11 resistors. DC analysis is first performed by SPICE to obtain the operation point, and then small-signal model, shown in Fig. 9, is used for DDD symbolic analysis and numerical evaluation. The AC analysis is performed using both CPU DDD evaluation and GPU parallel DDD evaluation proposed in our work. In Fig. 10 plots the frequency responses of the gain and the phase at the amplifier's output node from the two comparison methods. It can be observed that GPU parallel DDD evaluation provides the same result as it CPU serial counterpart does. We measured the run

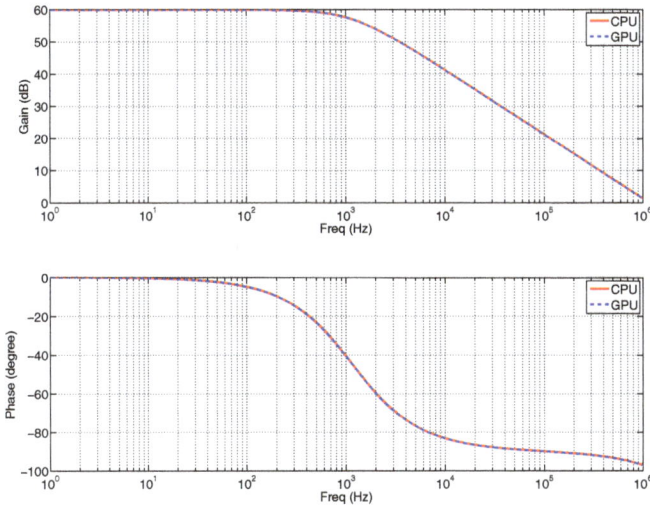

Fig. 10. Frequency response of μA741 amplifier. The red solid curve is the result of CPU DDD evaluation, while the blue dashed line is the result of GPU parallel DDD evaluation.

time of these two methods: the program of CPU evaluation costs 0.84 second, while the GPU parallel version only takes 0.029 second. For this benchmark circuit, we can judge that the parallel computation can easily achieve a speedup of 29 times. As the size of the circuit and the number of DDD nodes grow larger, more speedup can be expected.

From Table 1, we can make some observations. For a variety of circuits tested in the experiment, the GPU-accelerated version outperforms all of their counterparts. The maximum performance speedup is 38.33 times for *bigtst*. The time cost of the serial version is growing fast along the increasing of circuit size (nodes in the circuit). On the other side, however, the GPU-based parallel version performs much better for larger circuits. And more importantly, the larger the circuit is, the better performance improvement we can gain using GPU-acceleration. This trend is illustrated in Fig. 11. This result implies that the

circuit	# nodes	# elements	# DDD nodes	# terms	CPU time (s)	GPU time (s)	speedup
bigtst	32	112	642	2.68×10^{7}	9.21	0.240	38.33
cascode	14	76	2110	2.32×10^{5}	6.65	0.369	18.00
ccstest	9	35	109	260	0.32	0.014	23.40
ladder100	101	301	301	9.27×10^{20}	11.31	0.323	35.00
ladder21	22	64	64	28657	0.55	0.021	25.69
ladder7	8	22	22	34	0.08	0.007	10.86
rctree1	40	119	211	1.15×10^{8}	2.53	0.076	33.30
rctree2	53	158	302	4.89×10^{10}	4.76	0.134	35.51
rlctest	9	39	119	572	0.01	0.001	8.82
μA741	23	89	6205	363914	0.84	0.029	29.14
vcstst	12	46	121	536	0.28	0.013	20.74

Table 1. Performance comparison of CPU-serial and GPU-parallel DDD evaluation for a set of circuits

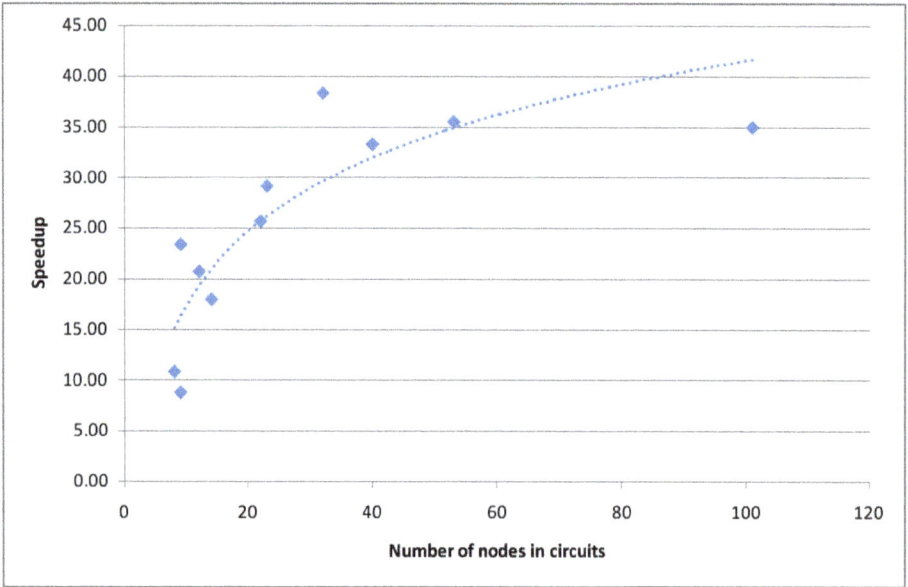

Fig. 11. The performance speedup of GPU-acceleration vs. circuits size (number of nodes)

GPU-acceleration is suitable to overcome the performance problem of DDD-based numerical evaluation for large circuits.

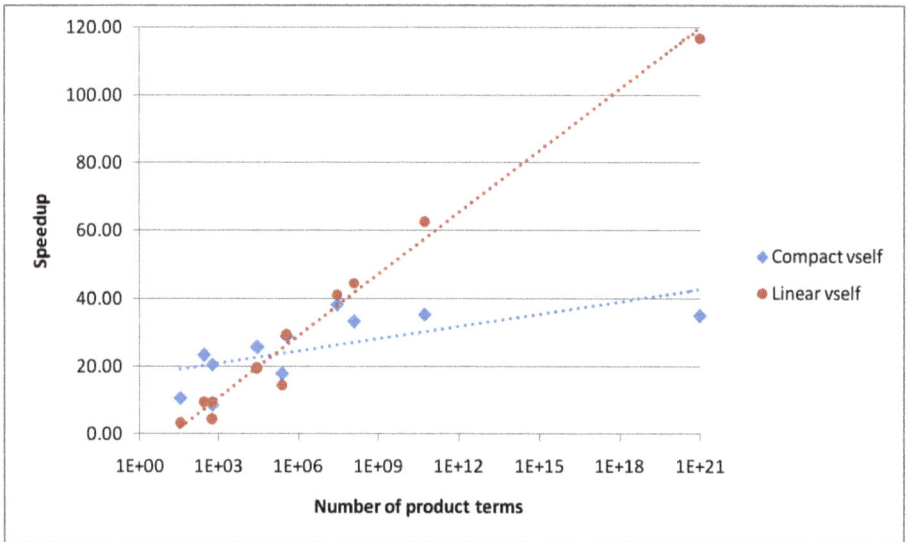

Fig. 12. Performance comparison for two approaches of vself data structure (the x-axis is in logarithm scale)

In this experiment, both of data structures for storing *vself* are implemented. The performance comparison is listed in Table 2. The GPU parallel version under both of the two data structures for *vself* outperforms the serial version. And the performance speedup is clearly related to the number of product terms in MNA matrix, as shown in Fig. 12. For small circuits with less MNA matrix product terms, the compact version of *vself* is more efficient due to the lowering of global memory traffic when calculating *vself*. However, for large circuits with bigger number of MNA matrix product terms, the linear version of *vself* outperforms the compact version *comp vself* owing to the effect of coalesced memory access as discussed in the prior section.

circuit	# terms	CPU (s)	GPU time (s)		speedup	
			w/ comp vself	w/ linear vself	w/ comp vself	w/ linear vself
bigtst	2.68×10^7	9.21	0.240	0.223	38.33	41.21
cascode	2.32×10^5	6.65	0.369	0.452	18.00	14.70
ccstest	260	0.32	0.014	0.033	23.40	9.65
ladder100	9.27×10^{20}	11.31	0.323	0.097	35.00	116.92
ladder21	28657	0.55	0.021	0.028	25.69	19.40
ladder7	34	0.08	0.007	0.025	10.86	3.20
rctree1	1.15×10^8	2.53	0.076	0.057	33.30	44.71
rctree2	4.89×10^{10}	4.76	0.134	0.076	35.51	62.93
rlctest	572	0.01	0.001	0.002	8.82	4.40
μA741	363914	0.84	0.029	0.029	29.14	29.27
vcstst	536	0.28	0.013	0.029	20.74	9.62

Table 2. Performance comparison for two implementations of *vself* data structure

6. Summary

In this chapter, a GPU-based graph-based parallel analysis method for large analog circuits has been presented. Two data structures have been designed to cater the favor of GPU computation and device memory access pattern. Both the CPU version and the GPU version's performance has been studied and compared for circuits with different number of product terms in MNA matrix. The GPU-based DDD evaluation performs much better than its CPU-based serial counterpart, especially for larger circuits. Experimental results from tests on a variety of industrial benchmark circuits show that the new evaluation algorithm can achieve about one to two order of magnitudes speedup over the serial CPU based evaluations on some large analog circuits. The presented parallel techniques can be also used for the parallelization of other decision diagrams based applications such as Binary Decision Diagrams (BDDs) for logic synthesis and formal verifications.

7. References

Akers, S. B. (1976). Binary decision diagrams, *IEEE Trans. on Computers* 27(6): 509–516.

AMD Inc. (2006). Multi-core processors—the next evolution in computing (White Paper). http://multicore.amd.com.

AMD Inc. (2011a). AMD developer center, http://developer.amd.com/GPU.

AMD Inc. (2011b). AMD Steam SDK, http://developer.amd.com/gpu/ATIStreamSDK.

Bryant, R. E. (1986). Graph-based algorithms for Boolean function manipulation, *IEEE Trans. on Computers* pp. 677–691.

Bryant, R. E. (1995). Binary decision diagrams and beyond: enabling technologies for formal verification, *Proc. Int. Conf. on Computer Aided Design (ICCAD)*.

Gielen, G., Wambacq, P. & Sansen, W. (1994). Symbolic analysis methods and applications for analog circuits: A tutorial overview, *Proc. of IEEE* 82(2): 287–304.

Göddeke, D. (2011). General-purpose computation using graphics harware, `http://www.gpgpu.org/`.

Hachtel, G. D. & Somenzi, F. (1996). *Logic Synthesis and Verification Algorithm*, Kluwer Academic Publishers.

Intel Corporation (2006). Intel multi-core processors, making the move to quad-core and beyond (White Paper). `http://www.intel.com/multi-core`.

Khronos Group (2011). Open Computing Language (OpenCL), `http://www.khronos.org/opencl`.

Kirk, D. B. & Hwu, W.-M. (2010). *Programming Massively Parallel Processors: A Hands-on Approach*, Morgan Kaufmann Publishers Inc., San Francisco, CA.

Minato, S. (1993). Zero-suppressed BDDs for set manipulation in combinatorial problems, *Proc. Design Automation Conf. (DAC)*, pp. 272–277.

Minato, S. (1996). *Binary Decision Diagrams and Application for VLSI CAD*, Kluwer Academic Publishers, Boston.

NVIDIA Corporation (2011a). http://www.nvidia.com.

NVIDIA Corporation (2011b). CUDA (Compute Unified Device Architecture). `http://www.nvidia.com/object/cuda_home.html`.

Shi, C.-J. & Tan, X.-D. (2000). Canonical symbolic analysis of large analog circuits with determinant decision diagrams, *IEEE Trans. on Computer-Aided Design of Integrated Circuits and Systems* 19(1): 1–18.

Shi, C.-J. & Tan, X.-D. (2001). Compact representation and efficient generation of s-expanded symbolic network functions for computer-aided analog circuit design, *IEEE Trans. on Computer-Aided Design of Integrated Circuits and Systems* 20(7): 813–827.

Shi, G. (2010a). Computational complexity analysis of determinant decision diagram, *IEEE Trans. on Circuits and Systems II: Analog and Digital Signal Processing* 57(10): 828 –832.

Shi, G. (2010b). A simple implementation of determinant decision diagram, *Proc. Int. Conf. on Computer Aided Design (ICCAD)*, pp. 70 –76.

Tan, S. X.-D., Guo, W. & Qi, Z. (2005). Hierarchical approach to exact symbolic analysis of large analog circuits, *IEEE Trans. on Computer-Aided Design of Integrated Circuits and Systems* 24(8): 1241–1250.

Tan, X.-D. & Shi, C.-J. (2000). Hierarchical symbolic analysis of large analog circuits via determinant decision diagrams, *IEEE Trans. on Computer-Aided Design of Integrated Circuits and Systems* 19(4): 401–412.

8

Switching Noise in 3D Power Distribution Networks: An Overview

Waqar Ahmad and Hannu Tenhunen
KTH Royal Institute of Technology KTH/ICT/ECS/ESD, Stockholm
Sweden

1. Introduction

The design and analysis of a power distribution network is one of the most important areas in high-speed digital systems. The main function of a power distribution network is to supply power to core logic and I/O circuits in any digital system. With increasing clock speeds accompanied by decreasing signal rise times and supply voltages, the transient currents injected into the power distribution planes can induce voltage fluctuations on the power distribution network (Tummala et al., 1997). This undesired voltage fluctuation on the power/ground planes is commonly known as switching noise or delta-I noise. Power supply noise leads to unwanted effects on the power distribution network (PDN) such as ground bounce, power supply compression, and electromagnetic interference. The digital switching noise propagates through the substrate and power distribution networks to analog circuits, degrading their performance in system-on-chip (SoC) applications (Iogra, 2007). The modern advances in process technology along with tremendous increase in number of on-chip devices make the design of on-chip power distribution network a major design challenge. Increased switching activity of high speed devices therefore causes large current derivatives or current transients. These current transients may cause unwanted potential drops in supply voltage due to parasitic resistance and inductance of power distribution network. Over and above, scaling of the supply voltage may cause a large degradation in the signal-to-noise ratio of high speed CMOS circuits (Bai & Hajj, 2002). When on-chip logic cells switch, either they draw current from supply network or inject current into the ground network. If a lot of logic cells switch simultaneously, then they may cause voltage variations within the supply network due to parasitic associated with the power distribution network. This voltage variation is nothing but core switching noise. It is called the voltage surge if variation is above the nominal voltage and is called the sag if variation is below the nominal supply voltage (Bobba & Hajj, 2002). This variation in supply voltage may cause logic errors thereby adversely affecting the circuit performance (Bai & Hajj, 2002). Excessive drop in power bus voltage or surge in ground bus voltage can cause following problems: decrease in the device drive capability, increase in the logic gate delay, and reduction of the noise margin. Hence, it is important to estimate these voltage variations in the power distribution network (Bai & Hajj, 2002). Simultaneous switching noise is mainly caused by the parasitic inductance associated with the power distribution network at high frequency. The power supply level goes down at different nodes in a PDN because of the inductive voltage drop at high frequency. This is because of the simultaneous switching of on-chip logic load as well as drivers connected to the output pins of a chip. The glitch in voltage caused this way is proportional to the number of circuit components switching at a clock edge, the steepness of the clock edge and effective inductance of the power distribution network at this moment. On-chip core switching noise as well as the noise caused by switching of the external drivers is equally important at high operating frequencies of the order of GHz. Scaling down of the technology node shrinks the minimum feature size which in turn increases average capacitive load offered by on-chip core logic. Therefore average charging and discharging currents for on-chip logic load increase. The increased circuit speed also raises di/dt. Therefore, on-chip currents may fluctuate by large amounts within short interval of times. Hence voltage fluctuations caused by switching of on-chip logic load known as core switching noise is very significant and important under these circumstance.

On-chip power distribution noise has become a determining factor in performance and reliability where the core voltage has dramatically dropped to 0.9V for 40nm technology node (Shishuang et al., 2010) and the trend continuous. At the same time jitter tolerance and timing margins are shrinking due to ever increasing clock frequency (Shishuang et al., 2010). The chip-package PDN should therefore be optimized at early design stages to meet I/O jitter specifications and core logic timing closure (Bai & Hajj, 2002). For the IC designers and the signal integrity engineers, the most important issue is to understand how the on-chip transient current load interacts with the entire PDN system, and how the on-chip PDN noise affects the circuit performance (Shishuang et al., 2010). The experimental results in (Shishuang et al., 2010) show that on-chip PDN noise may be much higher even if PCB level PDN noise is well under control.

On-chip noise margins decrease proportionally with the power supply with the scaling of the technology nodes (Coenen & Roermund, 2010). Generally the voltage fluctuation should be kept within 5-10% of the supply voltage in VLSI design (Dennis et al., 2008). When supply voltage is less than the nominal value in synchronous digital circuits, it causes timing violations in a register during a clock period (Dennis et al., 2008). This timing error caused by power supply noise may become permanent when stored in a register (Dennis et al., 2008). Core switching noise has been neglected in past due to higher package inductance as compared to on-chip inductance of the power distribution network. Therefore, switching noise was considered to be inductive voltage noise caused by fast switching I/O drivers (Bathy, 1996; Kabbani & Al-Khalili, 1999; Senthinathan & Prince, 1991; Vaidyanath, 1994;Yang & Brews, 1996). On the other hand today several folds increase in clock frequency as compared to I/O speed accompanied with higher integration densities and scaling of on-chip interconnects has made the core switching noise more critical than ever (Zheng & Tenhunnen, 2001). The supply-noise becomes more problematic when microprocessors share the same substrate as the analogue circuits like PLL (Stark et al., 2008) (Vonkaenel, 2002). Therefore core switching noise may cause jitter in clock frequency thereby reducing the usable cycle time and consequently causing critical path failure in the processor. If core switching noise is extended over several clock cycles, then jitter accumulation will take place thereby causing deviation of each subsequent clock edge more and more from the ideal location (Larsson, 2002). The noise accumulation therefore causes synchronization failure between different clock domains more than the critical path failure. The other side effect of switching noise is substrate noise which may modulate the threshold voltage of MOS devices.

Three-Dimensional (3D) integration is a key enabling technology in today's high speed digital design. The purpose of 3D integration is either to partition a single chip into multiple strata to reduce on-chip global interconnects length (Joyner, 2004) or stacking of chips together through TSVs. By increasing the number of strata from one to four reduces the length of the longest interconnect by 50% with 75% improvement in latency and 50% improvement in interconnect energy dissipation (Meindl, 2002). Using 3D integration the wire-limited clock frequency can be increased by 3.9x and wire-limited area and power can be reduced by 84% (Miendl, 2003) and 51% (Khan et al., 2011) respectively. The power delivery to a 3D stack of high power chips also presents many challenges and requires careful and appropriate resource allocation at the package level, die level, and interstratal interconnect level (Huang et al., 2007). Three-Dimensional (3D) integration provides the potential for tremendously increased level of integration per unit footprint as compared to its 2D counterpart (Xie et al., 2010). While the third dimension introduced this way is attractive for many applications but puts some stringent requirements and bottlenecks on 3D power delivery. The huge current requirements per package pin for 3D integration lead to significant complications in reliable power delivery. A k-tier 3D chip could use k times as much current as a single 2D chip of the same footprint under similar packaging technology (Xie et al., 2010). Through silicon vias used in 3D integration introduce extra resistance and inductance in the power distribution path. The power distribution network impedance has not been kept up with the scaling of technology node due limited wire resources, increased device density and current demands (Xie et al., 2010) and situation is further worsened by 3D integration. The increased IR and Ldi/dt supply noise in 3D chips may cause a larger variation in operating speed leading to more timing violations (Xie et al., 2010). The supply noise overshoot due to inductive parasitic may aggravate reliability issues such as oxide breakdown, hot carrier injection (HCI), and negative bias temperature instability (NBTI) (Sapatnekar, 2009).

Three-Dimensional (3D) integration increases integration density by increasing number of on-chip devices per unit footprint which has following effects with the scaling of technology nodes:

- Tremendous increase in current per unit foot print.
- Increase in power per unit foot print.
- Increase in inductance per unit foot print at high frequency of the order of GHz.
- Consequent rise in switching noise imposed on 3D power distribution network.

Each power distribution TSV pair has to supply a logic load with decoupling capacitance. In order to increase the switching speed, the time constant RC needs to be reduced which means to reduce TSV resistance which is difficult with the scaling of technology node. Even if the instantaneous voltage fluctuation is very small, the periodic nature of digital circuits can cause resonance (Larsson, 1998). The resonance frequency due to effective TSV inductance and decoupling capacitance is given as follows

$$f_r = \frac{1}{2\pi\sqrt{L_{TSV}^{eff}C_{dec}}}$$

In order to prevent oscillations through a TSV pair the resonance frequency should be higher or lower than the system clock frequency. The effective resistance of TSV produces IR-drop, whereas reduces the resonance oscillations by providing damping. If simultaneous switching noise is dominant, the decrease in TSV effective resistance may increase the total noise in a 3D power distribution network.

A three-dimensional (3D) stack of logic dies interconnected through TSVs has to connect the core logic circuits, IO circuits, and rest of the circuits from three-dimensional (3D) stack to the printed circuit board. Therefore, both the simultaneous switching noise and the core switching noise depend on high-speed switching currents through power distribution network in three-dimensional (3D) stack of dies, location and number of power distribution TSV pairs, vias and routing of various serial and parallel signal interfaces on signal layers of the package. The signal distribution TSV pairs share the IO power distribution environment. Three-dimensional (3D) power distribution network parasitic significantly account for core noise, significantly influence the simultaneous switching noise, crosstalk, signal propagation delay and skew between signal distribution TSVs. The transient currents drawn by core logic produce voltage droops across outputs of the core power distribution network thereby degrading the operating frequency performance of the microprocessor. Similar voltage droops are produced across IO power distribution network as a result of total transient current pulled through IO drivers and buffers. These droops weaken the signal driving capability of IO drivers thereby causing signal integrity issues.

2. Simultaneous switching in 3D power distribution network

On-chip simultaneous switching noise is caused by switching of the output buffers or drivers as shown by Figure 1. These drivers have to drive the off chip load. The noise on

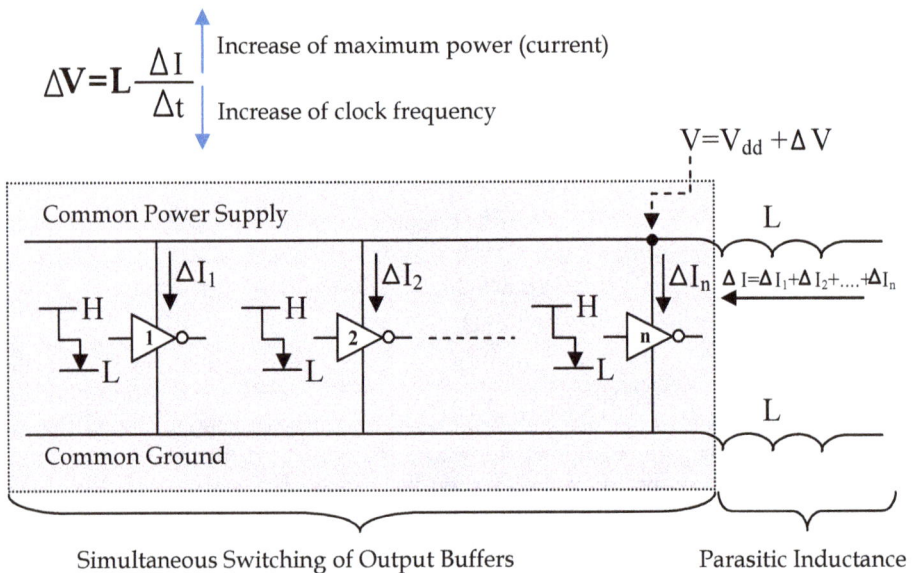

Fig. 1. Simultaneous Switching of Output Drivers

power distribution network depends on the parasitic inductance of the current path between the driver and the output load, the maximum current demand of the load, and the clock frequency.

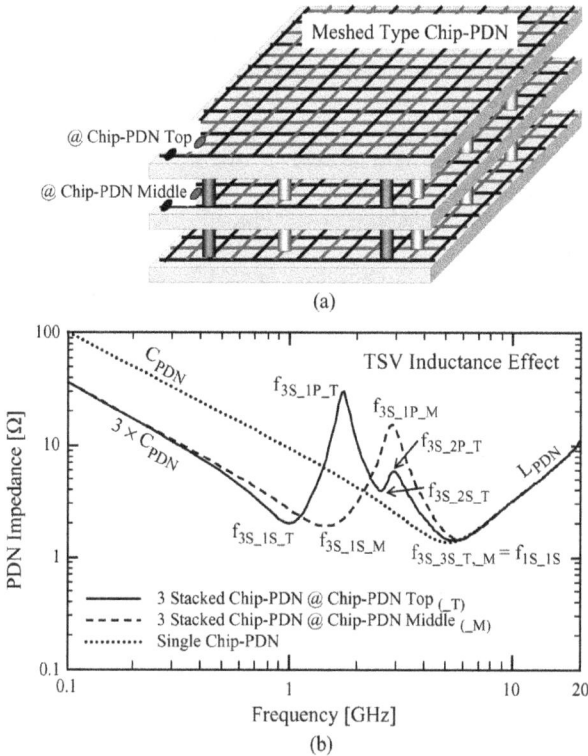

(a)

(b)

Fig. 2. (a) Configuration of three stacked chip-PDN connected by a multi-P/G TSV. (b) Simulated PDN impedances of a single chip-PDN (dotted line) and three stacked chip-PDN (solid and dashed lines) by the proposed separated P/G TSV and chip-PDN models. The Figure is taken from (Pak et al., 2011).

The impedance peaks for the power distribution network of a three-dimensional (3D) stack of chips as compared to a two-dimensional (2D) chip is shown by Figure 2 (Pak et al., 2011). The impedance peaks are in GHz range and are created due to faster switching on the power distribution network. A huge amount of decoupling capacitance is required to suppress these peaks for a three-dimensional (3D) power distribution network (Pak et al., 2011). The impedance peaks are mainly generated due to TSV inductance at high frequency.

Simultaneous switching noise causes the following problems:
- Reduction of the voltage margins.
- Failure of the logic.
- Noise coupling to sensitive circuits like RF and analogue circuits.
- Circuit reliability degradation like decrease in the signal to noise ratio or increase in the noise sensitivity.

3. Core switching noise in 3D power distribution network

Logic cells are connected between supply and ground TSVs for a three-dimensional (3D) power distribution network. Each logic cell has an equivalent capacitance as a load to the power distribution TSV pair. On-chip logic cells switch either low to high or high to low at different clock edges in a synchronous logic system. When on-chip logic cells switch, either they draw current from supply network or inject current into the ground network. If a lot of logic cells switch simultaneously, they may produce voltage variations within the supply network due to the parasitic associated with the power distribution network. This voltage variation is nothing but core switching noise. It is called voltage surge if variation is above the nominal voltage and is called the sag if variation is below the nominal supply voltage (Bobba & Hajj, 2002). The core switching noise depends on the on-chip power distribution network parasitic rather than the package parasitic because of the scaling of the interconnect in modern high speed ULSI design. In addition to that the core switching noise has become more on-chip centric as the package inductance is significantly less as compared to the on-chip inductance at high frequency because of the introduction of BGAs and TSVs in modern packaging. The core switching noise is a major part of the total simultaneous switching noise as the current drawn by the core logic load is generally much higher than the I/O driver's current (Radhakrishnan et al., 2007). The core switching noise in a three-dimensional (3D) stack of logic dies interconnected through TSVs is more significant as compared to 2D ICs due to extra parasitic introduced through vertical TSVs. A large switching noise is introduced in a three-dimensional (3D) power distribution network if various stacked dies switch simultaneously (Huang et al., 2007). The number of power distributions TSVs for a three-dimensional (3D) stack of dies is basically limited by the footprint of the die (Jain et al., 2008) and on top of that the power supply noise is further worsened with the addition of dies in vertical stack due to additional parasitic involved in the power distribution paths through TSVs. The core switching noise can be a critical issue in a system like three-dimensional (3D) multi-processor system-on-chip (3D MOPSoC) (Tao et al., 2010). The core switching noise may introduce common mode noise in mixed analog and digital design as well as increase radiation at resonant frequencies. Overall power consumption in a three-dimensional (3D) stack reduces due to less interconnect, however, power density is increased in parts of three-dimensional (3D) stack due to increase in the number of transistors per unit volume as compared to two-dimensional (2D) counterpart.

3.1 Effects of core switching noise

While the third dimension is attractive for many applications but puts some stringent requirements and bottlenecks on three-dimensional (3D) power delivery. Huge current requirements per package pin for three-dimensional (3D) integration lead to significant complications in reliable power delivery. A k-tier three-dimensional (3D) chip could use k times as much current as a single two-dimensional (2D) chip of the same footprint under similar packaging technology (Xie et al., 2010). Through-silicon-vias used in three-dimensional (3D) integration introduce extra resistance and inductance in the power distribution path. The power distribution network impedance has not been kept up with the scaling of the technology node due to limited wire resources, increased device density and current demands (Xie et al., 2010) and situation is further worsened by 3D integration. The increased IR and Ldi/dt supply noise in three-dimensional (3D) chips may cause a larger variation in operating speed leading to more timing violations (Xie et al., 2010). The supply noise overshoot due to inductive parasitic may aggravate reliability issues such as oxide breakdown, hot carrier injection (HCI), and negative bias temperature instability (NBTI) (Sapatnekar, 2009). The power delivery to a three-dimensional (3D) stack of high power chips also presents many challenges and requires careful and appropriate resource allocation at the package level, die level, and interstratal interconnect level (Huang et al., 2007). Any drop in the core supply voltage directly impacts the maximum operating frequency of the processor (Huang et al., 2007). Simultaneous switching noise originated from the internal logic circuitry has become a serious issue with the increase in speed and density of the internal circuitry.

3.1.1 Propagation delay

The voltage variations due to core switching noise are spread out to the diverse nodes of the power distribution network, thereby causing severe performance degradations in the form of propagation delays (Andarde et al., 2007). The timing violation in a register is produced when value of the supply voltage is less than the nominal value in a synchronous digital circuit. Due to slow variation of the supply voltage as compared to the clock period the value of the supply voltage may remain same for all the gates in a combinational path. Therefore, the value of supply voltage may vary period to period causing severe reliability issues is the logic. The logic cells are prone to more delays with the voltage scaling whereas keeping the threshold voltage relatively constant (Ajami et al., 2003). The propagation delay of the gates increases by 10% with 10% drop in the supply voltage for 180nm (Saleh et al., 2002), by 30% with 10% variation in the supply voltage for 130nm (Pant et al., 2004), and by 4% with 1% change in the supply voltage for 90nm technology node (Tirumuri et al., 2004).

3.1.2 Logic errors

The scaling of the threshold voltage with the scaling of the power supply voltage has reduced the noise margins, thereby making the CMOS circuits more vulnerable to the noise (Bobba & Hajj, 2002). The excessive drop in the power voltage or surge in the ground voltage may drop the noise margins of the circuits. Consequently, a circuit may erroneously latch up to a wrong value or switch at a wrong time if magnitude of the voltage surge/droop is greater than the noise margin of the circuit for a given clock edge (Pant et al., 2000). The problem is expected to grow for a three-dimensional (3D) stack of logic dies interconnected through TSVs, with the addition of each die in the vertical stack.

3.1.3 Impairing driving capabilities of a gate

The gates are becoming increasingly sensitive to the switching noise due to limited scaling of the threshold voltage as compared to the supply voltage scaling with each technology node (Junxia et al., 2009). The reduction in supply voltage not only reduces the noise immunity but also produces signal integrity as well as the performance and the reliability issues. The voltage spikes are droops produced in the power distribution network due to core switching because of the parasitic inductance and resistance associated with the power supply network. The excessive drop in power voltage or surge in the ground voltage, therefore, slows down the cell transition capability, thereby seriously compromising the cell driving capabilities. Consequently clock skews are produced as a result of violations in the setup and the hold times.

3.1.4 Gate oxide reliability issue

With the scaling of power supply voltage the thickness of the gate oxide in modern CMOS VLSI circuits is very thin in order to reduce the nominal supply voltage (Ming-Dou & Jung-Sheng, 2008). The excessive surge in power voltage or drop in ground voltage, therefore, may cause the transistor gate oxide reliability issue due to the electrical over-stress.

3.1.5 Hot carrier injection (HCI)

The channels of CMOS devices are already very short in length due to scaling with the technology nodes. Therefore, excessive surge in supply voltage or drop in ground voltage may cause the carriers to inject into the substrate or the gate oxide due to over voltage thereby depleting the drain-channel junction. It is called hot carrier injection and occurs when the transistor is in saturation (or switched). Consequently, it increases the switching time of an NMOS device and decreases the switching time of a PMOS device.

3.1.6 Cross coupling of through-silicon-vias (TSVs)

The results in (Liu et al., 2011) show that TSVs cause a significant coupling noise and timing problems even if TSV count is significantly less compared to the gate count. The core switching noise through power distribution TSVs may directly or through substrate couple to I/O drivers power supply network, signal links, clock lines, and analog components of the chip. In addition to that the core switching noise may couple to neighboring dies through TSVs for a three-dimensional (3D) stack of dies interconnected through TSVs as the substrates of different planes may essentially be biased through a common ground (Salman, 2011). Various inter-plane noise coupling paths have been identified in (Salman et al., 2010). The results in (Radhakrishna et al., 2007) show that I/O voltage is more sensitive to transient currents produced by switching of the core logic. Power distribution TSVs have significant capacitance due to larger in size as compared to signal TSVs and therefore, may produce significant noise coupling to substrate. The coupling noise from a power distribution TSV may cause path delay in signal line due to Miller effect. The coupling noise through power distribution TSVs may cause charge sharing to dynamic logic, thereby flipping the signal unintentionally or may change the state of a sequential element in static logic.

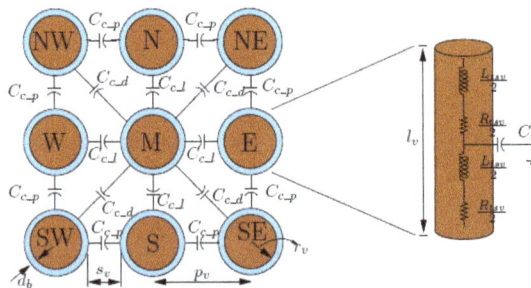

Fig. 3. Cross coupling of a TSV to nearby TSVs where M indicates the middle TSV and E, W, N, S, SE, SW, NW, and NE indicate the locations of closer TSVs to the middle TSV. The diagram is taken from (Roshan, 2008).

Figure 3 shows the capacitive coupling of a given TSV to all the six TSVs arount it. This coupling depends on the distance of a TSV from other TSVs as well as the size of the TSVs. The capacitive coupling can be much stronger in bulky TSVs having significant hight. The coupling also, increases by increasing the density of TSVs. In addition to that, the coupling is inversly proportional to the thickness of the barrier layer around a TSV.

4. How to overcome core switching noise

The core switching noise depends on the amount of logic load driven on rising/falling edge of the clock, sharpness of the clock edge (i.e. rise time), and nature of the network between power supply and logic load. The core switching noise can be reduced through different ways like placing on-chip decoupling capacitance close to the load, placing integrated decoupling capacitance with lower values of ESL and ESR into the substrate, keeping the output impedance across load as close to the target impedance as possible, and determining the optimum value of the damping factor for the power distribution network between supply and load. The rise time increases with the speed of the circuit and logic load increases with the integration density of transistors with each technology node and the problem is exacerbated for three-dimensional (3D) power distribution network.

4.1 Using on-chip decoupling capacitance

Decoupling capacitors are used as charge reservoirs to reduce the power supply noise during switching of the on-chip logic load. The decoupling capacitor is placed across the power and ground conductors to satisfy the target impedance that should be met at all the specified frequencies (Yamamoto & Davis, 2007). Practically, the decoupling capacitance is not a pure capacitance at high frequency because of the intrinsic effective series inductance and effective series resistance. Above the resonance frequency, the impedance of the decoupling capacitance appears inductive and the decoupling capacitance is therefore, not effective as desired above the self resonance frequency. Figure 4 (a) (Jakushokas et al., 2011) shows that the impedance of a power distribution network is resistive at low frequency, whereas it increases linearly with the frequency for higher frequencies due to the dominance of the inductive reactance of the network. There is a maximum frequency ω_{max} at which the network impedance exceeds the target impedance. Figure 4 (b) (Jakushokas et al., 2011) shows that the impedance of the network shoots up at the resonance frequency by using decoupling capacitance as compared to the no decoupling capacitance case. It is because of the parallel resonance produced by the LC tank circuit which produces the maximum impedance. However, above this frequency, the impedance starts increasing linearly with the frequency because of the dominance of the inductive reactance of the network at high frequency.

Figure 4 (Jakushokas et al., 2011) shows that the target impedance is reached at a higher frequency when using decoupling capacitance as compared to without the use of decoupling capacitance. Therefore, decoupling capacitance is used to increase the frequency at which the impedance of the power distribution network exceeds the target impedance. The impedance of a decoupling capacitance is equal to the effective series resistance of a capacitor at the resonance frequency. A logic gate that is not switching connects its output load capacitance to either the positive supply or ground (Dally & Poulton, 1998). Thus, most of the time these output loads serve as symbiotic bypass capacitors that help maintain the supply voltage during current transients (Dally & Poulton, 1998). The method of calculating the symbiotic bypass capacitance is also given by (Dally & Poulton, 1998). The symbiotic bypass capacitance, therefore, enhances the strength of the intentional on-chip decoupling capacitance. However, too large decoupling capacitance reduces the resonance frequency of the power distribution network. Therefore, there is always a tradeoff between the resonance frequency and the amount of decoupling capacitance.

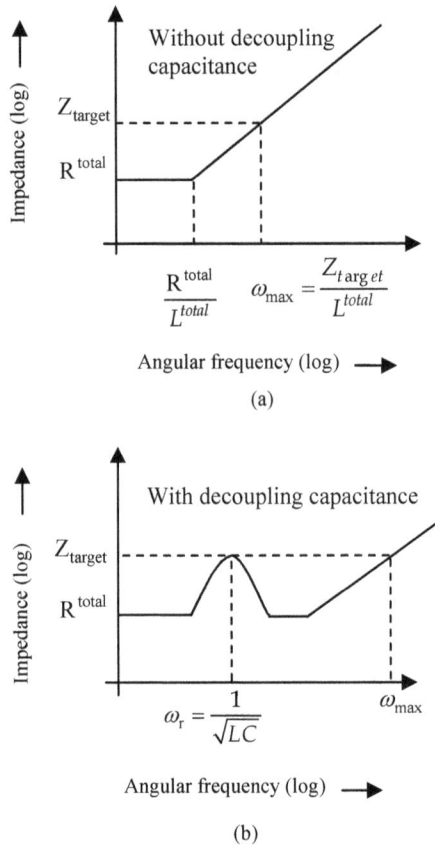

Fig. 4. (a) Frequency response of the impedance of a power distribution network without decoupling capacitance. (b) Frequency response of the impedance of a power distribution network with decoupling capacitance (Jakushokas et al., 2011).

4.2 Using integrated decoupling capacitance for 3D chip stack

The decoupling capacitors integrated into the Si substrate can provide high capacitance at low cost (Sharma et al., 2008). The integrated decoupling capacitors are famous for comparatively low effective series resistance and inductance at high frequency. The effective series resistance and inductance can be further brought down by inserting banks of small parallel decoupling capacitors in the substrate. The integrated decoupling capacitors provide noise immunity and improved power distribution for a monolithic three-dimensional (3D) chips stack (Dang et al., 2009). Therefore, the integrated decoupling capacitors are more attractive for high frequency three-dimensional (3D) integrated systems.

5. Resonance and damping in 3D power distribution network

There may be resonance oscillations in the power distribution network by adding on-chip decoupling capacitance (Bakoglu et al., 1990). A three-dimensional (3D) power distribution network has lower resonance frequency as compared to its two-dimensional (2D) counterpart (Jain et al., 2008). The resonance oscillations produced this way may cause worst case noise accumulation during subsequent clock cycles if not damped in a proper way. The on-chip decoupling capacitance should therefore be selected with a significant ESR (effective series resistance) in order to damp the resonance oscillations. The logic load on each die may have a resonance frequency as a result of the interaction between inductance of the power distribution TSV pairs and decoupling capacitance across the logic load. The damping is only required in the frequency domain around the resonance frequency, rather than at all the frequencies, therefore decoupling capacitance should be selected to have maximum ESR (effective series resistance) around the resonance frequency. The peak-to-peak ground noise on a power distribution TSV pair is given by (Larsson, 1998) through the following equation, assuming that TSV pair forms an under-damped system with damping factor less than one:

$$\Delta v_{pp} = \Delta v \left(1 + e^{-\frac{\pi \zeta}{\sqrt{1-\zeta^2}}} \right)$$

Where

Δv_{pp} = Peak-to-peak ground noise on TSV pair.

Δv = Peak ground noise on TSV pair.

$\zeta = \dfrac{R_{TSV}^{eff}}{2} \sqrt{\dfrac{C_{dec}}{L_{TSV}^{eff}}}$ = Damping factor for the power distribution TSV pair.

R_{TSV}^{eff} = Effective series resistance associated with a power distribution TSV pair.

L_{TSV}^{eff} = Effective series inductance associated with a power distribution TSV pair.

C_{dec} = On-Chip decoupling capacitance associated with a decoupling capacitance.

The performance and reliability of a three-dimensional (3D) power distribution network also depends on the magnitude and duration of the resonance oscillations. These oscillations must be controlled or significantly damped, otherwise noise accumulation will take place at subsequent clock cycles. The damping factor should have significant value in order to suppress the resonance oscillations. The effective resistance of a power distribution TSV pair should be kept much higher than the effective inductance of the power distribution TSV pair in order to increase the value of the damping factor.

6. TSV-induced substrate noise in 3D integrated circuits

Through-Silicon-Via (TSV) is a cylindrical metallic structure that is assumed to be used for power/signal distribution in a three-dimensional (3D) stack of dies. It has dielectric layer around it and normally passes through the Si-substrate in vertical direction. Figure 5 shows the cross section of a Si-Substrate with MOSFET transistor and TSV. Part of the Signal/logic switching transition though TSV can cross-through the barrier layer and may pass through the substrate and impact the performance of neighboring active devices and TSVs, known as TSV induced noise. It depends on TSV to device distance and substrate contacts. The transitions through TSVs may vary the body voltage V_B of the MOSFET device. TSVs-induced substrate noise is almost directly related to the density of TSVs.

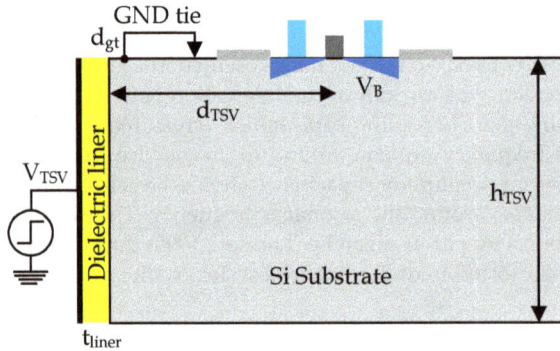

Fig. 5. Cross-section view of TSV-to-device coupling (Khan, 2009, 2011).

Figure 6 shows variations in the MOSFET device body voltage, V_B, for different distances from a TSV for the set of design parameters shown by this Figure. The transitions are very short lived with only 50ps transition time.

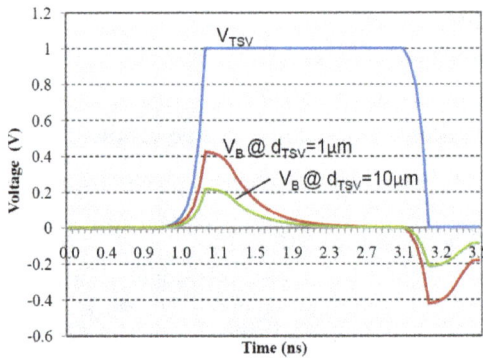

Fig. 6. Body voltage during TSV signal transition at different distances, d_{TSV}, for V_{TSV}=1V square wave, h_{TSV}=20um, t_{liner}=1um, d_{gt}=0.5um, signal transition time=50ps (Khan, 2009, 2011).

7. Summary and future work

On-chip switching noise for a three-dimensional (3D) power distribution network has deleterious effects on power distribution network itself as well the active devices. The extent of switching noise is related to the TSV density on one hand, whereas the integration density of on-chip devices on the other hand. Peaks of the switching noise largely depend on effective inductance of the power distribution network at high frequencies of the order of GHz. Therefore, efficient implementation of on-chip decoupling capacitance along with other on-chip inductance reduction techniques at high frequency is necessary to overcome the switching noise. In addition to that some accurate and efficient modeling techniques are also necessary for early estimation of the switching noise in order to lay down the rest of the design parameters for a three-dimensional (3D) power distribution network.

8. Acknowledgment

The author would like to acknowledge European Union research funding under grant FP7-ICT-215030 (ELITE) of the 7th framework program and Higher Education Commission (HEC), Pakistan, for the support of this work.

9. References

Ajami, A.H.; Banerjee, K.; Mehrotra, A. and Pedram, M. (2003). Analysis of IR-Drop Scaling with Implications for Deep Submicron P/G Network Designs, *Proceeding of IEEE International Symposium on Quality Electronic Design*, ISBN 0-7695-1881-8, pp. 35-40, San Jose, CA, USA, March 2003

Andrade, D.; Martorell, F.; Pons, M.; Moll, F. and Rubio, A. (2007). Power Supply Noise and Logic Error Probability, *Proceedings of IEEE European Conference on Circuit Theory and Design*, pp. 152-155, ISBN 978-1-4244-1341-6, University of Sevilla, Seville, Spain , August 2007

Gen Bai, and Hajj I.N. (2002), Simultaneous switching noise and resonance analysis of on-chip power distribution network. *Proceedings of International Symposium on Quality Electronics Design*, pp. 163-168, ISBN 0-7695-1561-4, Sanjose, California, USA, March 2002

Bakoglu H.B. (1990). *Circuits, Interconnections and Packaging for VLSI, Reading, MA: Addison-Wesley,1990*

Bathey, K.; Swaminathan, M.; Smith, L.D.; and Cockerill, T.J. (1996). Noise computation in single chip packages. *IEEE Transactions on Components, Packaging,and Manufacturing Technology*, Vol. 19, No. 2, (May 1996), pp. 350–360, ISSN 1070-9894

Bing Dang; Wright, S.L.; Andry, P.; Sprogis, E.; Ketkar, S.; Tsang, C.; Polastre, R.; and Knickerbocker, J. (2009). 3D Chip Stack with Integrated Decoupling Capacitors. *Proceedings of IEEE Electronic Components and Technology Conference*, pp. 1-5, ISBN 978-1-4244-4475-5, June 2009

Bobba, S., and Hajj, I.N. (1999). Simultaneous Switching Noise in CMOS VLSI Circuits, *Proceedings of IEEE Symposium on Mixed-Signal Design*, pp. 15-20, ISBN 0-7803-5510-5, August 2002.

Coenen, M.; and Roermund A.V. (2010). Noise Reduction in Nanometer CMOS. *Proceedings of IEEE Asia-Pacific International Symposium on Electromagnetic Compatibility*, pp. 1060-1063, ISBN 978-1-4244-5621-5, Beijing, China, April 2010.

Dally and Poulton (1998). *Digital System Engineering*. Cambridge UK: Cambridge Univ. Press, 1998, pp. 237-242.

Gang Huang; Bakir, M.; Naeemi, A.; Chen, H.; and Meindl, J.D. (2007). Power Delivery for 3D Chips Stack: Physical Modeling and Design Implication. *Proceedings of IEEE Electrical Performance of Electronic Packaging and Systems*, pp. 205-208, ISBN 978-1-4244-0883-2, Atlanta, GA, USA, November 2007

Jain, P.; Tae-Hyoung Kim; Keane, J.; and Kim, C.H. (2008). A Multi-Story Power Delivery Technique for 3D Integrated Circuits. *Proceedings of IEEE International Symposium on Low power Electronics and Design*, pp. 57-62, ISBN 978-1-4244-8634-2, Bangalore, India , August 2008

Jakushokas, R.; Popovich, M.; Mezhiba, A.V.; Köse, S.; and Friedman, E.G. (2011). *Power Distribution Networks with On-Chip Decoupling Capacitors*. Berlin, Germany: Springer-Verlag, 2011

Joyner, J.W.; Zarkesh-Ha, P.; and Meindl, J.D. 82004). Global Interconnect Design in a Three Dimensional System-on-a-Chip. *IEEE Transactions on Very Large Scale Integration Systems*,Vol. 12, No. 4, April 2004, pp. 367-372, ISSN 1063-8210

Junxia Ma; Lee, J.; and Tehranipoor, M. (2009). Layout-Aware Pattern Generation for Maximizing Supply Noise Effects on Critical Paths. *Proceedings of IEEE VLSI Test Symposium*, pp. 221-226, ISBN 978-0-7695-3598-2, Santa Cruz, California, USA, June 2009.

Khan, N.H.; Alam, S.M.; and Hassoun, S. (2009). Through-silicon Via (TSV)-induced Noise Characterization and Noise Mitigation using Coaxial TSVs. *Proceedings of IEEE International Conference on 3D System Integration*, pp. 1-7, ISBN 978-1-4244-4511-0, San Francisco, USA, September 2009

Khan, N.H.; Alam, S.M.; Hassoun, S. (2011). Mitigating TSV-induced Substrate Noise in 3-D ICs using GND plugs. *Proceedings of IEEE International Symposium on Quality Electronics Design*, pp. 1-6, ISSN 1948-3287, SANTA CLARA, CA, USA, March 2011

Larsson, P. (1998). Resonance and Damping in CMOS Circuits with On-Chip Decoupling Capacitance. *IEEE Transactions on Circuits and Systems I: Fundamental Theory and Applications*, Vol. 45, No. 8, ISSN 1057-7122, August 2002, pp. 849-858

Larsson, P. (1999). Power Supply Noise in Future IC's: A Crystal Ball Reading. *Proceedings of IEEE Custom Integrated Circuits*, pp. 467-474, ISBN 0-7803-5443-5, San Diego, CA, USA, August 2002

Liu, Chang; Song, Taigon; Cho, Jonghyun; Kim, Joohee; Kim, Joungho; Lim, Sung Kyu (2011). Full-Chip TSV-to-TSV Coupling Analysis and Optimization in 3D IC. *Proceedings of IEEE Design Automation Conference*, pp. 783-778, ISBN 978-1-4503-0636-2, San Diego, USA, August 2011

Meindl, J.D. (2002). The Evolution of Monolithic and Polylithic Interconnect Technology. *IEEE Symposium on VLSI Circuit Design*, pp. 2-5, ISBN 0-7803-7310-3, Honolulu, Hawaii, USA, August 2002

Ming-Dou Ker; and Jung-Sheng Chen (2008). Impact of MOSFET Gate-Oxide Reliability on CMOS Operational Amplifier in a 130-nm Low-Voltage Process. *IEEE Transactions on Device and Materials Reliability*, Vol. 8, No. 2, pp. 394-405, ISSN 1530-4388, June 2008

Jun So Pak; Joohee Kim; Jonghyun Cho; Kiyeong Kim; Taigon Song; Seungyoung Ahn; Junho Lee; Hyungdong Lee; Kunwoo Park; Joungho Kim. PDN Impedance Modeling and Analysis of 3D TSV IC by Using Proposed P/G TSV Array Model Based on Separated P/G TSV and Chip-PDN Models. *IEEE Transactions on Components, Packaging, and Manufacturing Technology*, Vol. 1, No. 2, pp. 208-219, ISSN 2156-3950, February 2011

Pant, M.D.; Pant, P.; Wills, D.S.; Tiwari, V. (2000). Inductive Noise reduction at the Architectural Level. *Proceedings of IEEE International Conference on VLSI Design*, pp. 162-167, ISBN 0-7695-0487-6, Calcutta, India, January 2000

Radhakrishnan, K.; Li, Y.-L.; and Pinello, W.P. (2001). Integrated Modeling Methodology for Core and I/O Power delivery. *Proceedings of IEEE Electronic Components and Technology Conference*, pp. 1107-1110, ISBN 0-7803-7038-4, Orlando, Florida, USA, May 2001

Roshan Weerasekera, *System interconnection design trade-offs in three dimensional integrated circuits*, PhD Dissertation: Royal Institute of Technology (KTH), Stockholm, Sweden, November 2008

Saleh, R.; Hussain, S.Z.; Rochel, S.; Overhauser, D. (2000). Clock skew verification in the presence of IR-Drop in the power distribution network. *IEEE Transactions on Computer-Aided Design of Integrated Circuits and Systems*, Vol. 19, No. 6, ISSN 0278-0070, August 2002

Salman, Emre; Doboli A.; and Stanacevic, M. (2010). Noise and Interference Management in 3-D Integrated Wireless Systems. *Proceedings of the International Conference and Expo on Emerging Technologies for a Smarter World*, September 2010

Salman, Emre (2011). Noise Coupling Due To Through Silicon Vias (TSVs) in 3-D Integrated Circuits. *Proceedings of IEEE International Symposium on Circuits and Systems*, pp. 1411-1414, ISNN 0271-4302, Rio de Janeiro, Brazil, July 2011

Sapatnekar, S.S. (2009). Addressing Thermal and Power Delivery Bottlenecks in 3D Circuits. *Proceedings of IEEE Asia and South Pacific Design Automation Conference*, pp. 423-428, 978-1-4244-2748-2, Taiwan, January 2009

Sharma, U.; Gee, H.; Liou, D.; Holland, P.; Rong Liu (2008). Integration of Precision Passive Components on Silicon for Performance Improvements and Miniaturization. *Proceedings of Electronics System-Integration Technology Conference*, pp. 485-490, ISBN 978-1-4244-2813-7, British Columbia, November 2008

Shishuang Sun, Larry D Smith, and Peter Boyle, *On-chip PDN Noise Characterization and Modeling*, DesignCon, January 2010

Strak, A.; Gothenberg, A.; Tenhunen, H. (2008). Power-supply and substrate-noise-induced timing jitter in non-overlapping clock generation circuits. *IEEE Transactions on Circuits and Systems*, Vol. 55, No. 4, pp. 1041-1054, ISSN 1549-8328, May 2008

Shuai Tao; Yu Wang; Jiang Xu; Yuchun Ma; Yuan Xie; Huazhong Yang (2010). Simulation and Analysis of P/G Noise in TSV Based 3D MPSoC. *Proceedings of IEEE International Conference on Green Circuits and Systems*, pp. 573-577, 978-1-4244-6876-8, Shanghai, China, June 2010

Tirumurti, C.; Kundu, S.; Sur-Kolay, S.; Yi-Shing Chang (2004). A Modeling Approach for Addressing Power Supply Switching Noise Related Failures of Integrated Circuits.

Proceedings of Design, Automation and Test in Europe Conference, pp. 1078-1083, Paris, France, March 2004

Tummala, R.R.; E.J. Rymaszewski, and A.G. Klopfenstein, *Microelectronics Packaging Handbook,* pt. I, second edn. New York: Chapman Hall, 1997

Vonkaenel, V.R. (2002). A High-Speed, Low-Power Clock Generator for a Microprocessor Application. *IEEE Journal of Solid-State Circuits,* Vol. 33, No. 11, pp. 1634-1639, August 2002

Xie, Yuan; Cong, Jason; Sapatnekar, Sachin, *Three-Dimensional Integrated Circuit Design,* Integrated Circuits and Systems, Springer: 2010

Yamamoto, H.; Davis, J.A. (2007). Decreased effectiveness of on-chip decoupling capacitance in high-frequency operation. *IEEE Transactions on Very Large Scale Integration System,* Vol. 15, No. 6, pp. 649-659, ISSN 1063-8210, June 2007

Zheng, L-R; Tenhunen, H. (2001). Fast Modeling of Core Switching Noise on Distributed LRC Power Grid in ULSI Circuits. *IEEE Transactions on Advanced Packaging,* Vol. 24, No. 3, pp. 245-254, August 2001

9

A Multilevel Approach Applied to Sat-Encoded Problems

Noureddine Bouhmala
Vestfold University College
Norway

1. Introduction

1.1 The satisfiability problem

The satisfiability problem (SAT) which is known to be NP-complete (7) plays a central role problem in many applications in the fields of VLSI Computer-Aided design, Computing Theory, and Artificial Intelligence. Generally, a SAT problem is defined as follows. A propositional formula $\Phi = \bigwedge_{j=1}^{m} C_j$ with m clauses and n Boolean variables is given. Each Boolean variable, $x_i, i \in \{1, \ldots, n\}$, takes one of the two values, *True* or *False*. A clause , in turn, is a disjunction of literals and a literal is a variable or its negation. Each clause C_j has the form:

$$C_j = \left(\bigvee_{k \in I_j} x_k \right) \vee \left(\bigvee_{l \in \bar{I}_j} \bar{x}_l \right),$$

where $I_j, \bar{I}_j \subseteq \{1, \ldots . n\}$, $I \cap \bar{I}_j = \varnothing$, and $\bar{x}_i$ denotes the negation of x_i. The task is to determine whether there exists an assignment of values to the variables under which Φ evaluates to *True*. Such an assignment, if it exists, is called a satisfying assignment for Φ, and Φ is called satisfiable. Otherwise, Φ is said to be unsatisfiable. Since we have two choices for each of the n Boolean variables, the size of the search space S becomes $|S| = 2^n$. That is, the size of the search space grows exponentially with the number of variables. Since most known combinatorial optimization problems can be reduced to SAT (8), the design of special methods for SAT can lead to general approaches for solving combinatorial optimization problems. Most SAT solvers use a Conjunctive Normal Form (CNF) representation of the formula Φ. In CNF, the formula is represented as a conjunction of clauses, with each clause being a disjunction of literals. For example, $P \vee Q$ is a clause containing the two literals P and Q. The clause $P \vee Q$ is satisfied if either P is *True* or Q is *True*. When each clause in Φ contains exactly k literals, the resulting SAT problem is called k-SAT.

The rest of the paper is organized as follows. Section 2 provides an overview of algorithms used for solving the satisfiability problem. Section 3 reviews some of the multilevel techniques that have been applied to other combinatorial optimization problems. Section 4 gives a general description of memetic algorithms. Section 5 introduces the multilevel memetic algorithm. Section 6 presents the results obtained from testing the multilevel memetic algorithm on large industrial instances. Finally, in Section 7 we present a summary and some guidelines for future work.

2. SAT solvers

One of the earliest local search algorithms for solving SAT is GSAT (36)(3)(38). Basically, GSAT begins with a random generated assignment of values to variables, and then uses the steepest descent heuristic to find the new variable-value assignment which best decreases the number of unsatisfied clauses. After a fixed number of moves, the search is restarted from a new random assignment. The search continues until a solution is found or a fixed number of restarts have been performed. Another widely used variant of GSAT is the Walksat algorithm and its variants in (37)(30)(17)(26)(27)(18). It first picks randomly an unsatisfied clause, and then, in a second step, one of the variables with the lowest *break count*, appearing in the selected clause, is randomly.

Other algorithms (11)(15) (9) (10) used history-based variable selection strategies in order to avoid flipping the same variable selected.In parallel to the development of more sophisticated versions of randomized improvement techniques, other methods based on the idea of modifying the evaluation function (47)(19)(40)(34)(35) in order to prevent the search from getting stuck in non attractive areas of the underlying search space have become increasingly popular in SAT solving. A new approach to clause weighting known as Divide and Distribute Fixed Weights (DDFW) (20) exploits the transfer of weights from neighboring satisfied clauses to unsatisfied clauses in order to break out from local minima. Recently, a strategy based on assigning weights to variables (33) instead of clauses greatly enhances the performance of the Walksat algorithm, leading to the best known results on some benchmarks.

Evolutionary algorithms are heuristic algorithms that have been applied to SAT and many other NP-complete problems. Unlike local search methods that work on a current single solution, evolutionary approaches evolve a set of solutions. GASAT (21)(24) is considered to be the best known genetic algorithm for SAT. GASAT is a hybrid algorithm that combines a specific crossover and a tabu search procedure. Experiments have shown that GASAT provides very competitive results compared with state-of-art SAT algorithms. Gottlieb at al. proposed several evolutionary algorithms for SAT (13). Results presented in that paper show that evolutionary algorithms compare favorably to Walksat. Finally, Boughaci et al. introduced a new selection strategy (5) based on both fitness and diversity to choose individuals to participate in the reproduction phase of a genetic algorithm. Experiments showed that the resulting genetic algorithm was able to find solutions of a higher quality than the scatter evolutionary algorithm (4).

Lacking the theoretical guidelines while being stochastic in nature, the deployment of several meta-heuristics involves extensive experiments to find the optimal noise or walk probability settings. To avoid manual parameter tuning, new methods have been designed to automatically adapt parameter settings during the search (25)(32), and results have shown their effectiveness for a wide range of problems.

3. Multilevel techniques

The multilevel paradigm is a simple technique which at its core involves recursive coarsening to produce smaller and smaller problems that are easier to solve than the original one. The pseudo-code of the multilevel generic algorithm is shown in Algorithm 1.

The multilevel paradigm consists of four phases: coarsening, initial solution, uncoarsening and refinement. The coarsening phase aims at merging the variables associated with the problem to form clusters. The clusters are used in a recursive manner to construct a

---

Algorithm 1: The Multilevel Generic Algorithm

---

input : Problem P_0
output: Solution $S_{final}(P_0)$
begin
    level := 0;
    While Not reached the desired number of levels $P_{level+1}$:= Coarsen (P_{level});
    level := level + 1;
    /* Initial Solution is computed at the lowest level */;
    $S(P_{level})$ =Initial Solution (P_{level}) ;
    While $(level > 0)$ $S_{start}(P_{level-1})$: = Uncoarsen $(S_{final}(P_{level}))$;
    $S_{final}(P_{level-1})$:= Refine $(S_{start}(P_{level-1}))$;
    level := level - 1;
end

---

hierarchy of problems each representing the original problem but with fewer degrees of freedom. The coarsest level can then be used to compute an initial solution. The solution found at the coarsest level is uncoarsened (extended to give an initial solution for the next level) and then improved using a chosen optimization algorithm. A common feature that characterizes multilevel algorithms, is that any solution in any of the coarsened problems is a legitimate solution to the original one. Optimization algorithms using the multilevel paradigm draw their strength from coupling the refinement process across different levels. Multilevel techniques were first introduced when dealing with the graph partitioning problem (GGP) (1) (14) (16) (22) (23) (44) and have proved to be effective in producing high quality solutions at a lower cost than single level techniques. The traveling salesman problem (TSP) was the second combinatorial optimization problem to which the multilevel paradigm was applied (45) (46) and has clearly shown a clear improvement in the asymptotic convergence of the solution quality. When the multilevel paradigm was applied to the graph coloring problem (42), the results do not seem to be in line with the general trend observed in GCP and TSP as its ability to enhance the convergence behavior of the local search algorithms was rather restricted to some class of problems. Graph drawing is another area where multilevel techniques gave a better global quality to the drawing and the author suggests its use to both accelerate and enhance force drawing placement algorithms (43).

4. Memetic Algorithms (MAs)

An important prerequisite for the multilevel paradigm is the use of an optimization search strategy in order to carry out the refinement during each level. In this work, we propose a memetic algorithm (MA) that we use for the refinement phase. Algorithm 2 provides a canonical memetic algorithm.

MAs represent the set of hybrid algorithms that combine genetic algorithms and local search. In general the genetic algorithm improves the solution while the local search fine tunes the solution. They are adaptive based search optimizations algorithms that take their inspiration from genetics and evolution process (31). Memetic algorithms simultaneously examine and manipulate a set of possible solution. Given a specific problem to solve, the input to MAs is an initial population of solutions called individuals or chromosomes. A gene is part of a chromosome, which is the smallest unit of genetic information. Every gene is able to assume different values called allele. All genes of an organism form a genomem which

---

Algorithm 2: A Canonical Memetic Algorithm

---

begin

    Generate initial population ;

    Evaluate the fitness of each individual in the population ;

    While (Not Convergence reached) Select individuals according to a scheme to reproduce ;

    Breed if necessary each selected pairs of individuals through crossover;

    Apply mutation if necessary to each offspring ;

    Apply local search to each chromosome ;

    Evaluate the fitness of the intermediate population ;

    Replace the parent population with a new generation; ;

end

---

affects the appearance of an organism called phenotype. The chromosomes are encoded using a chosen representation and each can be thought of as a point in the search space of candidate solutions. Each individual is assigned a score (fitness) value that allows assessing its quality. The members of the initial population may be randomly generated or by using sophisticated mechanisms by means of which an initial population of high quality chromosomes is produced.

The reproduction operator selects (randomly or based on the individual's fitness) chromosomes from the population to be parents and enters them in a mating pool. Parent individuals are drawn from the mating pool and combined so that information is exchanged and passed to offspring depending on the probability of the crossover operator. The new population is then subjected to mutation and entered into an intermediate population. The mutation operator acts as an element of diversity into the population and is generally applied with a low probability to avoid disrupting crossover results. The individuals from the intermediate population are then enhanced with a local search and evaluated.

Finally, a selection scheme is used to update the population giving rise to a new generation. The individuals from the set of solutions which is called population will evolve from generation to generation by repeated applications of an evaluation procedure that is based on genetic operators and a local search scheme. Over many generations, the population becomes increasingly uniform until it ultimately converges to optimal or near-optimal solutions.

5. The Multilevel Memetic Algorithm (MLVMA)

The implementation of a multilevel algorithm for the SAT problem requires four basic components: a coarsening algorithm, an initialization algorithm, an extension algorithm (which takes the solution on one problem and extends it to the parent problem), and a memetic algorithm which will be used during the refinement phase. In this section we describe all these components which are necessary for the derivation of a memetic algorithm operating in a multilevel context. This process, is graphically illustrated in Figure 1 using an example with 10 variables. The coarsening phase uses two levels to coarsen the problem down to three clusters. $Level_0$ corresponds to the original problem. A random coarsening procedure is used to merge randomly the variables in pairs leading to a coarser problem with 5 clusters. This process is repeated leading to the coarsest problem with 3 clusters. An initial solution is generated where the first cluster is assigned the value of true and the remaining two clusters are assigned the value false. At the coarsest level, our MA wil generate an initial population and then improves

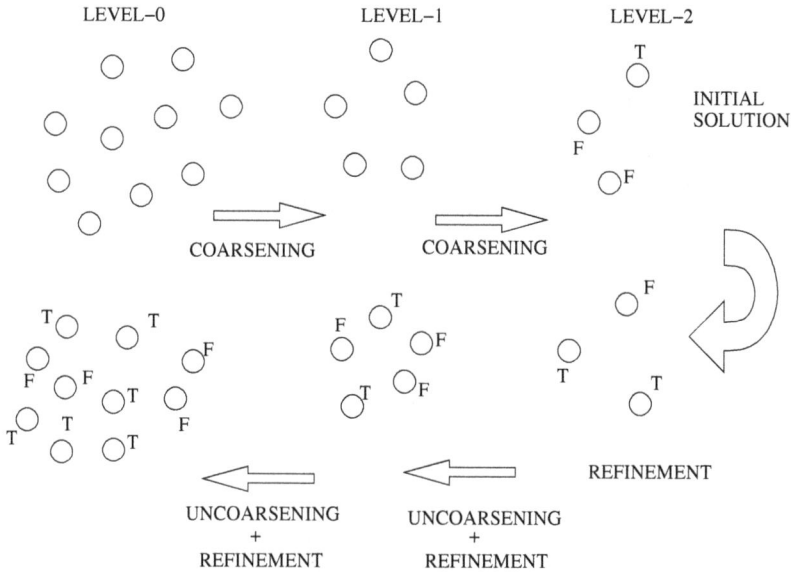

Fig. 1. The various phases of the multilevel memetic algorithm.

it. As soon as the convergence criteria is reached at $Level_2$, the uncoarsening phase takes the solution from that level and extends it to give an initial solution for $Level_1$ and then proceed with the refinement. This iteration process ends when MA reaches the stop criteria that is met at $Level_0$.

5.1 Coarsening

The coarsening procedure has been implemented so that each coarse problem P_{l+1} is created from its parent problem P_l by merging variables and representing each merged pair v_i and v_j with a child variable that we call a cluster in P_{l+1}. The coarsening scheme uses a simple randomized algorithm similar to (16). The variables are visited in a random order. If a variable v_i has not been merged yet, then we randomly select one randomly unmerged variable v_j, and a cluster consisting of these two variables is created. Unmatched variables are simply copied to the next level. The new formed clusters are used to define a new and smaller problem and recursively iterate the coarsening process until the size of the problem reaches some desired threshold.

5.2 Initial solution & refinement

As soon as the coarsening phase is ended, a memetic algorithm is used at different levels. The next subsections describes the main features of the memetic algorithm used in this work.

5.2.1 Fitness function

The notion of fitness is fundamental to the application of memetic algorithms. It is a numerical value that expresses the performance of an individual (solution) so that different individuals can be compared. The fitness of a chromosome (individual) is equal to the number of clauses that are unsatisfied by the truth assignment represented by the chromosome.

5.2.2 Representation

A representation is a mapping from the state space of possible solutions to a state of encoded solutions within a particular data structure. The chromosomes (individuals) which are assignments of values to the variables are encoded as strings of bits, the length of which is the number of variables (or clusters if MA is operating on a coarse level). The values *True* and *False* are represented by 1 and 0 respectively. In this representation , an individual X corresponds to a truth assignment and the search space is the set $S = \{0,1\}^n$.

5.2.3 Initial population

A initial solution is generated using a population consisting of 50 individuals. According to our computational experience, larger populations do not bring effective improvements on the quality of the results. At the coarsest level, MA will randomly generate an initial population of 50 individuals in which each gene's allele is assigned the value 0 or 1.

5.2.4 Crossover

The task of the crossover operator is to reach regions of the search space with higher average quality. New solutions are created by combining pairs of individuals in the population and then applying a crossover operator to each chosen pair. Combining pairs of individuals can be viewed as a matching process. The individuals are visited in random order. An unmatched individual i_k is matched randomly with an unmatched individual i_l. Thereafter, the two-point crossover operator is applied using a crossover probability to each matched pair of individuals. The two-point crossover selects two randomly points within a chromosome and then interchanges the two parent chromosomes between these points to generate two new offspring. Recombination can be defined as a process in which a set of configurations (solutions referred as parents) undergoes a transformation to create a set of configurations (referred as offspring). The creation of these descendants involves the location and combinations of features extracted from the parents. The reason behind choosing the two point crossover are the results presented in (41) where the difference between the different crossovers are not significant when the problem to be solved is hard. The work conducted in (39) shows that the two-point crossover is more effective when the problem at hand is difficult to solve. In addition, the author propose an adaptive mechanism in order to have evolutionary algorithms choose which forms of crossover to use and how often to use them, as it solves a problem.

5.2.5 Mutation

The purpose of mutation which is the secondary search operator used in this work, is to generate modified individuals by introducing new features in the population. By mutation, the alleles of the produced child have a chance to be modified, which enables further exploration of the search space. The mutation operator takes a single parameter p_m, which specifies the probability of performing a possible mutation. Let $C = c_1, c_2, \ldots c_m$ be a chromosome represented by a binary chain where each of whose gene c_i is either 0 or 1. In our mutation operator, each gene c_i is mutated through flipping this gene's allele from 0 to 1 or vice versa if the probability test is passed. The mutation probability ensures that, theoretically, every region of the search space is explored. If on the other hand, mutation is applied to all genes, the evolutionary process will degenerate into a random search with no benefits of the information gathered in preceding generations. The mutation operator prevents the searching process form being trapped into local optimum while adding to the diversity of the population and thereby increasing the likelihood that the algorithm will generate individuals with better fitness values.

5.2.6 Selection

The selection operator acts on individuals in the current population. During this phase, the search for the global solution gets a clearer direction, whereby the optimization process is gradually focused on the relevant areas of the search space. Based on each individual quality (fitness), it determines the next population. In the roulette method, the selection is stochastic and biased toward the best individuals. The first step is to calculate the cumulative fitness of the whole population through the sum of the fitness of all individuals. After that, the probability of selection is calculated for each individual as being $P_{Selection_i} = f_i / \sum_1^N fi$, where f_i is the fitness of individual i.

5.2.7 Local search

Algorithm 3: local-search

---

input : $Chromosome_i$
output: A possibly improved $Chromosome_i$;
begin
    PossFlips ← a randomly selected variable with the largest decrease (or smallest increase) in unsatisfied clauses;;
    v ← Pick (PossFlips);;
    $Chromosome_i$ ← $Chromosome_i$ with v flipped ;
    If $Chromosome_i$ satisfies Φ return $Chromosome_i$;
end

Finally, the last component of our MA is the use of local improvers. By introducing local search at this level, the search within promising areas is intensified. This local search should be able to quickly improve the quality of a solution produced by the crossover operator, without diversifying it into other areas of the search space. In the context of optimization, this rises a number of questions regarding how best to take advantage of both aspects of the whole algorithm. With regard to local search there are issues of which individuals will undergo local improvement and to what degree of intensity. However care should be made in order to balance the evolution component (exploration) against exploitation (local search component). Bearing this thought in mind, the strategy adopted in this regard is to let each chromosome go through a low rate intensity local improvement. Algorithm 3 shows the local search algorithm used. This heuristic is used for one iteration during which it seeks for the variable-value assignment with the largest decrease or the smallest increase in the number of unsatisfied clauses. Random tie breaking strategy is used between variables with identical score.

5.2.8 Convergence criteria

As soon as the population tends to loose its diversity, premature convergence occurs and all individuals in the population tend to be identical with almost the same fitness value. During each level, the proposed memetic algorithm is assumed to reach convergence when no further improvement of the best solution (the fittest chromosome) has not been made during two consecutive generations.

5.3 Uncoarsening

Having improved the assignment at the level L_{m+1}, the assignment must be projected onto its parent level L_m. The uncoarsening process is trivial; if a cluster $C_i \in L_{m+1}$ is assigned the value of true then the matched pair of clusters that it represents, C_j and $C_k \in L_m$ are also assigned the value true. The idea of refinement is to use the projected population from L_{m+1} onto L_m as the initial population for further improvement using the proposed memetic algorithm. Even though the population at L_{m+1} is at local minimum, the projected population at level L_m may not be at a local optimum. The projected population is already a good solution and contains individuals with high fitness value, MA will converge quicker within a few generation to a better assignment.

6. Experimental results

6.1 Boundary model checking

The instances used in our experiments arise from model checking (6) which is considered to be one among many real-world problems that are often characterized by large and complex search spaces. Model checking is an automatic procedure for verifying finite-state concurrent systems. Given a model of a design and a specification in temporal logic, one is interested to check whether the model satisfies the specification. Methods for automatic model checking of complex hardware design systems are gaining wide industrial acceptance compared to traditional techniques based on simulation. The most widely used of these methods is called Bounded Model Checking (2) (BMC). In BMC the design to be validated is represented as a finite state machine, and the specification is formalized by writing temporal logic properties. The reachable states of the design are then traversed in order to verify the properties. The basic idea in BMC is to find bugs or counterexamples of length k.

In practice, one looks for longer counterexamples by incrementing the bound k, and if no counterexample exists after a certain number of iterations , one may conclude that the correctness of the specification holds. The main drawback with model checking real systems is the so-called state-explosion problem: as the size of of the system being verified increases, the total state space of the system increases exponentially. This problem makes exhaustive search exploration intractable.

In recent years, there has been a growing interest in applying methods based on propositional satisfiability (SAT) (29)(12) in order to improve the scalability of model checking. The BMC problem can be reduced to a propositional satisfiability problem, and can therefore be solved by SAT solvers. Essentially, there are two phases in BMC. In the first phase, the behavior of the system to be verified is encoded as a propositional formula. In the second phase, that formula is given to a propositional decision algorithm, i.e., a satisfiability solver, to either obtain a satisfying assignment or to prove there is none. If the formula is satisfiable, a bug has been located in the design, otherwise one cannot in general conclude that there is no bug; one must increase the bound, and search for "larger bugs".

6.2 Test suite

We evaluated the performance of the multilevel memetic algorithm on a set of large problem instances taken from real industrial bounded model checking hardware designs. This set is taken from the SATLIB website (http://www.informatik.tu-darmstadt.de/AI/SATLIB). All the benchmark instances used in this experiment are satisfiable instances. Due to the randomization nature of the algorithms, each problem instance was run 20 times with a cutoff parameter (max-time) set to (300sec). We use $|.|$ to denote the number of elements in a set, e.g., $|V|$ is the number of variables, while $|C|$ denotes the number of clauses. Table shows the instances used in the experiment. The tests were carried out on a a DELL machine with 800 MHz CPU and 2 GB of memory. The code was written in C and compiled with the GNU C compiler version 4.6. The parameters used in the experiment are listed below:

- Crossover probability = 0.85.
- Mutation probability = 0.1.
- Population size = 50 .
- Stopping criteria for the coarsening phase: The coarsening stops as soon as the size of the coarsest problem reaches 100 variables (clusters). At this level, MA generates an initial population.
- Convergence during the refinement phase: If no improvement of the fitness function of the best individual has not been observed during 10 consecutive generations, MA is assumed to have reached convergence and moves to a higher level.

6.3 Experimental results

Figures 2-9 show how the best assignment (fittest chromosome) progresses during the search. The plots show immediately the dramatic improvement obtained using the multilevel paradigm. The performance of MA is unsatisfactory and is getting even far more dramatic for larger problems as the percentage excess over the solution is higher compared to that of MLVMA. The curves show no cross-over implying that MLVMA dominates MA. The plots suggest that problem solving with MLVMA happens in two phases. The first phase which corresponds to the early part of the search, MLVMA behaves as a hill-climbing method. This phase which can be described as a long one, up to 85% of the clauses are satisfied. The best assignment improves rapidly at first, and then flattens off as we mount the plateau, marking the start of the second phase. The plateau spans a region in the search space where flips typically leave the best assignment unchanged, and occurs more specifically once the refinement reaches the finest level. Comparing the multilevel version with the single level version, MLVMA is far better than MA, making it the clear leading algorithm. The key success behind the efficiency of MLVMA relies on the multilevel paradigm. MLVMA uses the multilevel paradigm and draw its strength from coupling the refinement process across different levels. This paradigm offers two main advantages which enables MA to become much more powerful in the multilevel context:

- During the refinement phase MA applies a local a transformation (i.e, a move) within the neighborhood (i.e, the set of solutions that can be reached from the current one) of the current solution to generate a new one. The coarsening process offers a better mechanism for performing diversification (i.e, the ability to visit many and different regions of the search space) and intensification (i.e, the ability to obtain high quality solutions within those regions).
- By allowing MA to view a cluster of variables as a single entity, the search becomes guided and restricted to only those configurations in the solution space in which the variables grouped within a cluster are assigned the same value. As the size of the clusters varies from one level to another, the size of the neighborhood becomes adaptive and allows the possibility of exploring different regions in the search space while intensifying the search by exploiting the solutions from previous levels in order to reach better solutions.

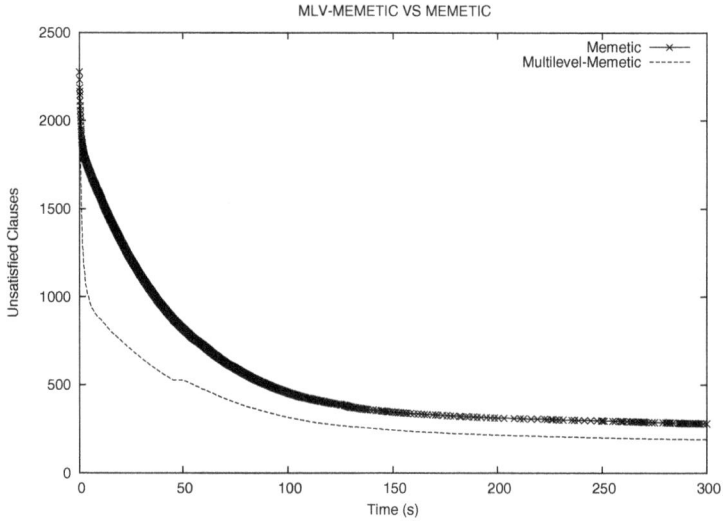

Fig. 2. bmc-ibm-2.cnf: $|V| = 3628$, $|C| = 14468$. Along the horizontal axis we give the time in seconds, and along the vertical axis the number of unsatisfied clauses.

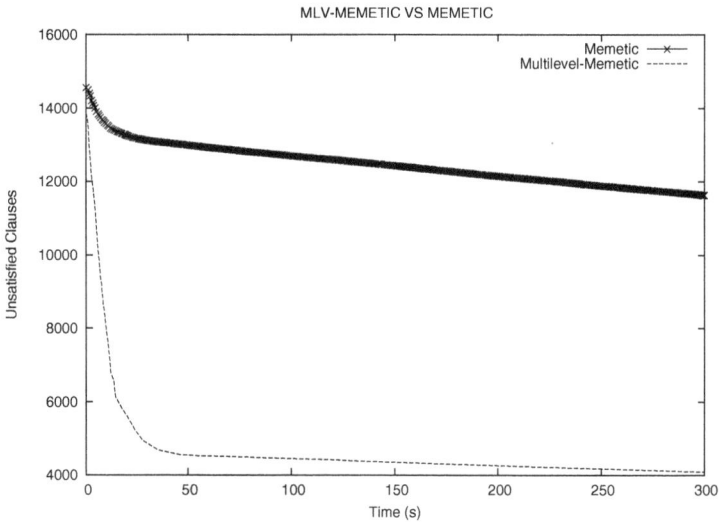

Fig. 3. bmc-ibm-3.cnf: $|V| = 14930$, $|C| = 72106..$ Along the horizontal axis we give the time in seconds , and along the vertical axis the number of unsatisfied clauses.

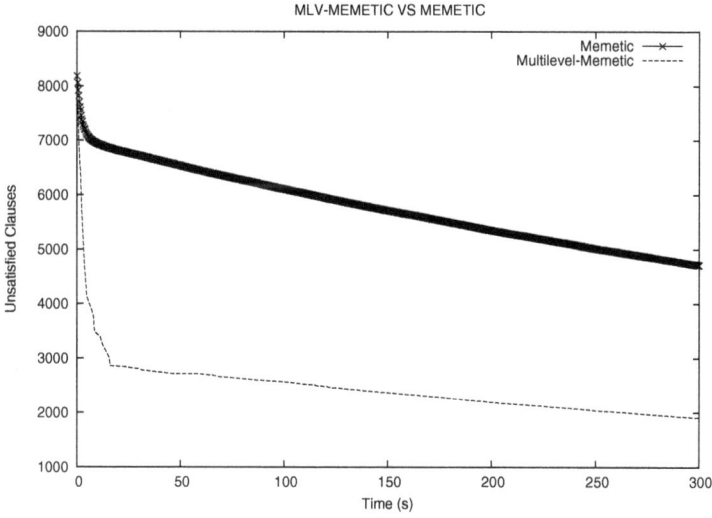

Fig. 4. bmc-ibm-5: $|V| = 9396$, $|C| = 41207$. Along the horizontal axis we give the time in seconds , and along the vertical axis the number of unsatisfied clauses.

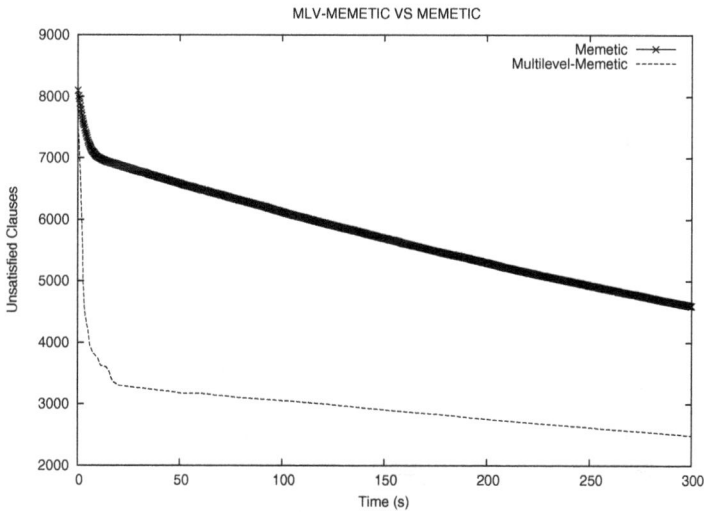

Fig. 5. bmc-ibm-7.cnf: $|V| = 8710$, $|C| = 39774$. Along the horizontal axis we give the time in seconds, and along the vertical axis the number of unsatisfied clauses.

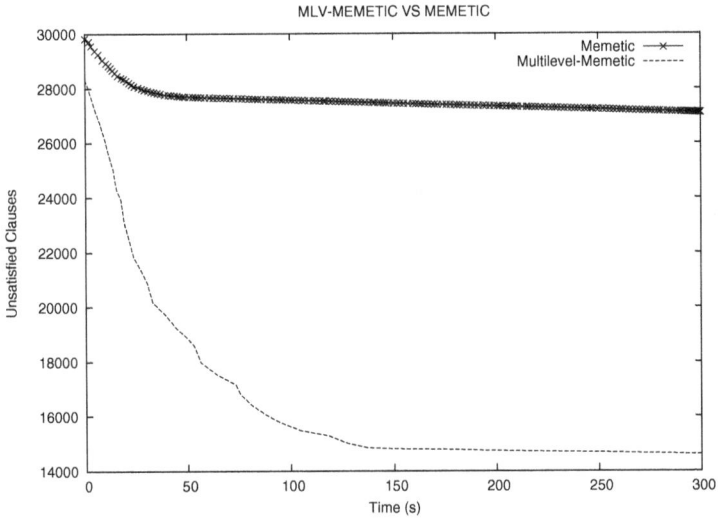

Fig. 6. bmc-ibm-11.cnf: $|V| = 32109$, $|C| = 150027$. Along the horizontal axis we give the time in seconds, and along the vertical axis the number of unsatisfied clauses.

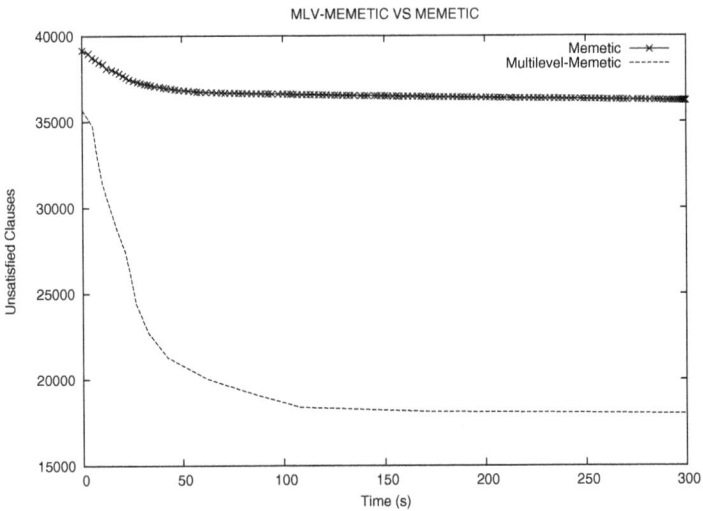

Fig. 7. bmc-ibm-12.cnf: $|V| = 39598$, $|C| = 19477$. Along the horizontal axis we give the time in seconds, and along the vertical axis the number of unsatisfied clauses.

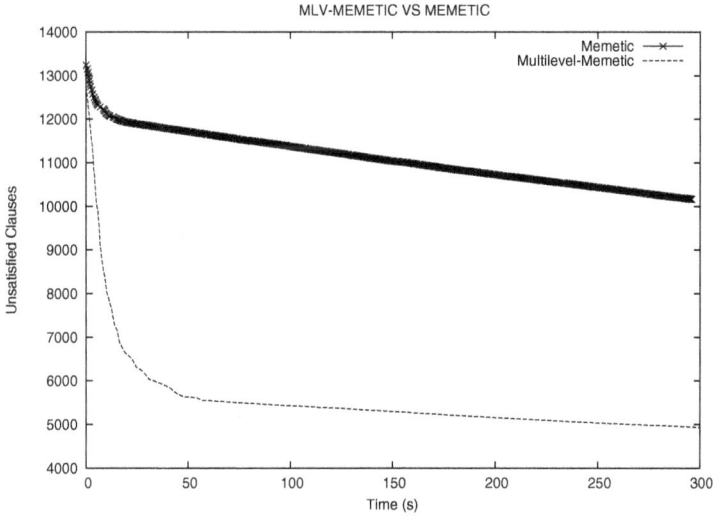

Fig. 8. bmc-ibm-13.cnf: $|V| = 13215$, $|C| = 6572$. Along the horizontal axis we give the time in seconds, and along the vertical axis the number of unsatisfied clauses.

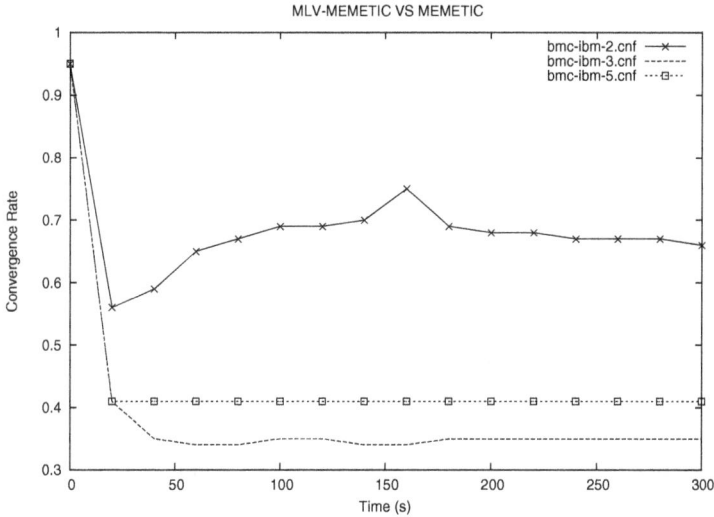

Fig. 9. Results on the convergence behavior for bmc-ibm-2.cnf, bmc-ibm-3.cnf, bmc-ibm-5.cnf. Along the horizontal axis we give the time (in seconds), and along the vertical axis the convergence rate.

Figures 9 shows the convergence behavior expressed as the ratio between the best chromosome of the two algorithms as a function of time. The plots show that the curves are below the value 1 leading to conclude that MLVMA is faster compared to MA. The asymptotic performance offered by MLVMA is impressive, and dramatically improves on MA. In some cases, The difference in performance reaches 30% during the first seconds, and maintains it during the whole search process. However, on other cases, the difference in performance continues to increase as the search progresses.

7. Conclusion

In this work, we have described a new approach for addressing the satisfiability problem. which combines the multilevel paradigm with a simple memetic algorithm. Thus, in order to get a comprehensive picture of the new algorithm's performance, we used a set of benchmark instances drawn from Bounded Model Checking. The experiments have shown that MLVMA works quite well with a random coarsening scheme combined with a simple MA used as a refinement algorithm. The random coarsening provided a good global view of the problem, while MA used during the refinement phase provided a good local view. It can be seen from the results that the multilevel paradigm greatly improves the MA and always returns a better solution for the equivalent runtime. The quality of the solution provided by MLVMA can get as high as 77%. A scale up test shows that the difference in performances between the two algorithms increases with larger problems. Our future work aims at investigating other coarsening schemes and study other parameters which may influence the interaction between the memetic algorithm and the multilevel paradigm.

8. References

[1] S.T. Barnard and H.D. Simon. A fast multilevel implementation of recursive spectral bisection for partitioning unstructured problems. Concurrency: Practice and Experience, 6(2) pages: 101-117, 1994.

[2] A.Biere at al. Bounded model checking. Advances in Computers,2003.

[3] C. Blum and A. Roli. Metaheuristics in combinatorial optimization: Overview and conceptual comparison. ACM Computing Surveys, 35(3) pages: 268-308, September 2003.

[4] D. Boughaci, H. Drias. Efficient and experimental meta-heuristics for MAX-SAT problems. In Lecture Notes in Computer Sciences, WEA 2005, vol. 3503/2005, pages: 501-512, 2005.

[5] D. Boughaci, B. Benhamou, and H. Drias. Scatter Search and Genetic Algorithms for MAX-SAT Problems. J.Math.Model.Algorithms, pages: 101-124, 2008.

[6] E.M. Clark, E.A. Emerson, and A.P.Sista. Automatic verification of finite state on current systems using temporal logic specifications. ACM Transactions on Programming Languages and Systems 8 , pages: 244-263, 1996.

[7] S.A. Cook. The complexity of theorem-proving procedures. Proceedings of the Third ACM Symposium on Theory of Computing, pages: 151-158, 1971.

[8] M.R. Gary and D.S. Johnson. Computers and intractability: A guide to the theory of NP-completeness. W.H. Freeman and Company, New York, 1979.

[9] I. Gent and T. Walsh. Unsatisfied Variables in Local Search. In J. Hallam, editor, Hybrid Problems, Hybrid Solutions, pages: 73-85. IOS Press, 1995.

[10] L.P. Gent and T. Walsh. Towards an Understanding of Hill-Climbing Procedures for SAT. Proceedings of AAAI'93, pages: 28-33. MIT Press, 1993.

[11] F. Glover. Tabu Search-Part1. ORSA Journal on Computing, 1(3), pages: 190-206, 1989.

[12] E. Goldberg, Y. Novikov. Bermin: a Fast and Robust SAT-solver. Proc.of the Design, Automation and Test in Europe, IEEE Computer Society, 2002.

[13] J. Gottleib, E. Marchiori, and C. Rossi. Evolutionary Algorithms for the satisfiability problem. Evolutionary Computation, 10(1), pages: 35-50, 2002. Satisfiability Problem Using Finite Learning Automata. International Journal of Computer Science and Applications, Volume 4, Issue 3, pages: 15-29, 2007.

[14] R. Hadany and D. Harel. A multi-scale algorithm for drawing graphs nicely. Tech.Rep.CS99-01, Weizmann Inst.Sci., Faculty Maths.Comp.Sci, 1999.

[15] P. Hansen and B. Jaumand. Algorithms for the Maximum Satisfiability Problem. Computing, 44, pages: 279-303, 1990.

[16] B. Hendrickson and R. Leland. A multilevel algorithm for partitioning graphs. In S. Karin, editor, Proc. Supercomputing'95, San Diego, 1995. ACM Press, New York.

[17] H. Hoos. On the run-time behavior of stochastic local search algorithms for SAT. In. Proceedings of AAAI-99, pages: 661-666, 1999.

[18] H.Hoos. An adaptive noise mechanism for Walksat. In Proceedings of the Eighteen National Conference in Artificial Intelligence (AAAI-02), pages: 655-660, 2002.

[19] F. Hutter, D. Tompkins, H. Hoos. Scaling and probabilistic smoothing: Efficient dynamic local search for SAT. In Proceedings of the Eight International Conference of the Principles and Practice of Constraint Programming (CP'02), pages: 233-248, 2002.

[20] A. Ishtaiwi, J. Thornton, A. Sattar, and D.N.Pham. Neighborhood clause weight redistribution in local search for SAT. Proceedings of the Eleventh International Conference on Principles and Practice Programming(CP-05), volume 3709 of Lecture Notes in Computer Science, pages: 772-776, 2005.

[21] H. Jin-Kao, F. Lardeux, and F. Saubion. Evolutionary computing for the satisfiability problem. In Applications of Evolutionary Computing, volume 2611 of LNCS, pages: 258-267, University of Essex, England, UK, April 2003.

[22] G. Karypis and V. Kumar. A fast and high quality multilevel scheme for partitioning irregular graphs. SIAM J. Sci. Comput., 20(1) pages: 359-392, 1998.

[23] G. Karypis and V. Kumar. Multilevel k-way partitioning scheme for irregular graphs. J. Par. Dist. Comput., 48(1), pages: 96-129, 1998.

[24] F. Lardeux, F. Saubion, and Jin-Kao. GASAT: A Genetic Local Search Algorithm for the Satisfiability Problem. Evolutionary Computation, volume 14 (2), MIT Press, 2006.

[25] C.M. Li, W. Wei, and H. Zhang. Combining adaptive noise and look-ahead in local search for SAT. Proceedings of the Tenth International Conference on Theory and Applications of Satisfiability Testing(SAT-07), volume 4501 of Lecture Notes in Computer Science, pages: 121-133, 2007.

[26] C.M. Li, W.Q. Huang. Diversification and determinism in local search for satisfiability. In proceedings of the Eight International Conference on Theory and Applications of Satisfiability Testing (SAT-05), volume 3569 of Lecture Notes in Computer Science, pages: 158-172, 2005.

[27] C.M Li, W. Wei, and H. Zhang. Combining adaptive noise and look-ahead in local search for SAT. In Proceedings of the Tenth International Conference on Theory and Applications of Satisfiability Testing (SAT-07), volume 4501 of Lecture Notes in Computer Science, pages: 121-133, 2007.

[28] M. Lozano and C.G. Martinez. Hybrid metaheuristics with evolutionary algorithms specializing in intensification and diversification: Overview and progress report. Computers and operations Research, 37, pages: 481-497, 2010.

[29] Y. Mahajan, Z. Fu, S. Malik. Zchaff2004: An efficient SAT solver. To appear in SAT 2004 Special Volume
[30] D. McAllester, B. Selman, and H.Kautz. Evidence for Invariants in Local Search. Proceedings of AAAI'97, pages: 321-326. MIT Press, 1997.
[31] P.A. Moscato. On evolution search, optimization, genetic algorithms and martial arts: Towards memetic algorithms. Technical Report Caltech Concurrent Computation Program, Caltech,Pasadena, California, 1989.
[32] D.J. Patterson and H. Kautz. Auto-Walksat: A Self-Tuning Implementation of Walksat. Electronic Notes on Discrete Mathematics 9, 2001.
[33] S. Prestwich. Random walk with continuously smoothed variable weights. Proceedings of the Eight International Conference on Theory and Applications of Satisfiability Testing(SAT-05), volume 3569 of Lecture Notes, pages: 203-215, 2005.
[34] D. Schuurmans, and F. Southey. Local search characteristics of incomplete SAT procedures. In Proc.AAAI-2000, pages: 297-302, AAAI Press, 2000.
[35] D. Schuurmans, F.Southey, and R.C. Holte. The exponentiated sub-gradient algorithm for heuristic Boolean programming. In Proc. IJCAI-01, pages: 334-341, Morgan Kaufman Publishers, 2001.
[36] B. Selman, H. Levesque, and D. Mitchell. A New Method for Solving Hard Satisfiability Problems. Proceedings of AAA'92,pages: 440-446, MIT Press, 1992.
[37] B. Selman, H.A. Kautz, and B. Cohen. Noise Strategies for Improving Local Search. Proceedings of AAAI'94, pages: 337-343. MIT Press, 1994.
[38] B. Selman and H.A. Kautz. Domain-Independent extensions to GSAT: Solving large structured satisfiability problems. In R.Bajcsy,editor, Proceedings of the international Joint Conference on Artificial Intelligence, volume 1, pages: 290-295. Morgan Kaufmann Publishers Inc., 1993.
[39] W. Spears. Adapting Crossover in Evolutionary Algorithms. Proc of the Fourth Annual Conference on Evolutionary Programming, MIT Press, pages: 367-384, 1995.
[40] J. Thornton, D.N. Pham, S. Bain, and V. Ferreira Jr. Additive versus multiplicative clause weighting for SAT. Proceedings of the Nineteenth National Conference of Artificial Intelligence (AAAI-04), pages: 191-196, 2004.
[41] D. Vrajitoru. Genetic programming operators applied to genetic algorithms. In Proceedings of the Genetic and Evolutionary Computation Conference, pages: 686-693,Orlando (FL). Morgan Kaufmann Publishers, 1999,
[42] C. Walsahaw. A Multilevel Approach to the Graph Colouring Problem. Tech.Rep. 01/IM/69, Comp.Math. Sci., Univ.Greenwich, London SE10 9LS, UK, May 2001.
[43] C. Walshaw: A Multilevel Algorithm for Forced-Directed Graph-Drawing. Journal of Graph Algorithms and Applications, 7(3) pages: 253-285, 2003.
[44] C. Walshaw and M. Cross. Mesh partitioning: A multilevel balancing and refinement algorithm. SIAM J. Sci. Comput., 22(1) pages: 63-80,2000.
[45] C. Walshaw. A Multilevel Approach to the Traveling Salesman Problem. Oper. Res., 50(5) pages: 862-877, 2002.
[46] C. Walshaw. A Multilevel Lin-Kerninghan-Helsgaun Algorithm for the Traveling Salesman Problem. Tech. Rep. 01/IM/80, Comp. Math. Sci., Univ. Greenwich, 2001.
[47] Z.Wu., and B. Wah. An efficient global-search strategy in discrete Lagrangian methods for solving hard satisfiability problems. In Proceedings of the Seventeenth National Conference on Artificial Intelligence (AAAI-00), pages: 310-315, 2000.

10

Library-Based Gate-Level Current Waveform Modeling for Dynamic Supply Noise Analysis

Mu-Shun Matt Lee and Chien-Nan Jimmy Liu

National Central University
Taiwan (ROC)

1. Introduction

As the VLSI technology goes into the nanometer era, the device sizes and supply voltages are continually decreased. The smaller supply voltage reduces the power dissipation but also decreases the noise margin of devices. Therefore, the power integrity problem has become one of the critical issues that limit the design performance (Blakiewicz & Chrzaniwska-Jeske, 2007; kawa, 2008 & Michael et al., 2008). Most of the power supply noises (PSNs) come from two primary sources. One is the IR-drop and the other is the simultaneous switching noise (SSN). Figure 1(a) illustrates a typical RLC model for power supply networks, which is the combination of on-chip power grids and off-chip power pins. The IR-drop is a power supply noise when the supply current goes through those non-zero resistors and results in a I·R voltage drop. The simultaneous switching noise (SSN) is the supply noise which happens when large instantaneous current goes through those non-zero inductors on power networks and generates a L·(di/dt) voltage drop. When the supply voltage is reduced , the noise margin of devices also decreases as shown in Fig.1(b). It may induce worse performance because the driving capability of devices becomes week due to smaller supply voltage. If serious power supply noise occurs, the logic level may be changed, which causes function error in the circuit. The worst situation is the electron-migration (EM) effects. Supply wires are shorten or broken because a large current travels through the small supply wires. Therefore, the power supply noise analysis is reguired at design stages to evaluate the effects caused by power supply noise.

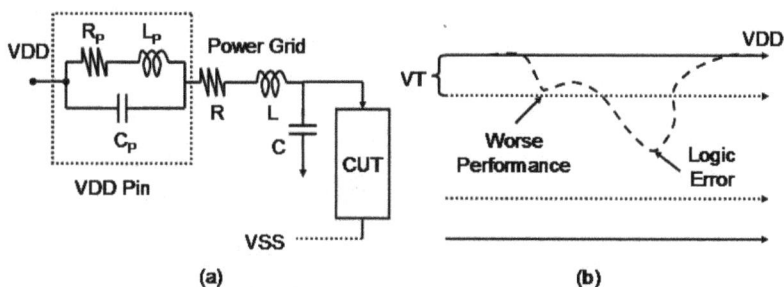

Fig. 1. (a)RLC model for power supply (b) Supply Voltage over Time at Silicon Device.

While esimating the power supply noise, both the magnitude and slope of supply currents are required. Traditionally, accurate supply current waveforms can only be obtained from the transistor-level simulation. Therefore, in the present design flow, the power supply noise (PSN) check is mostly performed at very late design stage. Although the analysis results are accurate at transistor level, this approach may be impractical for large designs because simulating the entire design at transistor level requires great computation resources. If any problem is found, the designers often tune the width of the supply lines or add another current path to fit the specification. However, if the supply current waveforms are obtained at early stage, more efficient low-power technologies, like multiple supply voltages and power-gating, can be used to reduce the supply power and noise (Chen et al., 2005; Juan et al., 2010; Kawa, 2008; Michael et al., 2008; Popovich et al., 2008; Xu et at., 2011 & Zhao et al. 2002). The primary reason of lacking tools for checking the power integrity problems at gate level or higher levels is the limited design information, that current cannot provide waveforms directly. In this research, we propose the gate-level IR-drop analysis method with limited design information to build the missing link of the traditional design flow.

The most popular format to store the gate-level information is the liberty format (LIB) (Synopsys, 2003). The LIB file of a cell library keeps the information of all cells and is widely used in the synthesis and timing analysis at gate level and RT level. However, due to the format limitation, only timing information and average energy consumption are kept in LIB files. They cannot provide instantaneous supply current information directly. One straightforward approach is to approximate the instantaneous supply current using the average power divided by the user-given time interval as illustrated in Fig. 2(a). However, even if the average power is the same, the waveforms can be quite different with different time intervals. It may not be accurate enough to estimate real instantaneous supply current.

Several advanced library formats have been proposed for recording voltage waveforms (ECSM) (Candence, 2006) or current waveforms (CCSM) (Synopsys, 2008) to provide the more accurate timing and power information. These formats need large storage space to record these piece-wise-linear waveforms. Therefore, those new formats are only used in very advanced process, like 65 nm technology. Typically, the libraries with new formats are used to support the static timing analysis to obtain more accurate estimation. It may also support the gate-level power estimation to obtain more accurate peak power. However, because the peak power is often evaluated in the cycle-accurate basis at gate level, it will suffer the same time-interval issue.

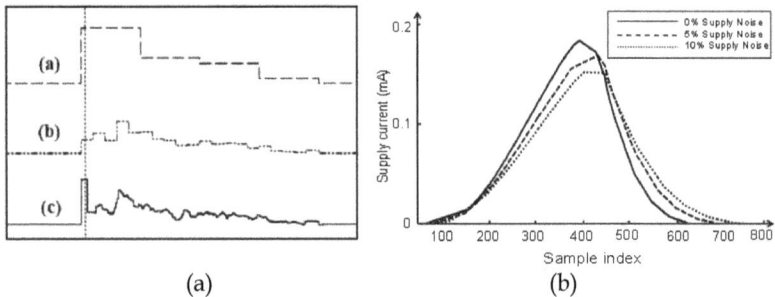

(a)   (b)

Fig. 2. (a) The power waveforms with different time intervals (b) The current waveforms with different supply noises

In the literature, the authors in (Boliolo et al., 1997) propose an approach to estimate power supply noise at gate level. In their approach, the capacitance of each internal node in a cell, the energy consumption of each transition, and several regression equations representing the timing behavior, are required to estimate the supply current waveforms. Given an input pattern to a cell, its supply current will be approximated as a simple triangle, whose area is the total energy. The base and the height of this triangle are obtained from the regression equations. Then, combining all triangles of every changed cell in time obtains the overall supply current waveform. This approach is a practical solution that can be combined with logic simulation tools. The results shown in the paper are also accurate. However, the required timing behaviors of supply current waveforms are not available in standard library files. Extra characterization efforts for different cell libraries are still required before using this approach, which is a very time-consuming process.

In anthor work (Shimazaki et al., 2000), the authors propose an EMI-noise analysis approach based on a rough supply current waveform. Although their approach also uses standard library infotmation, their current waveform estimaiton approach is too simple to provide accurate supply current waveforms. Most importantly, their approach can be used in combination caircuits only, which is not feasibal for modern complex designs. Therefore, an accurate gate-level supply current model using standard library information, even for sequential circuits, is propsed to avoid addtional charcterization process (Lee et al., 2008).

The proposed current model has provided the solutions to estimate the ideal supply current waveforms without noise effects. However, the estimated waveforms cannot be directly used to analyze IR-drop effects because the supply currents will have significant difference with non-zero resistance on the supply lines. Figure 2(b) shows an example obtained from the c432 circuit suffering from different supply noises. In typical cases, the current with supply noise is less than the ideal current. If the ideal supply current waveforms are used to calculate the IR-drop, the results are often overestimated. The direct solution to consider the effects of IR-drop is to extend the libraries with different supply resistors. However, this approach will greatly increase the storage space and characterization efforts for library information, which may be not a good solution. Therefore, a library adjustment method is also proposed to consider the IR-drop effect on supply current modeling with standard library information (Lee et al., 2010).

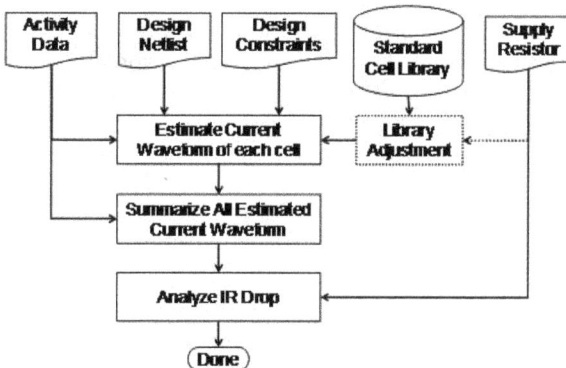

Fig. 3. The proposed gate-level IR-drop analysis flow

The proposed gate-level IR-drop analysis flow is illustrated in Fig.3. According to the cell switching from gate-level activity files, the corresponding supply current waveform of each cell can be constructed by using standard library information. The supply current waveforms obtained from the original standard libraries are then modified to consider IR-drop effects. Second, the estimated supply current waveforms of all switching cells are summarized in time to obtain the supply current waveforms of the whole circuit. Finally, the IR-drop voltage caused from the supply resistor can be derived from the current waveform.

The rest of this article is organized as follow. In Section 2, the most popular library format, the liberty format, is presented. A gate-level supply current waveform estimation method using standard library information is proposed in Section 3. A correction method of the library information is also proposed to modify the IR-drop effect in Section 4. The experimental results of this work are demonstrated in Section 5 and a simple conclusion is presented in Section 6.

2. Standard library: Liberty format (LIB)

Liberty format (LIB) (Synosys, 2003) is the most popular library format at gate level to store the timing information and the average energy consumption of each cell in the standard library. Those data are stored using some look-up tables. The definitions of some commonly used variables are listed as follows. They will be used later to derive the proposed current waveform model.

Transition Time: This is defined as the duration time of a signal from 10% to 90% VDD in the rising case and from 90% to 10% VDD in the falling case. $TR(X)$ is defined as the transition time of the node X in the rising case. $TF(X)$ is defined as the transition time of the node X in the falling case.

Propagation Time: This is defined as the duration time from the input signal crossing 50% VDD to the output signal crossing 50% VDD. $TDR(X{\rightarrow}Y)$ is defined as the propagation delay from the related pin X to the output Y when the output Y is rising. D represents the propagation delay and R represents the rising case. $TDF(X{\rightarrow}Y)$ is defined as the propagation delay from the related pin X to the output Y when the output Y is falling. F represents the falling case.

Setup Time: This is a timing constraint of the sequential cell, which is defined as the minimum time that the data input D must remain stable before the active edge of the clock CK to ensure correct functioning of the cell. In other words, it is the duration from D crossing 50% VDD to CK crossing 50% VDD if the output value can be evaluated successfully. $TSR(D)$ is defined as the setup time when the data input D is rising. S represents the setup time and R represents the rising case. $TSF(D)$ is the setup time when the data input D is falling. F represents the falling case.

Load: This is the total capacitance at a node. $Load(Y)$ is defined as the capacitance at the node Y.

Internal Power: This is the internal energy consumption of a cell without the energy consumed on its output loading. E_{INT} is defined as the internal energy consumption of the cell.

Changing Time: $T(X)$ is defined as the time that the signal X is crossing 50% VDD, which is the signal transition point in logic simulators recorded in VCD (Value Changed Dump) files.

Voltage Definitions: VDD is defined as the supply voltage. VT is defined as the threshold voltage of the transistor.

3. Current waveform estimation using library information

In order to avoid extra characterization efforts while migrating to new cell libraries, a supply current model is proposed based on standard library information. The key idea is using a triangular waveform to approximate the real supply current waveform generated by a cell switching as shown in Fig. 4. Then, the parameters of the triangle are calculated by standard library information only. Finally, the overall supply current waveform can be obtained by combining all triangles of every changed cells in time. Before presenting the proposed approach, some variables must be defined first. For each triangle shown in Fig. 4, four variables, T_{START}, T_{END}, T_{PEAK} and I_{PEAK}, are defined to represent the triangular waveform. T_{START} and T_{END} are the start/end time of the supply current waveform. These two variables define the duration of the waveform. T_{PEAK} and I_{PEAK} are the location and current value when the maximum supply current occurs.

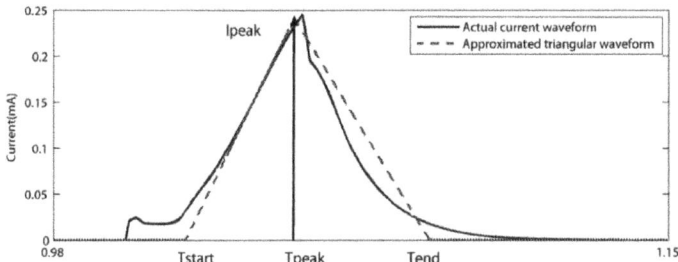

Fig. 4. The definition of the triangular current waveform

Although there are a lot of cells in a cell library, most of them can be classified into three categories in our approach. In the following sections, the formulas to construct the current waveform model in each category will be presented. During the formula construction, this work assumes that only the LIB file is available. Therefore, the transistor-level netlist and detailed device sizes are avoided. If some general structures are required to build the formulas, only the information provided in the library data sheet will be used. While applying the proposed methodology to different libraries, users can make necessary adjustment easily from that public information.

3.1 Simple logic cells

If the CMOS implementation of a cell is a single layer structure, it is called a simple logic cell in this work, such as **INVERTER, NAND, NOR** as shown in Fig. 5. Those cells can be modeled as an equivalent inverter with two parts, the equivalent PMOS and NMOS. Therefore, in the following discussion, an inverter is used as an example to discuss its supply current model in the charging period (the output signal is rising) and the discharging period (the output signal is falling).

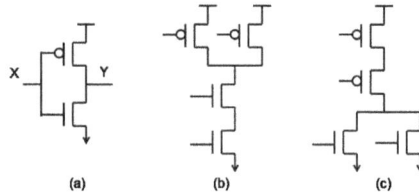

(a)   (b)   (c)

Fig. 5. The structures of simple logic cells (a) INVERTER (b) NAND (c) NOR

3.1.1 Charging period

In the charging period, the relationship between the input signal X, the output signal Y and the timing parameters of the triangular waveform can be illustrated in Fig. 6. T_{START} is defined as the time that the input voltage achieves (VDD-VT) because the equivalent PMOS turns on at this time. The corollary of T_{START} is shown as follows.

$$\frac{T_{START}-[T(X)-0.625\times TF(X)]}{VT}=\frac{1.25\times TF(X)}{VDD} \qquad (1)$$

$$\Rightarrow T_{START}=T(X)-1.2\times TF(X)\times\frac{0.5\times VDD-VT}{VDD}$$

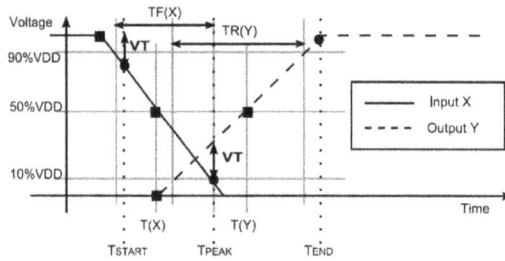

Fig. 6. The parameters of a simple cell in the charging period

In typical cases, the shape of the charging current for a simple logic cell is similar to a RC charging behavior. Therefore, the exponential RC charging function is used to approximate this behavior. Theoretically, T_{END} is defined as the time when the output loading is charged to VDD. However, due to the long tail of the RC charging curve, T_{END} is defined as the time that the output loading is charged to 95% VDD in this work to reduce the error while the waveform is simplified to a triangle. The corollary of T_{END} is shown as follows, where τ is the RC time constant.

$$V(t)=V_0\times(1-e^{(\frac{-t}{\tau})})\Rightarrow t=\ln(\frac{V_0-V(t)}{V_0})\times\tau$$

$$\begin{cases} t_{10\%VDD}=\ln(\dfrac{VDD-0.1\times VDD}{VDD})\times\tau \\ t_{90\%VDD}=\ln(\dfrac{VDD-0.9\times VDD}{VDD})\times\tau \end{cases} \Rightarrow TR(Y)=[\ln(0.9)-\ln(0.1)]\times\tau\Rightarrow\tau=\frac{TR(Y)}{\ln(9)}$$

$$\left\{ \begin{array}{l} t_{0\%VDD} = \ln(\dfrac{VDD}{VDD}) \times \tau \\[3mm] t_{50\%VDD} = \ln(\dfrac{VDD - 0.5 \times VDD}{VDD}) \times \tau \end{array} \right.$$

$$\Rightarrow TD[V(Y = 0\%VDD) \to V(Y = 50\%VDD)] = \ln(0.5) \times \tau$$

$$
\begin{aligned}
T_{END} &= T(Y) - TD[V(Y = 0\%VDD) \to V(Y = 50\%VDD)] \\
&\quad + TD[V(Y = 0\%VDD) \to V(Y = 95\%VDD)] \\
&= T(Y) - \ln(0.5) \times \tau + \ln(1 - 0.95) \times \tau
\end{aligned}
\tag{2}
$$

In this paper, two points (X1, Y1) and (X2, Y2) on a plane are used to define a line. Then, the slope (a) and intercept (b) can be calculated as follows.

$$X(t) = \{(X_1, Y_1), (X_2, Y_2)\} = a \times t + b$$
$$Slope(a) \Rightarrow a = \frac{Y_2 - Y_1}{X_2 - X_1}$$
$$Intercept(b) \Rightarrow b = Y_1 - a \times X_1$$

Under this definition, the time t that the equation Y(t) is larger than the equation X(t) with VT can be calculated as follows.

$$\left\{ \begin{array}{l} X(t) = a_X \times t + b_X \\ Y(t) = a_Y \times t + b_Y \end{array} \right.$$
$$If \quad (Y(t) - X(t) = VT)$$
$$Then \quad t = \frac{VT - (b_X - b_Y)}{a_X - a_Y} = T(Y(t) - X(t) = VT)$$

In the charging period, T_{PEAK} is defined as the time that the operation mode of NMOS is in the saturation mode and the operation mode of PMOS is changing from the saturation mode to the linear mode, which is the point that allows most current to flow through PMOS. In other words, T_{PEAK} happens at the time when the voltage difference between the output Y and the input X is equal to VT (VSG=VT). Therefore, T_{PEAK} can be obtained when Y(t) − X(t) = VT. Because the definitions of TF(X) and TR(Y) are the signal duration from 10% to 90% VDD, using them to calculate the signal duration from 0% to 50% VDD should be multiplied by 0.625(=0.5/(90% − 10%)) instead of 0.5. Finally, the corollary of T_{PEAK} is shown as follows.

$$\left\{ \begin{array}{l} X(t) = \{(T(X), 0.5 \times VDD), (T(X) - 0.625 \times TF(X), VDD)\} \\ Y(t) = \{(T(Y), 0.5 \times VDD), (T(Y) - 0.625 \times TR(Y), 0)\} \end{array} \right.$$
$$\Rightarrow T_{PEAK} = T(Y(t) - X(t)) = VT)$$

(3)

If the total consumed energy is used as the area of this triangle and the base of this triangle is (T_{END}-T_{START}), I_{PEAK} can be obtained from the formula of the triangle area. Please note that the energy stored in the LIB file is the internal energy consumption (E_{INT}) of the cell only.

The energy consumed on the output loading (E_{LOAD}) should be added to obtain the correct area of the triangle. The corollary of I_{PEAK} is shown as follows.

$$E_{INT} + E_{LOAD} = \frac{1}{2} \times (T_{END} - T_{START}) \times I_{PEAK} \qquad (4)$$

$$\Rightarrow I_{PEAK} = 2 \times \frac{E_{INT} + E_{LOAD}}{T_{END} - T_{START}}$$

3.1.2 Discharging period

Because the supply current does not charge the output loading in the discharging period, most of the supply current can appear only when NMOS is turned on but PMOS is not completely turned off yet. Therefore, in this case, T_{START} is defined as the time that input voltage achieves VT because NMOS is turned on at this time. T_{END} is defined as the time that the input voltage achieves (VDD-VT) when PMOS is turned off. Using these definitions, the duration of the supply current waveform in the discharging period can be decided. Following the same assumption in Section 3.1.1, T_{PEAK} is still defined as the time that the operation mode of PMOS is changed from linear to saturation. Figure 7 shows their relationship to the input/output waveforms. Because there is no current charging the output loading, the E_{INT} obtained in the LIB file can be used as the triangle area in the discharging period to obtain the T_{PEAK} value. The corollary of T_{END} is shown as follows.

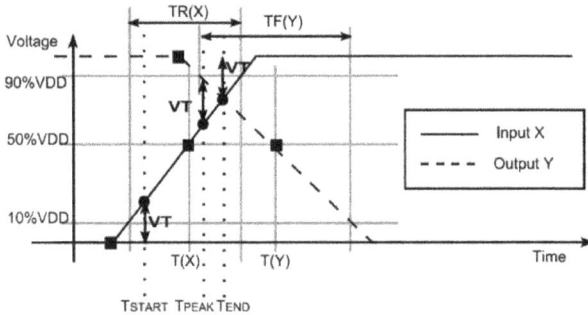

Fig. 7. The parameters of a simple cell in the discharging period

$$\frac{T_{START} - [T(X) - 0.625 \times TR(X)]}{VT} = \frac{1.25 \times TR(X)}{VDD} \qquad (5)$$

$$\Rightarrow T_{START} = T(X) - 1.25 \times TR(X) \times \frac{0.5 \times VDD - VT}{VDD}$$

$$\frac{[T(X) + 0.625 \times TR(X)] - T_{END}}{VT} = \frac{1.25 \times TR(X)}{VDD} \qquad (6)$$

$$\Rightarrow T_{END} = T(X) + 1.25 \times TR(X) \times \frac{0.5 \times VDD - VT}{VDD}$$

$$\begin{cases} X(t) = \{(T(X), 0.5 \times VDD), (T(X) - 0.625 \times TR(X), 0)\} \\ Y(t) = \{(T(Y), 0.5 \times VDD), (T(Y) - 0.625 \times TF(Y), VDD)\} \end{cases} \qquad (7)$$

$$\Rightarrow T_{PEAK} = T(Y(t) - X(t)) = VT)$$

$$E_{INT} = \frac{1}{2} \times (T_{END} - T_{START}) \times I_{PEAK} \Rightarrow I_{PEAK} = 2 \times \frac{E_{INT}}{T_{END} - T_{START}} \tag{8}$$

3.2 Composite logic cells

As shown in Fig. 8, some cells are composed of two or more simple logic cells, such as **BUFFER, AND,** and **OR** cells. Those cells are called "composite logic cells" in this work. In the following descriptions, a **BUFFER** is used as an example to explain the proposed approach for those cells. Because the information of the internal signal I in Fig. 8(a) cannot be obtained in the LIB file, an assumption is made in this work that the input signal of the second stage in a composite cell will start rising/falling when the output voltage of its first stage achieves 50% VDD. With this assumption, the internal signal I can be rebuilt using existing library information as shown in Fig. 9. Since the timing information of the internal node can be estimated, the methods proposed in Sect. 3.1 can be used to handle the two simple cells respectively and the total current waveform of this composite cell can be estimated.

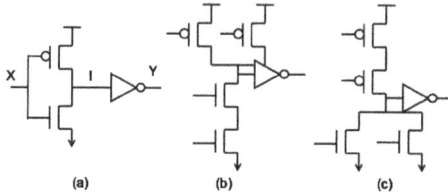

Fig. 8. The structure of the composite logic cell (a) BUFFER (b) AND (c) OR

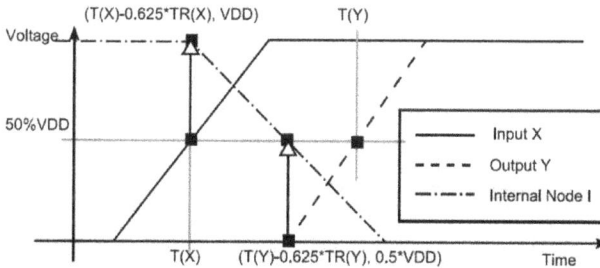

Fig. 9. The internal voltage waveform in a composite cell

3.2.1 Charging period

If the output of the composite cell is rising, the internal node I will be in the falling case as shown in Fig. 10. Therefore, the simple-cell methods in the discharging period are used to calculate T_{START_1stF}, T_{END_1stF} and T_{PEAK_1stF} of the first stage. Then, the simple-cell methods in the charging period are used to calculate T_{START_2ndR}, T_{END_2ndR} and T_{PEAK_2ndR} of the second stage. Because there is only one energy value in the library and no proper method to split it into two parts, an assumption is made that the transition of the two stages are very close such that the composition of the two triangles still approximates to a triangle. While combining the triangles of the two stages, the T_{START}, T_{PEAK} and T_{END} of the composed triangle are defined as the average values of the two triangles in this work for easier calculation. Then, I_{PEAK} can be obtained in the same way from those timing information and the stored energy information. The detailed formulas to construct the current waveforms in this case are summarized as follows.

$$X(t) = \{(T(X), 0.5 \times VDD), (T(X) - 0.625 \times TR(X), 0)\}$$
$$I(t) = \{(T(X), VDD), (T(X) - 0.625 \times TR(Y), 0.5 \times VDD)\} \quad (9)$$
$$Y(t) = \{(T(Y), 0.5 \times VDD), (T(Y) - 0.625 \times TR(Y), 0)\}$$

$$\Rightarrow T_{START} = avg\{T_{START\_1stF}, T_{START\_2ndR}\} \quad (10)$$

$$\Rightarrow T_{PEAK} = avg\{T_{PEAK\_1stF}, T_{PEAK\_2ndR}\}$$

$$\Rightarrow T_{END} = avg\{T_{END\_1stF}, T_{END\_2ndR}\} \quad (11)$$

$$\Rightarrow I_{PEAK} = \frac{2 \times (\dfrac{E_{INT}}{VDD} + Load(Y) \times VDD)}{T_{END} - T_{START}} \quad (12)$$

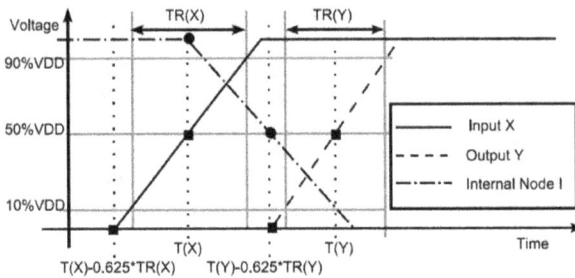

Fig. 10. The rebuilding voltage waveform of a composite logic cell in the charging period

3.2.2 Discharging period

If the output of a buffer is falling, the internal node I will be in the rising case. Therefore, the simple-cell formulas in the charging period are used to handle the first stage. The simple-cell formulas in the discharging period are used to handle the second stage. Then, using the similar approach for the case in charging period, T_{START}, T_{PEAK} and T_{END} can be obtained from the average values of the two triangles. The rebuilt voltage waveforms and timing parameters are shown in Fig. 11. Because the energy of the reversed supply current at the second stage can be eliminated by the energy of the first stage, the internal power in the discharging period can be used directly to estimate the I_{PEAK} of this cell. The detailed formulas to construct the current waveform in this case are listed as follows.

$$X(t) = \{(T(X), 0.5 \times VDD), (T(X) - 0.625 \times TF(X), VDD)\}$$
$$I(t) = \{(T(X), VDD), (T(X) - 0.625 \times TF(Y), 0.5 \times VDD)\} \quad (13)$$
$$Y(t) = \{(T(Y), 0.5 \times VDD), (T(Y) - 0.625 \times TF(Y), VDD)\}$$

$$\Rightarrow T_{START} = avg\{T_{START\_1stR}, T_{START\_2ndF}\}$$

$$\Rightarrow T_{PEAK} = avg\{T_{PEAK\_1stR}, T_{PEAK\_2ndF}\} \quad (14)$$

$$\Rightarrow T_{END} = avg\{T_{END\_1stR}, T_{END\_2ndF}\} \tag{15}$$

$$\Rightarrow I_{PEAK} = \frac{2 \times \dfrac{E_{INT}}{VDD}}{T_{END} - T_{START}} \tag{16}$$

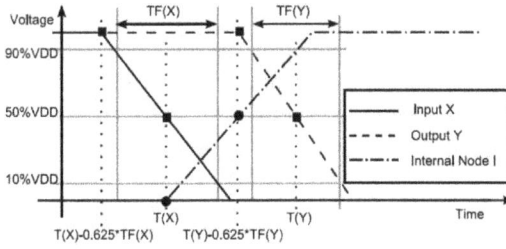

Fig. 11. The rebuilding voltage waveform of a composite logic cell in the discharging period

3.3 Sequential elements

In real applications, most circuits contain sequential cells. For a feasible solution, it is important to develop proper approaches to handle sequential cells. Like composite cells, sequential cells are often composed of several simple cells. In a standard library, the information of the internal nodes in a sequential cell is not stored, either. In order not to use extra information, some assumptions are made to rebuild the internal signals of a sequential cell. In the following descriptions, a positive-edge-triggered D-flip-flop (**DFF**) is used as an example to explain the proposed approach on sequential cells. Other flip-flops in the standard cell library, such as the flip-flops with set/reset, can be handled by using similar methods for their normal operations. The special set/reset behaviors can be characterized as a special case since they do not appear very often.

Figure 12 shows the typical architecture of a DFF. It can be divided into three blocks, which are clock generator, setup block and evaluation block. The total supply current waveform of the DFF is the summation of the waveforms from the three blocks. Since the operation modes of a DFF are more complex, its current waveform model is discussed in three cases.

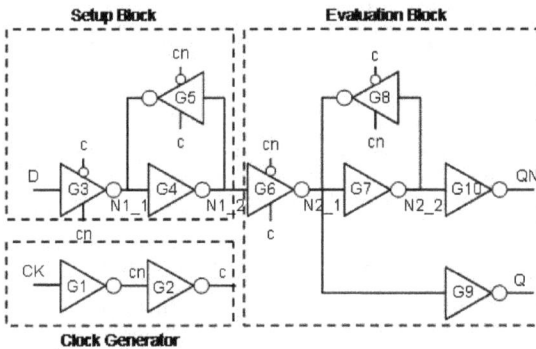

Fig. 12. The architecture of a typical DFF

3.3.1 Only clock pin is changed

In this case, the data pin D is stable and its value is the same as the output Q. In most cases, the internal signals, N1_1, N1_2, N2_1 and N2_2, are stable, too. Therefore, a supply current only occurs in the clock generator when only the clock pin is changing. The clock generator is often composed of two inverters to generate two inverse signals, c and cn, as shown in Fig. 13.

First, the case of CK rising (active edge) is discussed. Using the same idea for composite logic cells, the voltage waveforms of CK, cn and c will be rebuilt first. Then, the formulas of composite logic cells in the charging period can be used directly to decide T_{START}, T_{PEAK}, T_{END} and I_{PEAK}. However, there is still no timing information for the internal nodes of flip-flops in the LIB file. In order to solve this problem, two assumptions are made to rebuild the internal signals (cn and c) with approximate timing information.

The first assumption is that the maximum current of the tri-state inverter (G6) occurs when its output voltage (N2_1) reaches 50% VDD, as illustrated in Fig. 13. Then, following the T_{PEAK} definition of simple cells, the maximum current happens when the difference between the gate voltage of c and the drain voltage of N2_1 is equal to VT. The time that the voltage of c reaches [0.5×VDD+VT] can be implied with [T(CK) + TDR(CK→Q) − 0.625×TR(Q)].

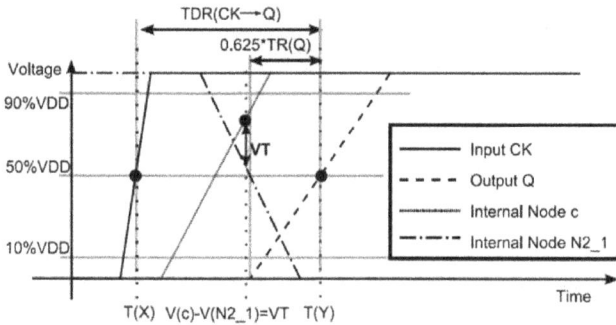

Fig. 13. The illustration of the first assumption to imply the timing information of internal node (c).

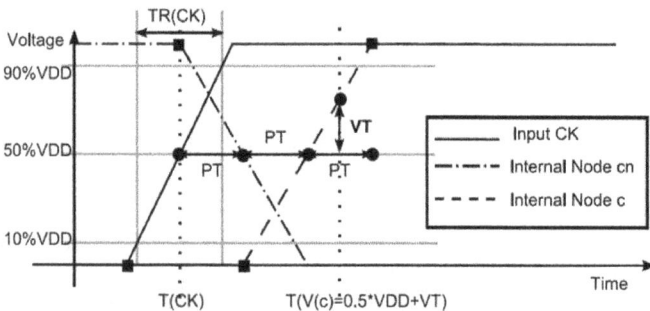

Fig. 14. The illustration of the second assumption to imply the timing information of internal nodes (cn) and (c).

The second assumption is that the rising and falling times of the nodes cn and c are very similar because most clock buffers are designed to have similar rising and falling time. From the first assumption, the time when V(c)=0.5 × VDD + VT can be obtained. Following the same assumption for composite cells, the input signal of the second stage will start rising/falling when the output voltage of the previous stage achieves 50% VDD. In order to simplify the explanation, a time interval (PT) is defined in Fig. 14. Since PT can be obtained with these two assumptions, the times that c and cn reach 0.5×VDD can be expressed with PT. Then, the internal voltage waveforms can be rebuilt as shown in Fig. 14. The detailed corollary is listed as follows.

$$Assume \quad X = [T(V(c) = 0.5 \times VDD + VT] - [T(CK) + PT]$$

$$T(V(c) = 0.5 \times VDD + VT)$$

$$= T(CK) + TDR(CK \to Q) - 0.625 \times TR(Q)$$

$$\frac{X}{0.5 \times VDD + VT} = \frac{2 \times PT}{VDD} \Rightarrow PT = \frac{VDD}{2 \times (VDD + VT)} \times [TDR(CK \to Q) - 0.625 \times TR(Q)]$$

$$\begin{aligned} CK(t) &= \{(T(CK), 0.5 \times VDD), (T(CK) - 0.625 \times TR(CK), 0)\} \\ cn(t) &= \{(T(CK), VDD), (T(CK) + PT, 0.5 \times VDD)\} \\ c(t) &= \{(T(CK) + PT, 0), (T(CK) + 3 \times PT, VDD)\} \end{aligned} \tag{17}$$

$$\Rightarrow T_{START} = avg\{T_{START\_1stF}, T_{START\_2ndR}\}$$

$$\Rightarrow T_{PEAK} = avg\{T_{PEAK\_1stF}, T_{PEAK\_2ndR}\} \tag{18}$$

$$\Rightarrow T_{END} = avg\{T_{END\_1stF}, T_{END\_2ndR}\} \tag{19}$$

$$\Rightarrow I_{PEAK} = \frac{2 \times \dfrac{E_{INT}}{VDD}}{T_{END} - T_{START}} \tag{20}$$

As to the CK falling case, no outputs change and no timing information is stored in the library because it is not the active edge. Although the internal nodes might change in this case, there is no information to make any reasoning. Therefore, the same timing information in the CK rising case is used to be the T_{START}, T_{PEAK} and T_{END} when only CK is falling. The internal energy consumption when only CK is falling is available in the library. It can be used to calculate a different I_{PEAK} for CK falling case.

3.3.2 Only data pin is changed

In this case, the clock pin CK is stable and only the data pin D is changed. The supply current is generated by the setup block only. If CK is logic-1, the gate G3 is turned off such that the whole cell has no switching current. When CK is logic-0, the current waveform is determined by whether the data pin D is rising or falling. Because the timing information of the internal nodes N1_1 and N1_2 are not stored in the library, two assumptions are made in this case to rebuild the approximate voltage waveforms of N1_1 and N1_2.

The first assumption is that the data propagation time from the input D to the internal node N1_1 equals to the setup time of this DFF. Because the definition of setup time is the minimum time that input data must be stable before clock arriving, it can be viewed as the time that the data has been propagated to N1_1 to enter the first latch.

The second assumption is that the node N1_2 will become stable before the gate G6 is turned on to allow the data to enter the second latch successfully. Because N2_1 is discharging in the D rising case, N1_2 must reach VDD when the voltage of the node c achieves VT. TC(VT) is defined to express the duration time between V(CK)=0.5×VDD and V(c)=VT. Following these assumptions, the time that N1_1 reaches 50% VDD and the time that N1_2 reaches VDD can be obtained. Then, following the same assumption of composite cells, the time that N1_1 reaches 50% VDD is the time that N1_2 reaches 0. The voltage waveforms of N1_1 andN1_2 can be rebuilt as shown in Fig. 15.

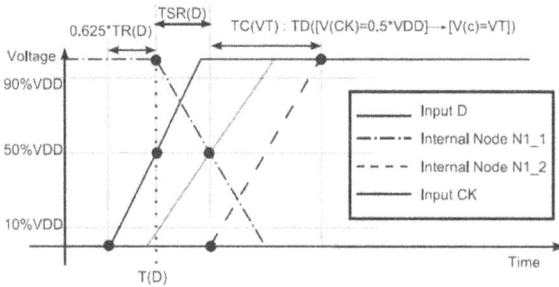

Fig. 15. The parameters of a DFF when only D is rising.

In the D falling case, TR(D) and TSR(D) are changed to TF(D) and TSF(D), respectively. E_{INT} is changed from the rising energy to the falling one. With the two internal waveforms of N1_1 and N1_2, the triangle parameters can be determined by the same approach for composite cells. Finally, the detailed corollary is shown as follows.

$$TC(VT):TD([V(CK)=0.5\times VDD]\rightarrow[V(c)=VT])$$

D Rising Case

$$D(t)=\{(T(D),0.5\times VDD),(T(D)-0.625\times TR(D),0)\}$$
$$N1\_1(t)=\{(T(D),VDD),(T(D)+TSR(D),0.5\times VDD)\}$$
$$N1\_2(t)=\{(T(D)+TSR(D),0),(T(D)+TSR(D)+TC(VT),VDD)\}$$

D Falling Case

$$D(t)=\{(T(D),0.5\times VDD),(T(D)-0.625\times TF(D),VDD)\}$$
$$N1\_1(t)=\{(T(D),0),(T(D)+TSF(D),0.5\times VDD)\}$$
$$N1\_2(t)=\{(T(D)+TSF(D),VDD),(T(D)+TSF(D)+TC(VT),0)\}$$

$$\Rightarrow T_{START}=avg\{T_{START\_1st},T_{START\_2nd}\} \tag{21}$$

$$\Rightarrow T_{PEAK}=avg\{T_{PEAK\_1st},T_{PEAK\_2nd}\} \tag{22}$$

$$\Rightarrow T_{END} = avg\{T_{END\_1st}, T_{END\_2nd}\}   \tag{23}$$

$$\Rightarrow I_{PEAK} = \frac{2 \times \dfrac{E_{INT}}{VDD}}{T_{END} - T_{START}}   \tag{24}$$

3.3.3 Output changed with clock active edge

In this case, the clock pin has an active edge, the data pin is stable, and the output Q is evaluated. Both the clock generator and the evaluation block generate supply currents. Therefore, the current waveform is composed of two triangular waveforms in this case. The first current waveform of the clock generator is discussed in Sect. 3.3.1. It is focus on how to estimate the second triangular waveform of the evaluation block in this section.

Figure 16 illustrates the rebuilt signals of the evaluation block when output Q is rising. First, using the rebuilt internal signal c in Sect. 3.1.1, the time that N2_1 starts to discharge can be obtained when the voltage of node c reaches VT. Second, T(Q) - 0.625 × TR(Q) implies the time that N2_1 reaches 0.5×VDD by the assumption of composite logic cells. Then, the internal waveform of N2_1 can be rebuilt. Third, T(QN) - 0.625 × TF(QN) implies the time that N2_2 reaches 0.5×VDD by the assumption of composite logic cells, which helps to rebuild the internal waveform of N2_2. After rebuilding the internal signals of the evaluation block, the similar approach for composite logic cells can be used to generate the composite triangular waveform of this DFF.

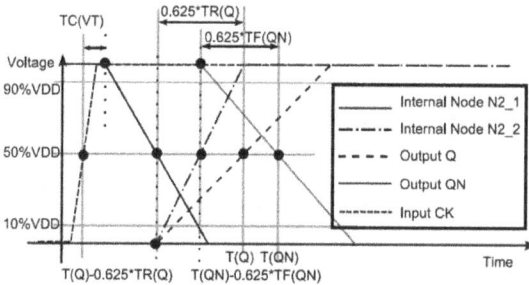

Fig. 16. The signals in a DFF when Q is rising with active clock edge.

When the output Q is falling, the time when c reaches VT is defined as the start time of N2_1 because the gate G6 starts to transition when c reaches VT. Then, changing TR(Q) and TF(QN) to TF(Q) and TR(QN) respectively, the same approach for the Q rising case can be used to rebuild the internal signals when the output Q is falling.

With the two internal waveforms of N2_1 and N2_2, T_{START} of the evaluation block is defined as the earliest start time of N2_1 and N2_2. T_{END} of the evaluation block is defined as the time that both Q and QN complete their transitions. T_{PEAK} can be calculated by the waveforms of internal nodes. The consumed internal energy of the evaluation block is the internal energy of total DFF minus the internal energy of the clock generator obtained in Sect. 3.3.1. After adding the energy of the output loading, the total triangle area of the evaluation block and the I_{PEAK} of this block can be obtained. Finally, combining the waveform of the evaluation block with the waveform of the clock generator calculated in

Sect. 3.3.1, the supply current waveform of the DFF in this case is obtained. The detailed formulas to construct the current waveform in this case are summarized as follows.

$$TC(VT) : TD([V(CK) = 0.5 \times VDD] \rightarrow [V(c) = VT])$$

Q Rising Case

$$N2\_1(t) = \{(T(CK) + TC(VT), VDD), (T(Q) - 0.625 \times TR(Q), 0.5 \times VDD)\}$$
$$N2\_2(t) = \{(T(Q) - 0.625 \times TR(Q), 0), (T(QN) - 0.625 \times TF(QN), 0.5 \times VDD)\}$$
$$Q(t) = \{(T(Q), 0.5 \times VDD), (T(Q) - 0.625 \times TR(Q), 0)\}$$
$$QN(t) = \{(T(QN), 0.5 \times VDD), (T(QN) - 0.625 \times TF(QN), VDD)\}$$

Q Falling Case

$$N2\_1(t) = \{(TCN(VDD - VT), 0), (T(Q) - 0.625 \times TF(Q), 0.5 \times VDD)\}$$
$$N2\_2(t) = \{(T(Q) - 0.625 \times TF(Q), VDD), (T(QN) - 0.625 \times TR(QN), 0.5 \times VDD)\}$$
$$Q(t) = \{(T(Q), 0.5 \times VDD), (T(Q) + 0.625 \times TF(Q), 0)\}$$
$$QN(t) = \{(T(QN), 0.5 \times VDD), (T(QN) + 0.625 \times TR(QN), VDD)\}$$

$$\Rightarrow T_{START} = TC(VT) \tag{25}$$

$$\Rightarrow T_{PEAK} = avg\{T_{PEAK\_(G2\_1)}, T_{PEAK\_(G2\_2)}, T_{PEAK\_(G2\_5)}\} \tag{26}$$

$$\Rightarrow T_{END} = avg\{T_{END\_(G2\_4)}, T_{END\_(G2\_5)}\} \tag{27}$$

$$\Rightarrow I_{PEAK} = \frac{2 \times (\frac{E_{INT} - E_{ClockGenerator}}{VDD} + Load(Q) \times VDD)}{T_{END} - T_{START}} \tag{28}$$

4. IR-Drop aware library adjustment methods

In this section, an analytical library adjustment approach is proposed to consider the effects of the supply resistors without extra characterization. The timing and power information stored in LIB file can be modified to reflect the effect of the supply resistor by the proposed equations. Therefore, the proposed gate-level supply current estimation method can obtain the accurate waveforms with IR-drop effects. Most importantly, this method can be easily embedded into present design flow to improve the accuracy of gate-level IR-drop analysis and provide designers a fast solution to consider IR-drop effect at early design stages. In this section, the adjustment methods of combination cells, simple logic and composite logic cells are discussed first in Section 4.1. Then, in Section 4.2, the methods of sequential cell are presented. Finally, the adjustment methods of activity files (VCD) are explained in Section 4.3.

4.1 Timing and power adjustment of combination cells

4.1.1 Output transition time

Figure 17 illustrates a simple cell with a supply resistor. In the output rising case, the supply current flows through the supply resistor, which increases the transition time due to the increased total resistance. Therefore, the RC charging model is used to calculate the increased transition time caused by the supply resistor.

R_{EFF} represents the effective resistance of the cell. C_{EFF} represents the effective capacitance of the cell. E_{INT} and E_{LOAD} represent the energy consumption caused by the cell and its output loading. In the output rising case, the C_{EFF} is approximated by the total energy divided by supply voltage. Assume $TR(Y)_{ORG}$ represents the original transition time in LIB files. $TR(Y)_{ADJ}$ represents the adjusted transition time in the output rising case. The detailed corollary and the adjustment formula can be derived as follows, in which the increased term is related to the known variables (R_{WIRE}, C_{EFF}) only. In the output falling case, the transition time is not changed because the current does not flow through the supply resistor. If there is a resistor in current path to ground, similar approach can be used to adjust $TF(Y)$.

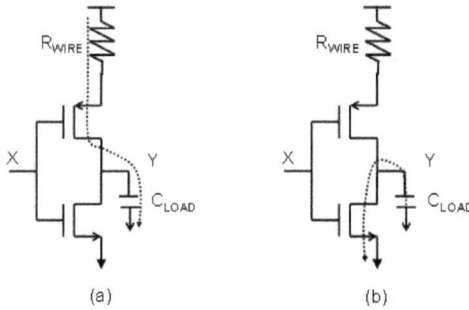

Fig. 17. The circuit structure of a simple cell (**INVETER**) with a supply resistor in (a) the output rising case (b) the output falling case

$$\begin{cases} C_{EFF} = \dfrac{E_{INT} + E_{LOAD}}{VDD} \\ TR(Y)_{ORG} = \ln 9 \times R_{EFF} \times C_{EFF} \end{cases}$$

$$\begin{aligned} TR(T)_{ADJ} &= \ln 9 \times (R_{EFF} + R_{WIRE}) \times C_{EFF} \\ &= \underline{\ln 9 \times R_{EFF} \times C_{EFF}} + \ln 9 \times R_{WIRE} \times C_{EFF} = TR(Y)_{ORG} + \ln 9 \times R_{WIRE} \times C_{EFF} \end{aligned} \quad (29)$$

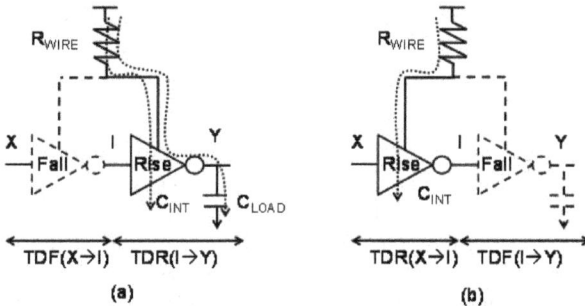

Fig. 18. The circuit structure of a composite cell (**BUFFER**) with a supply resistor in (a) the output rising case (b) the output falling case

Figure 18 illustrates a composite cell with a supply resistor. Typically, this kind of cells is composed of multiple stages of simple cells. In Fig.18(a), the supply current flows through the second stage in the output rising case. The first stage is in the output falling case. Therefore, only the increased transition time of the second stage should be considered in the output rising case. Applying the same method for the simple logic cells on the second stage can obtain the increased transition time. In the output falling case, the output transition time is still not changed because the current does not flow through the second stage. Only the propagation delay may be changed in such case, which is discussed in the next section.

4.1.2 Propagation delay time

According to the same model shown in Fig.18, the adjustment method of the propagation time for simple logic cells can be derived. Similarly, only the increased propagation time in the output rising case should be considered to adjust the original timing information. The adjustment formulas are listed as follows, in which the increased term is related to the known variables (R_{WIRE}, C_{EFF}) only.

$$\begin{cases} C_{EFF} = \dfrac{E_{INT} + E_{LOAD}}{VDD} \\ TDR(Y)_{ORG} = -\ln 0.5 \times R_{EFF} \times C_{EFF} \end{cases}$$

$$\begin{aligned} TDR(T)_{ADJ} &= (-\ln 0.5) \times (R_{EFF} + R_{WIRE}) \times C_{EFF} \\ &= \underline{(-0.5) \times R_{EFF} \times C_{EFF}} + (-\ln 0.5) \times R_{WIRE} \times C_{EFF} = TDR(Y)_{ORG} + (-\ln 0.5) \times R_{WIRE} \times C_{EFF} \end{aligned} \tag{30}$$

For composite logic cells, the adjustment of the propagation time in the output rising case is the same with the simple logic cell as shown in Fig.18 (a). In the output falling case shown in Fig.18(b), C_{EFF} is the internal capacitance C_{INT}. This internal capacitance can be approximated as the E_{INT} divided by the supply voltage because the operation current flows through the cell only. The adjustment formulas are listed as follows, in which the increased term is related the known variables only.

$$C_{EFF} = \frac{E_{INT} + E_{LOAD}}{VDD} \Rightarrow TDR(X \rightarrow Y)_{ADJ} = TDR(X \rightarrow Y)_{ORG} + (-\ln 0.5) \times R_{WIRE} \times C_{EFF} \tag{31}$$

$$C_{EFF} = \frac{E_{INT}}{VDD} \Rightarrow TDR(X \rightarrow Y)_{ADJ} = TDR(X \rightarrow Y)_{ORG} + (-\ln 0.5) \times R_{WIRE} \times C_{EFF} \tag{32}$$

4.1.3 Internal energy

Assume $E_{INT(ORG)}$ represents the internal energy stored in standard libraries, and $E_{INT(ADJ)}$ represents the modified internal energy. This internal energy can be viewed as the short-circuit energy by ignoring the effect of internal capacitances. Therefore, $E_{INT(OLD)}$ can be expressed as the short-circuit current (I_{SC}) times the duration of the short-circuit current (T_{SC}). Since I_{SC} can be rewritten as VDD/ R_{INT}, the $E_{INT(ADJ)}$ can be derived by the ratio of R_{EFF} and R_{ADJ}, as shown in the following equations. Please be noted that the R_{EFF} can be calculated from the original propagation time because the short-circuit current happens at the logic transition period.

$$E_{INT(ORG)} = I_{SC} \times T_{SC} = \frac{VDD}{R_{EFF}} \times T_{SC} \tag{33}$$

$$E_{INT(ADJ)} = \frac{VDD}{(R_{EFF} + R_{WIRE})} \times T_{SC} = \frac{R_{EFF}}{R_{EFF} + R_{WIRE}} \times E_{INT(ORG)}$$

4.2 Timing and power adjustment of sequential elements

Only the output Q rising case is to explain the adjusted formulas because the formulas fir other cases can be derived by similar ways. One difficulty of the adjustment of DFF cases is to estimate the effective capacitance of the gate because the internal capacitance is unavailable. In this work, the internal energy is used to approximate the effective capacitance. The other difficulty is the adjustment of effective supply resistance because more than one gates switch in the DFF. Therefore, the simple parallel connection formula is applied first to approximate the effective supply resistance seen by each switching gate. The details of the adjusted formulas in the timing and internal energy are discussed in the following subsections.

4.2.1 Output transition time

Only the increased transition time caused by the output stage (G9) should be added to adjust the output transition time of output Q. The E_{INT} (CK$_{RISE}$→Q$_{RISE}$) represents the internal energy consumption stored in the library for the output Q rising case when CK actives, which is composed of the energy of G1, G2, G6, G7, G9 and G10. The E_{INT}(CK$_{RISE}$) represents the internal energy consumption of G1 and G2 when only CK actives. It implies that the energy consumption of G6, G7, G10 and G9 in Fig.19 can be calculated by E_{INT} (CK$_{RISE}$→Q$_{RISE}$)-E_{INT}(CK$_{RISE}$). Therefore, the C_{EFF} of the path through G9 can be approximated as a half of E_{INT} (CK$_{RISE}$→Q$_{RISE}$)- E_{INT}(CK$_{RISE}$) divided by VDD because the energy are separated into two rising gates (G7 and G9).

When measuring the output transition time, three current paths travel through the R$_{WIRE}$. Assume the three inverters G2, G7 and G9 have similar sizes, the equivalent supply resistor of each cell must be three times the lumped supply resistor (R$_{WIRE}$) according to the parallel connection formula. Therefore, the adjusted formula is modified a little bit as follows. The falling time of QN (TF(QN)) is not necessary to be adjusted because it is a falling gate. The adjustment formulas for the output transition time in output Q falling case are also listed as follows, which can be derived by similar way as in the output Q rising case.

$$\begin{cases} C_{EFF} = \dfrac{E_{INT}(CK_{RISE} \rightarrow Q_{RISE}) - E_{INT}(CK_{RISE})}{2 \times VDD} + C_{LOAD} \\ R_{EFF} = 3 \times R_{WIRE} \end{cases}$$
$$TR(Q)_{ADJ} = TR(Q)_{ORG} + \ln 9 \times R_{WIRE} \times C_{EFF} \tag{34}$$
$$TF(QN)_{ADJ} = TF(QN)_{ORG}$$

$$\begin{cases} C_{EFF} = \dfrac{E_{INT}(CK_{RISE} \rightarrow Q_{FALL}) - E_{INT}(CK_{RISE})}{2 \times VDD} + C_{LOAD} \\ R_{EFF} = 3 \times R_{WIRE} \end{cases}$$
$$TF(Q)_{ADJ} = TF(Q)_{ORG} \tag{35}$$
$$TR(QN)_{ADJ} = TR(QN)_{ORG} + \ln 9 \times R_{WIRE} \times C_{EFF}$$

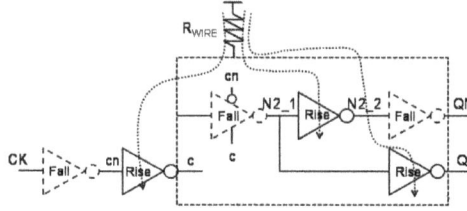

Fig. 19. The current flows of the DFF in the output rising case

4.2.2 Propagation delay time

In the signal propagation path from CK to Q, only G2 and G9 are rising in the output Q rising case. Therefore, the increased delay time of the two gates are added by the similar method shown in Section 4.1.2 to adjust the CK to Q delay of DFF circuits. The effective capacitance of G2, $C_{EFF}(G2)$, can be approximated as the $E_{INT}(CK_{RISE})$ divided by VDD. The effective capacitance of G9, $C_{EFF}(G9)$, and the R_{EFF} of G2 and G9 are obtained by the same approach for the output transition time. The adjusted propagation delay time $TDR(CK \rightarrow Q)_{ADJ}$ can be calculated by the following formulas, in which the increased term is related to known variables only.

$$\begin{cases} C_{EFF}(G2) = \dfrac{E_{INT}(CK_{RISE})}{VDD} \\ C_{EFF}(G9) = \dfrac{E_{INT}(CK_{RISE} \rightarrow Q_{RISE}) - E_{INT}(CK_{RISE})}{2 \times VDD} + C_{LOAD} \\ R_{EFF} = 3 \times R_{WIRE} \end{cases} \tag{36}$$

$$TDR(CK \rightarrow Q)_{ADJ} = TDR(CK \rightarrow Q)_{ORG} + \underline{(-\ln 0.5) \times R_{EFF} \times (C_{EFF}(G2) + C_{EFF}(G9))}$$

The propagation delay time of the output QN $TDF(CK \rightarrow QN)_{ADJ}$ can be calculated by the similar approach of $TDR(CK \rightarrow Q)_{ADJ}$. The adjustment formula is listed as follows, except that G7 is used instead of G9 for different output.

$$\begin{cases} C_{EFF}(G2) = \dfrac{E_{INT}(CK_{RISE})}{VDD} \\ C_{EFF}(G7) = \dfrac{E_{INT}(CK_{RISE} \rightarrow Q_{RISE}) - E_{INT}(CK_{RISE})}{2 \times VDD} + C_{LOAD} \\ R_{EFF} = 3 \times R_{WIRE} \end{cases} \tag{37}$$

$$TDF(CK \rightarrow Q)_{ADJ} = TDF(CK \rightarrow Q)_{ORG} + \underline{(-\ln 0.5) \times R_{EFF} \times (C_{EFF}(G2) + C_{EFF}(G7))}$$

4.2.3 Setup time

Figure 20 illustrates the internal status of a DFF when data is setting up. Following the same assumption of the setup time in Section 3, the data must reach N1_2 before the internal node c rises to ensure that the data can enter the next stage successfully. Therefore, the setup time of the D rising case TSR(D) can be expressed as the following formula.

$$TSR(D)_{ORG} = TDF(D \rightarrow N1\_2)_{ORG} - TDR(CK \rightarrow c)_{ORG} \tag{38}$$

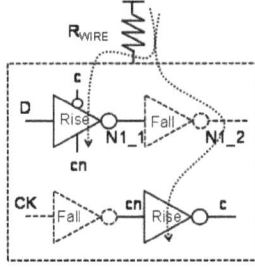

Fig. 20. The internal status of a DFF when D is setting up.

The propagation delay time TDF(D→N1_2)$_{ADJ}$ and TDR(CK→c)$_{ADJ}$ can be calculated with the similar method for the propagation delay time. Because there are two current paths in this case as shown in Fig.20, the equivalent supply resistor of each cell is two times R$_{WIRE}$. The formulas are listed as follows.

$$\left.\begin{cases} C_{EFF}(G3) = \dfrac{E_{INT}(D_{RISE})}{VDD} \\[2mm] C_{EFF}(G2) == \dfrac{E_{INT}(CK_{RISE})}{VDD} \\[2mm] R_{EFF} = 2 \times R_{WIRE} \end{cases}\right. \tag{39}$$

$$TDF(D \rightarrow N1\_2)_{ADJ} = TDF(D \rightarrow N1\_2)_{ORG} + (-\ln 0.5) \times R_{EFF} \times C_{EFF}(G3))$$

$$TDR(CK \rightarrow c)_{ADJ} = TDR(CK \rightarrow c)_{ORG} + (-\ln 0.5) \times R_{EFF} \times C_{EFF}(G2))$$

Therefore, the formula of the adjusted setup time TSR(D)$_{ADJ}$ can be obtained as follows. The setup time in the D falling case can be obtained by the similar way. The formula is also listed as follows.

$$\begin{aligned} TSR(D)_{ADJ} &= TDF(D \rightarrow N1\_2)_{ADJ} - TDR(CK \rightarrow c)_{ADJ} \\ &= TSR(D) + (-\ln 0.5) \times R_{EFF} \times [C_{EFF}(G3) - C_{EFF}(G2)] \\ TSF(D)_{ADJ} &= TDR(D \rightarrow N1\_2)_{ADJ} - TDR(CK \rightarrow c)_{ADJ} \\ &= TSF(D) + (-\ln 0.5) \times R_{EFF} \times [C_{EFF}(G4) - C_{EFF}(G2)] \end{aligned} \tag{40}$$

4.2.4 Internal energy

The internal energy of DFF cannot be separated to each cell. Therefore, the entire DFF is viewed as a super-gate to adjust its internal energy. The same formulas of the composite logic cells are used directly to adjust the internal energy of DFF.

4.3 Timing correction of cell switching activities

During gate-level simulation, the signal events are recorded in activity files (.vcd). Figure 21(a) shows the ideal timing diagram of four events, T(A), T(B), T(C), T(Y). With non-ideal supply lines, these events will occur at different time thus incurring different current waveforms. Therefore, the modification of activity files is also proposed in this paper, as illustrated in Fig.21(b). First, the modified propagation delay time $TD(G1)_{ADJ}$ can be obtained by the modification method of the signal cell. Then, Diff(G1) can be implied by $TD(G1)_{ADJ}$-$TD(G1)_{ORG}$ and be propagated to next event T(B). $T(B)_{ADJ}$ is derived by the summation of T(B) and Diff(G1). The other events can be modified in the similar way. After the timing errors are corrected, the accuracy of the constructed waveforms based on those events can be further improved.

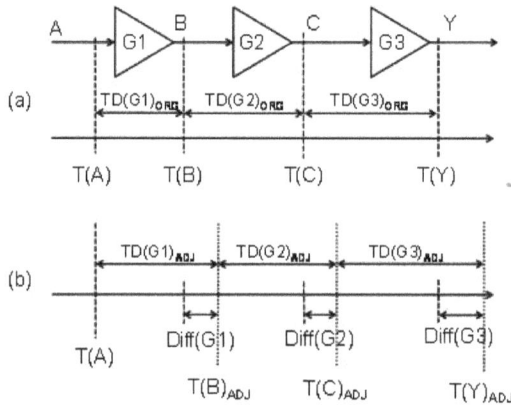

Fig. 21. Illustration of VCD events with (a) ideal (b) non-ideal supply lines

5. Experimental result

5.1 Experimental result of supply current waveform estimation method

We have implemented a supply current waveform estimation tool in C/C++. Given an input pattern, this tool can calculate the triangle that simulates the supply current waveform of each cell. The overall supply current waveform is then obtained by combining all triangles of every changed cell in time. All the input files of this tool follow standard formats, which are Verilog netlist file of the gate-level design, value changed dump (VCD) file of the design under given input patterns, and the LIB file of the standard cell library. The output format is a (time, voltage) pair that can be used to plot the dynamic supply current waveform. Those input/output files are compatible with current EDA tools. It allows our solution to be plugged into the existing EDA flow smoothly.

Very few commercial tools can provide the current waveform information at gate level. We choose PrimeTime- PX (Synopsys, 2009) for comparison, which can be used to estimate cycle-accurate peak power at gate level. Divided by the supply voltage, the peak power can be transformed to the peak current. Besides traditional LIB format, this tool also supports CCSM library format, which can be used to demonstrate the help from new library format. The results with and without CCSM data are shown in the rows "CCSM" and "LIB" of Table 1 respectively. Because 0.13 μm library does not have CCSM data yet, those CCSM data are characterized from HSPICE simulation by ourselves. Therefore, only combinational cells are characterized in our preliminary experiments. The previous approach (Shimazaki et al., 2000) is also rebuilt in our environment and tested in the same experiments to show our improvements on accuracy. Because they did not mention how to apply their approach on sequential cells, only combinational circuits are compared.

In the experiments, ISCAS' 85 and ISCAS' 89 benchmark circuits, which are implemented with TSMC 0.13um process, are used to test the accuracy. For each benchmark circuit, 200 random patterns are generated to trigger the circuit. After all, the average errors of the peak current and position with 200 pattern-pairs are shown in the row "eI_p" of Tables 1 and 2. The standard deviation of the peak current and position with the 200 results is shown in the row "sI_p" of Tables 1 and 2. The last column "AVG" in Tables 1 and 2 shows the average values of all cases. Figure 22 shows the estimated current waveforms of c7552 and s9234 as examples, which are very similar to HSPICE results.

According to the results estimated by PrimeTime-PX, the CCSM libraries significantly improve the accuracy of peak current estimation. However, the cycle-accurate results are still not accurate enough for analyzing the peak power or the IR-drop noise. The estimation results of the proposed methods, which are listed in the row "GCM" of Tables 2 and 3, demonstrates that the proposed approach can provide accurate estimations on the supply current waveforms by using the same information provided in traditional LIB libraries. The average estimation errors on eI_{PEAK} and eT_{PEAK} are about 10% with small standard deviation. The correlation between the estimated waveforms and HSPICE waveforms is higher than 0.97, which shows the similarity between the two waveforms. Compared to the rough estimation in (Shimazaki et al., 2000), the proposed approach does have a significant improvement on the estimation accuracy. Most importantly, the proposed approach can deal with sequential circuits, which enables this approach to be applied to modern designs.

The run time of the current waveform estimation for each benchmark circuit is provided in Table 3, which is measured on a XEON 3G machine with 2G RAM. The row "GCM" shows the run time of the proposed approach in seconds. The row "HSPICE" shows the run time of HSPICE simulation with the same patterns in hours. The row "Ratio" shows the ratio of the run time between HSPICE and GCM, which demonstrates a significant speed improvement.

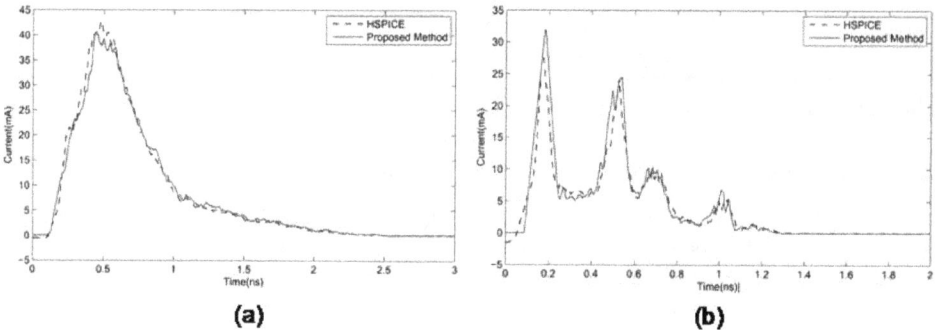

Fig. 22. The estimation supply current waveforms of (a)c7552 (b)s9234.

Circuit		c432	c499	c880	c1355	c1908	c2670	c3540	c5315	c6288	c7552	AVG
LIB	$eI_{P(\%)}$	60.63	203.96	44.51	111.75	64.72	812.42	69.66	1396.90	46.67	537.70	281.12
	$sI_{P(\%)}$	0.62	0.46	0.22	0.18	0.34	2.09	0.25	4.04	0.06	1.75	1.00
CCS	$eI_{P(\%)}$	25.25	38.41	33.49	42.43	52.22	255.50	50.68	299.00	51.94	73.05	92.29
	$sI_{P(\%)}$	0.16	0.11	0.10	0.07	0.13	0.91	0.14	0.94	0.08	0.26	0.29
(Shimaza	$eI_{P(\%)}$	42.16	35.81	59.85	81.11	64.88	41.01	51.61	38.48	27.73	31.67	47.43
ki et al.,	$sI_{P(\%)}$	15.92	14.58	19.68	16.47	16.99	10.21	15.70	9.96	18.56	10.02	14.80
2000)	$eT_{P(\%)}$	2.80	1.04	4.48	1.58	5.25	1.48	6.00	4.88	11.02	7.11	4.56
	$sT_{P(\%)}$	7.50	1.18	4.98	1.19	4.50	2.42	7.72	5.82	8.13	10.14	5.36
	Corr	0.959	0.977	0.961	0.928	0.973	0.981	0.971	0.988	0.993	0.980	0.971
GCM	$eI_{P(\%)}$	12.87	7.42	6.24	9.95	8.80	9.47	6.17	5.20	6.06	3.97	7.61
	$sI_{P(\%)}$	8.18	4.99	4.96	4.69	5.21	5.82	4.77	3.98	2.55	2.99	4.81
	$eT_{P(\%)}$	6.52	1.79	5.09	4.21	4.52	5.34	5.51	5.64	2.31	3.04	4.39
	$sT_{P(\%)}$	13.16	1.14	7.02	2.45	2.45	6.67	7.80	6.29	3.16	5.31	5.64
	Corr	0.964	0.985	0.985	0.976	0.987	0.977	0.982	0.989	0.992	0.988	0.983

Table 1. Experimental results of ISCAS85 benchmark circuits

Circuit		s298	s444	s526	s820	s1196	s1238	s1494	s5378	s9234	s15850	AVG
GCM	$eI_{P(\%)}$	8.96	12.52	10.96	12.96	8.92	9.32	2.84	10.81	13.77	13.04	10.40
	$sI_{P(\%)}$	6.11	1.51	8.56	10.60	5.23	6.63	3.57	2.01	1.14	0.97	4.63
	$eT_{P(\%)}$	10.99	4.12	7.25	6.97	5.87	5.12	8.25	2.12	1.52	3.29	5.55
	$sT_{P(\%)}$	12.38	1.78	7.53	5.98	6.79	5.54	6.96	0.39	0.26	1.36	4.89
	Corr	0.967	0.976	0.966	0.968	0.977	0.977	0.975	0.973	0.982	0.979	0.974

Table 2. Experimental results of ISCAS89 benchmark circuits

Circuit	c432	c499	c880	c1355	c1908	c2670	c3540	c5315	c6288	c7552	AVG
GCM (sec)	13.17	23.05	20.05	43.79	55.39	46.43	90.60	179.77	1083.94	414.68	-
HSPICE(hr)	9.73	17.58	16.44	27.45	26.54	40.61	54.82	86.76	69.16	128.38	-
Ratio	2661	2746	2951	2256	1725	3419	2178	1737	230	1115	2074
Circuit	s298	s444	s526	s820	s1196	s1238	s1494	s5378	s9234	s15850	AVG
GCM(sec)	4.38	1.78	3.15	2.76	6.89	6.12	4.88	28.55	35.35	71.75	-
HSPICE(hr)	13.49	10.33	12.05	14.50	23.03	23.88	28.68	107.83	183.15	495.78	-
Ratio	11089	20883	13770	18914	12034	14048	21161	13596	18652	24876	16902

Table 3. Experimental results of run time

5.2 Experimental result of library adjustment method

In order to demonstrate the accuracy of the IR-drop-aware adjustment approach, the same ISCAS85 and ISCAS89 benchmark circuits are used to perform some experiments. For each benchmark circuit, 200 random pattern pairs are generated to trigger the circuit. The average results of all circuits are illustrated in Fig.23. The average peak current errors using the method without adjustment the library information is draw with dash lines (w/o). The proposed library method is draw with bold line(w). According to the results, the proposed method can reduce the estimation errors successfully. Figure 24 shows the estimated supply current waveforms of c7552 circuit as example, which also confirm the accuracy of the proposed approach (GCM$_{(ADJ)}$). The waveforms obtained without IR-drop consideration (GCM) are also used to estimate the IR-drop directly with the same input pattern. The results show that estimation without considering R$_{wire}$ effects suffers large errors when the resistance on supply lines is getting larger. The proposed adjustment can consider the R$_{wire}$ effects and have a significant improvement on accuracy.

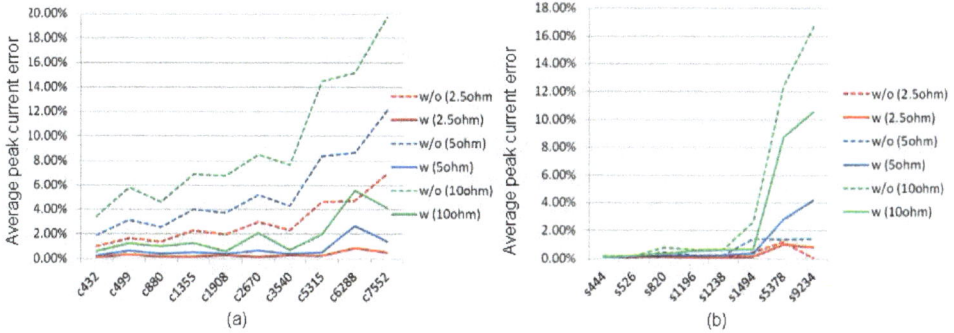

Fig. 23. The experimental results of library-adjustment methods on (a) ISCAS'85 (b) ISCAS'89.

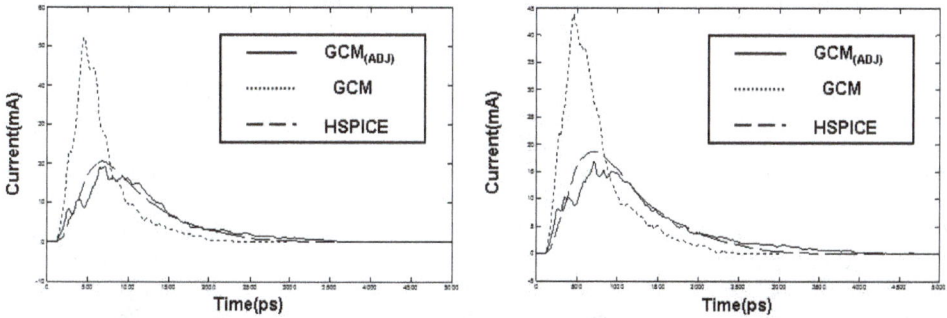

Fig. 24. The estimation supply current waveforms of c7552 with R$_{WIRE}$=10ohm

6. Conclusion

In this article, a library-based IR-drop estimation method is presented. This method concludes two parts, one is a gate-level supply current waveform estimation method using standard library information and the other is an analytical library adjustment method with IR-drop effect consideration. Extra characterization efforts and regression cost can be avoided to obtain accurate IR-drop estimation with less overhead. As shown in the experimental results, such an efficient modification method can provided good accuracy on IR-drop estimation with limited information. The estimation errors of our approach are about 5% compared with HSPICE results.

7. Acknowledgment

This work was partially supported by R.O.C National Science Council under Grant NSC99-2221-E-008-104. Their support are greatly appreciated.

8. References

Blakiewicz, G. & Chrzanowska-Jeske, M. (2007). Supply current spectrum estimation of digital cores at early design. *IET Circuits Devices Syst.*, vol.1, no.3 (Jun. 2007), pp. 233-240, ISSN: 1751-858X

Boliolo, A.; Benini, L.; de Micheli, G. & Ricco, B.(1997). Gate-level power and current simulation of CMOS integrated circuits. *IEEE Trans. Very Large Scale Integr.(VLSI) Syst.* , vol.5, no.4 (Dec. 1997), pp.473-488, ISSN: 1063-8210

Cadence (2006), *Open Source ECSM Format Specification Version 2.1*, Cadence

Chen, H.-M.; Huang, L.-D.; Liu, I-M. & Wong, M.D.F (2005). Simultaneous power supply planning and noise avoidance in floorplan design. *IEEE Trans. Comput.-Aided Des. Integr. Circuits Syst.*, vol. 24, no.4 (Apr. 2005), pp. 578-587, ISSN: 0278-0070

Juan, D.-C.; Chen, Y.-T.; Lee, M.-C. & Chang, S.-C. (2010). An Efficient Wake-Up Strategy Considering Spurious Glitches Phenomenon for Power Gating Designs. *IEEE Trans. Very Large Scale Integr.(VLSI) Syst.*, vol. 18, no. 2 (Feb. 2010), pp. 246-255, ISSN: 1063-8210

Kawa, J. (2008). Low Power and Power Management for CMOS-An EDA Perspective. *IEEE Trans. Electron Devices*, vol. 55, no. 1 (Jan. 2008), pp.186-196, ISSN: 0018-9383

Lee, M.-S.; Lin, C.-H.; Liu, C.-N.J., & Lin, S.-C. (2008). Quick supply current waveform estimation at gate level using existed cell library information. *Proceedings of ACM Great Lakes Symp. on VLSI*, pp. 135-138, Orlando, FL, May 4-6 2008, ISBN: 978-1-59593-999-9

Lee, M.-S.; Lai, K.-S.; Hsu; C.-L., & Liu, C.-N.J. (2010). Dynamic IR drop estimation at gate level with standard library information. *Proceedings of IEEE Intl. Symp. Circuits and Systems*, pp. 2606-2609, Paris, France, May 2010, ISBN: 978-1-4244-5308-5

Michael, K.; David, F.; Rob, A.; Alan G. and Shi, K. (2008). *Low Power Methodology Manual for System-on-Chip Design*. Springer, ISBN-10: 0387718184, New York, USA

Popovich, M.; Sotman, M.; Kolodny, A. & Friedman, E. (2008). Effective radii of on-chip decoupling capacitors. *IEEE Trans. Very Large Scale Integr.(VLSI) Syst.*, vol. 16, no. 7 (July 2008), pp. 894-907, ISSN: 1063-8210

Shimazaki, K.; Tsujikawa, H.; Kojima, S. & Hirano, S. (2000). LEMINGS: LSI's EMI-noise analysis with gate level simulator. *Proceedings of IEEE Int. Symp. Quality Electronic Design*, pp. 129-136, San Jose, CA, Mar. 2000, ISBN: 0-7695-0525-2

Synopsys (2003), Library *Compiler User Guide: Modeling Timing and Power Technology Libraries*, Synopsys

Synopsys (2008), *CCS Timing Library Characterization Guidelines Version 3.2*, Synopsys

Synopsys (2009), *PrimeTime PX User Guide Version C-2009.06*, Synopsys

Xu, H. ; Vemuri, R. & Jone, W. (2011). Dynamic Characteristics of Power Gating During Mode Transition. *IEEE Trans. Very Large Scale Integr.(VLSI) Syst.*, vol. 19, no. 2 (Feb. 2011), pp. 237-249, ISSN: 1063-8210

Zhao, S.; Roy, K. & Koh, C.-K. (2002). Decoupling capacitance allocation and its application to power-supply noise-aware floorplanning. *IEEE Trans. Comput.-Aided Des. Integr. Circuits Syst.*, vol. 21, no.1 (Jan. 2002.), pp. 81-92, ISSN: 0278-0070

11

Study on Low-Power Image Processing for Gastrointestinal Endoscopy

Meng-Chun Lin

Department of Computer National Chengchi University
Taiwan

1. Introduction

Gastrointestinal (GI) endoscopy has been popularly applied for the diagnosis of diseases of the alimentary canal including Crohn's Disease, Celiac disease and other malabsorption disorders, benign and malignant tumors of the small intestine, vascular disorders and medication related small bowel injury. There are two classes of GI endoscopy; wired active endoscopy and wireless passive capsule endoscopy. The wired active endoscopy can enable efficient diagnosis based on real images and biopsy samples; however, it causes discomfort for the patients to push flexible, relatively bulky cables into the digestive tube. To relief the patients' discomfort, wireless passive capsule endoscopes are being developed worldwide (1)-(6).

The capsule moves passively through the internal GI tract with the aid of peristalsis and transmits images of the intestine wirelessly. Developed by Given Imaging Ltd., the PillCam capsule is a state-of-the-art commercial wireless capsule endoscope product. The PillCam capsule transmits the GI images at a resolution of 256-by-256 8-bit pixels and the frame rate of 2 frames/sec (or fps). Because of its high mobility, it has been successfully utilized to diagnose diseases of the small intestine and alleviate the discomfort and pain of patients. However, based on clinical experience; the PillCam still has some drawbacks. First, the PillCam cannot control its heading and moving direction itself. This drawback may cause image oversights and overlook a disease. Second, the resolution of demosaicked image is still low, and some interesting spots may be unintentionally omitted. Therefore, the images will be severely distorted when physicians zoom images in for detailed diagnosis. The first drawback is the nature of passive endoscopy. Some papers have presented approaches for the autonomous moving function (7; 8; 21; 22; 25). Very few papers address solutions for the second drawback. Increasing resolution may alleviate the second problem; however, it will result in significant power consumption in RF transmitter. Hence, applying image compression is necessary for saving the power dissipation of RF transmitter (9)-(14), (23), (24), (26).

Our previous work (11) has presented an ultra-low-power image compressor for wireless capsule endoscope. It helps the endoscope to deliver a compressed 512-by-512 image, while the RF transmission rate is at 1 megabits ($(256 \times 256 \times 2 \times 8)/1024^2$) per second. No any references can clearly define how much compression is allowed in capsule endoscope application. We define that the minimum compression rate is 75% according to two considerations for our capsule endoscope project. The first consideration is that the new image resolution (512-by-512) that is four times the one (256-by-256) of the PillCam can be an assistant to promote the diagnosis of diseases for doctors. The other one is that we do not significantly increase the power consumption for the RF circuit after increasing the image resolution from the sensor. Instead of applying state-of-the-art video compression techniques, we proposed a simplified image compression algorithm, called GICam, in which the memory size and computational load can be significantly reduced. The experimental results shows that the GICam image compressor only costs 31K gates at 2 frames per second, consumes 14.92 mW, and reduces the image size by at least 75% .

In applications of capsule endoscopy, it is imperative to consider the tradeoffs between battery life and performance. To further extend the battery life of a capsule endoscope, we herein present a subsample-based GICam image compressor, called GICam-II. The proposed compression technique is motivated by the reddish feature of GI image. We have previously proposed the GICam-II image compressor in paper (20). However, the color importance of primary colors in GI images has no quantitative analysis in detail because of limited pages. Therefore, in this paper, we completely propose a series of mathematical statistics to systematically analyze the color sensitivity in GI images from the RGB color space domain to the 2-D DCT spatial frequency domain in order to make up for a deficiency in our previous work (20). This paper also refines the experimental results to analyze the performance about the compression rate, the quality degradation and the ability of power saving individually.

As per the analysis of color sensitivity, the sensitivity of GI image sharpness to red component is at the same level as the sensitivity to green component. This result shows that the GI image is cardinal and different from the general image, whose sharpness sensitivity to the green component is much higher than the sharpness sensitivity to the red component. Because the GICam-II starts compressing the GI image from the Bayer-patterned image, the GICam-II technique subsamples the green component to make the weighting of red and green components the same. Besides, since the sharpness sensitivity to the blue component is as low as 7%, the blue component is down-sampled by four. As shown in experimental results, with the compression ratio as high as 4:1, the GICam-II can significantly save the power dissipation by 38.5% when compared with previous GICam work (11) and 98.95% when compared with JPEG compression, while the average PSNRY is 40.73 dB. The rest of the paper is organized as follows. Section II introduces fundamentals of GICam compression and briefs the previous GICam work. Section III presents the sensitivity analysis of GICam image and shows the importance of red component in GI image. In Section IV, the GICam-II compression will be described in details. Then, Section V illustrates the experimental results in terms of compression ratio, image quality and power consumption. Finally, Section VI concludes our contribution and merits of this work.

Except using novel ultra-low-power compression techniques to save the power dissipation of RF transmitter in high-resolution wireless gastrointestinal endoscope systems. How to efficiently eliminate annoying impulsive noise caused by a fault sensor and enhance the sharpness is necessary for gastrointestinal (GI) images in wired/wireless gastrointestinal endoscope systems. To overcome these problems, the LUM filter is the most suitable candidate because it simultaneously has the characteristics of smoothing and sharpening. In the operational procedure of LUM filter, the mainly operational core is the rank-order filtering (ROF) and the LUM filter itself needs to use different kind of rank values to accomplish the task of smoothing or sharpening. Therefore, we need a flexible ROF hardware to arbitrarily select wanted rank values into the operation procedure of LUM filter and we have proposed an architecture based on a maskable memory for rank-order filtering. The maskable memory structure, called dual-cell random-access memory (DCRAM), is an extended SRAM structure with maskable registers and dual cells. This dissertation is the first literature using maskable memory to realize ROF. Driving by the generic rank-order filtering algorithm, the memory-based architecture features high degree of flexibility and regularity while the cost is low and the performance is high. This architecture can be applied for arbitrary ranks and a variety of ROF applications, including recursive and non-recursive algorithms. Except efficiently eliminating annoying impulsive noises and enhance sharpness for GI images, the processing speed of ROF can also meet the real-time image applications.

2. GICam image compressor

2.1 The review of GICam image compression algorithm

Instead of applying state-of-the-art video compression techniques, we proposed a simplified image compression algorithm, called GICam. Traditional compression algorithms employ the YCbCr quantization to earn a good compression ratio while the visual distortion is minimized, based on the factors related to the sensitivity of the human visual system (HVS). However, for the sake of power saving, our compression rather uses the RGB quantization (15) to save the computation of demosaicking and color space transformation. As mentioned above, the advantage of applying RGB quantization is two-fold: saving the power dissipation on preprocessing steps and reducing the computing load of 2-D DCT and quantization. Moreover, to reduce the hardware cost and quantization power dissipation, we have modified the RGB quantization tables and the quantization multipliers are power of two's. In GICam, the Lempel-Ziv (LZ) coding (18) is employed for the entropy coding. The reason we adopted LZ coding as the entropy coding, is because the LZ encoding does not need look-up tables and complex computation. Thus, the LZ encoding consumes less power and uses smaller silicon size than the other candidates, such as the Huffman encoding and the arithmetic coding. The target compression performance of the GICam image compression is to reduce image size by at least 75%. To meet the specification, given the quantization tables, we exploited the cost-optimal LZ coding parameters to meet the compression ratio requirement by simulating with twelve tested endoscopic pictures shown in Fig.3.

When comparing the proposed image compression with the traditional one in (11), the power consumption of GICam image compressor can save 98.2% because of the reduction of memory requirement. However, extending the utilization of battery life for a capsule endoscope remains an important issue. The memory access dissipates the most power in GICam image compression. Therefore, in order to achieve the target of extending the battery life, it is necessary to consider how to efficiently reduce the memory access.

2.2 Analysis of sharpness sensitivity in gastrointestinal images

2.2.1 The distributions of primary colors in the RGB color space

In the modern color theory (16; 17), most color spaces in used today are oriented either toward hardware design or toward product applications. Among these color spaces, the

RGB(red, green, blue) space is the most commonly used in the category of digital image processing; especially, broad class of color video cameras and we consequently adopt the RGB color space to analyze the importance of primary colors in the GI images. In the RGB color space, each color appears in its primary spectral components of red, green and blue. The RGB color space is based on a Cartesian coordinate system, in which, the differ colors of pixels are points on or inside the cube based on the triplet of values (R, G, B). Due to this project was supported in part by Chung-Shan Institute of Science and Technology, Taiwan, under the project BV94G10P. The responsibility of Chung-Shan Institute of Science and Technology mainly designs a 512-by-512 raw image sensor. The block-based image data can be sequentially outputted via the proposed locally-raster-scanning mechanism for this raw image sensor. The reason for adopting a novel image sensor without using generally conventional ones is to efficiently save the size of buffer memory. Conventional raw image sensors adopt the raster-scanning mechanism to output the image pixels sequentially, but they need large buffer memory to form each block-based image data before executing the block-based compression. However, we only need a small ping-pong type memory structure to directly save the block-based image data from the proposed locally-raster-scanning raw image sensor. The structure of this raw image sensor is shown in Fig.1 (a) and the pixel sensor architecture for the proposed image sensor is shown in Fig.1 (b). In order to prove the validity for this novel image sensor before the fabrication via the Chung-Shan Institute of Science and Technology, the chip of the 32-by-32 locally-raster-scanning raw image sensor was designed by full-custom CMOS technology and this chip is submitted to Chip Implementation Center (CIC), Taiwan, for the fabrication. Fig.2 (a) and Fig.2 (b) respectively shows the chip layout and the package layout with the chip specification. The advantage of this novel CMOS image sensor can save the large area of buffer memory. The size of buffer memory can be as a simple ping-pong memory structure shown in Fig.9 while executing the proposed image algorithm, a novel block coding.   Our research only focuses on developing the proposed image compressor and other components are implemented by other research department for the GICam-II capsule endocopy. Therefore, the format of the GI image used in the simulation belongs to a raw image from the 512-by-512 sensor designed by Chung-Shan Institute of Science and Technology. In this work, we applied twelve GI images captured shown in Fig.3 for testcases to evaluate the compression technique. The distribution of GI image pixels in the RGB color space is non-uniform. Obviously, the GI image is reddish and the pixels are amassed to the red region. Based on the observation in the RGB color space, the majority of red values are distributed between 0.5 and 1 while most of the green and blue values are distributed between 0 and 0.5 for all tested GI images. To further analyze the chrominance distributions and variations in the RGB color space for each tested GI images, two quantitative indexes are used to quantify these effects. The first index is to calculate the average distances between total pixels and the maximum primary colors in each GI image, and the calculations are formulated as Eq.1, Eq.2 and Eq.3. First, Eq.1 defines the the average distance between total pixels and the most red color $(\overline{R})$, in which, $R(i, j)$ means the value of red component of one GI image at (i, j) position and the value of most red color (R_{max}) is 255. In addition, M and N represent the width and length for one GI image, respectively. The M is 512 and the N is 512 for twelve tesed GI images in this work. Next, Eq.2 also defines the average distance between total pixels and the most green color $(\overline{G})$ and the value of most green one (G_{max}) is 255. Finally, Eq.3 defines the average distance between total pixels and the most blue color $(\overline{B})$ and the value of most blue color (B_{max}) is 255. Table 1 shows the statistical results of $\overline{R}$, $\overline{G}$ and $\overline{B}$

for all tested GI images. From Table 1, the results clearly show that $\overline{R}$ has the shortest average distance. Therefore, human eyes can be very sensitive to the obvious cardinal ingredient on all surfaces of tested GI images. Moreover, comparing $\overline{G}$ with $\overline{B}$, $\overline{G}$ is shorter than $\overline{B}$ because $\overline{G}$ contributes larger proportion in luminance.

$$\overline{R} = E[(1 - \frac{R(i,j)}{R_{max}})]$$

$$= (\frac{1}{M \times N}) \sum_{i=0}^{M-1} \sum_{j=0}^{N-1} (1 - \frac{R(i,j)}{R_{max}}) \tag{1}$$

$$\overline{G} = E[(1 - \frac{G(i,j)}{G_{max}})]$$

$$= (\frac{1}{M \times N}) \sum_{i=0}^{M-1} \sum_{j=0}^{N-1} (1 - \frac{G(i,j)}{G_{max}}) \tag{2}$$

$$\overline{B} = E[(1 - \frac{B(i,j)}{B_{max}})]$$

$$= (\frac{1}{M \times N}) \sum_{i=0}^{M-1} \sum_{j=0}^{N-1} (1 - \frac{B(i,j)}{B_{max}}) \tag{3}$$

The first index has particularly quantified the chrominance distributions through the concept of average distance, and the statistical results have also shown the reason the human eyes can sense the obvious cardinal ingredient for all tested GI images. Next, the second index is to calculate the variance between total pixels and average distance, in order to further observe

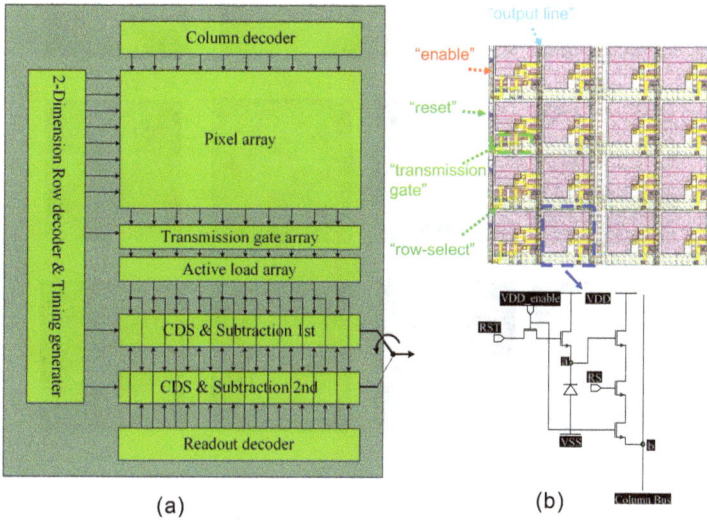

(a)   (b)

Fig. 1. (a) The structure of locally-raster-scanning raw image sensor (b)The pixel sensor architecture for the locally-raster-scanning raw image sensor.

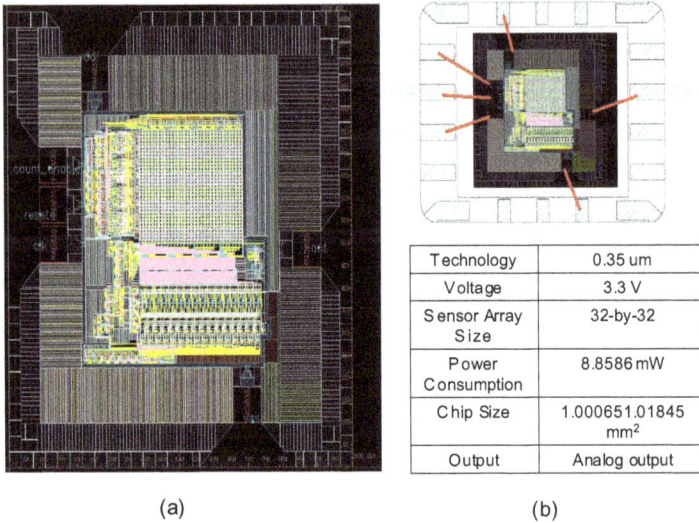

Technology	0.35 um
Voltage	3.3 V
Sensor Array Size	32-by-32
Power Consumption	8.8586 mW
Chip Size	1.000651.01845 mm^2
Output	Analog output

(a)   (b)

Fig. 2. (a)The chip layout of the locally-raster-scanning raw image sensor (b)The package layout and the chip specification of the locally-raster-scanning raw image sensor.

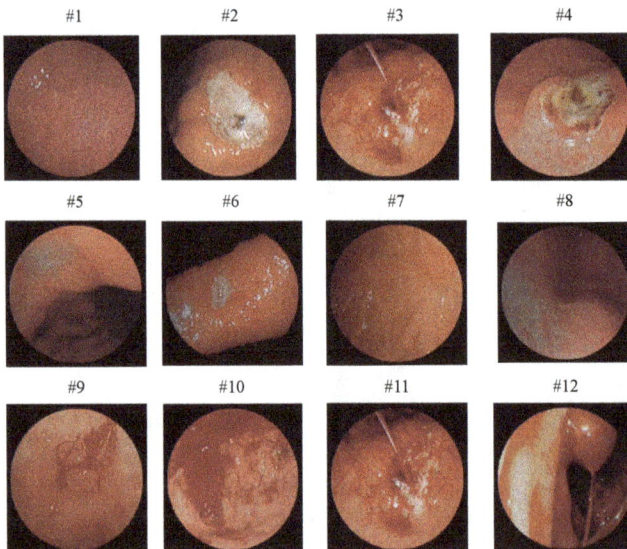

Fig. 3. The twelve tested GI images.

the color variations in GI images, and the calculations are formulated as Eq.4, Eq.7 and Eq.10. The Table 2 shows that the average variation of red signal is 0.09, the average variance of green one is 0.03, and the average variance of blue one is 0.02. It signifies that the color information of red signal must be preserved carefully more than other two primary colors; green and blue,

Average Distance			
Test Picture ID	$\bar{R}$	$\bar{G}$	$\bar{B}$
1	0.58	0.80	0.82
2	0.55	0.74	0.79
3	0.54	0.81	0.86
4	0.55	0.76	0.81
5	0.66	0.82	0.85
6	0.66	0.84	0.87
7	0.59	0.82	0.88
8	0.68	0.81	0.83
9	0.55	0.80	0.85
10	0.53	0.81	0.84
11	0.53	0.81	0.86
12	0.62	0.80	0.85
Average	0.59	0.80	0.84

Table 1. The Analysis of Average Distance.

for GI images because the dynamic range of red signal is broader than green and blue ones. In addition, the secondary is green signal and the last is blue signal.

$$VAR_R = E[(1 - \frac{R(i,j)}{R_{max}})^2] - \{E[(1 - \frac{R(i,j)}{R_{max}})]\}^2 \tag{4}$$

$$= (\frac{1}{M \times N}) \sum_{i=0}^{M-1} \sum_{j=0}^{N-1} [1 - \frac{R(i,j)}{R_{max}}]^2 - \tag{5}$$

$$[(\frac{1}{M \times N}) \sum_{i=0}^{M-1} \sum_{j=0}^{N-1} (1 - \frac{R(i,j)}{R_{max}})]^2 \tag{6}$$

$$VAR_G = E[(1 - \frac{G(i,j)}{G_{max}})^2] - \{E[(1 - \frac{G(i,j)}{G_{max}})]\}^2 \tag{7}$$

$$= (\frac{1}{M \times N}) \sum_{i=0}^{M-1} \sum_{j=0}^{N-1} [1 - \frac{G(i,j)}{G_{max}}]^2 - \tag{8}$$

$$[(\frac{1}{M \times N}) \sum_{i=0}^{M-1} \sum_{j=0}^{N-1} (1 - \frac{G(i,j)}{G_{max}})]^2 \tag{9}$$

$$VAR_B = E[(1 - \frac{B(i,j)}{B_{max}})^2] - \{E[(1 - \frac{B(i,j)}{B_{max}})]\}^2 \tag{10}$$

$$= (\frac{1}{M \times N}) \sum_{i=0}^{M-1} \sum_{j=0}^{N-1} [1 - \frac{B(i,j)}{B_{max}}]^2 - \tag{11}$$

$$[(\frac{1}{M \times N}) \sum_{i=0}^{M-1} \sum_{j=0}^{N-1} (1 - \frac{B(i,j)}{B_{max}})]^2 \tag{12}$$

Variance of Distance			
Test Picture ID	VAR_R	VAR_G	VAR_B
1	0.08	0.02	0.02
2	0.11	0.05	0.03
3	0.10	0.03	0.02
4	0.10	0.04	0.02
5	0.07	0.02	0.01
6	0.08	0.02	0.01
7	0.09	0.02	0.01
8	0.06	0.02	0.02
9	0.09	0.03	0.01
10	0.10	0.03	0.02
11	0.10	0.03	0.02
12	0.10	0.04	0.02
Average	0.09	0.03	0.02

Table 2. The Analysis of Variance.

2.2.2 The analysis of sharpness sensitivity to primary colors for gastrointestinal images

Based on the analysis of RGB color space, the importance of chrominance is quantitatively demonstrated for GI images. Except for the chrominance, the luminance is another important index because it can efficiently represent the sharpness of an object. Eq.13 is the formula of luminance (Y) and the parameters ; a1, a2 and a3 are 0.299, 0.587 and 0.114 respectively.

$$Y = a1 \times R + a2 \times G + a3 \times B \qquad (13)$$

To efficiently analyze the importance of primary colors in the luminance, the analysis of sensitivity is applied. Through the analysis of sensitivity, the variation of luminance can actually reflect the influence of each primary colors. Eq.14, Eq.15 and Eq.16 define the sensitivity of red ($S_{Y_{i,j}}^{R_{i,j}}$), the sensitivity of green ($S_{Y_{i,j}}^{G_{i,j}}$), and the sensitivity of blue ($S_{Y_{i,j}}^{B_{i,j}}$) at position (i,j), respectively for a color pixel of a GI image.

$$S_{Y_{i,j}}^{R_{i,j}} = \frac{\Delta Y_{i,j}/Y_{i,j}}{\Delta R_{i,j}/R_{i,j}} = \frac{R_{i,j}}{Y_{i,j}} \times \frac{\Delta Y_{i,j}}{\Delta R_{i,j}} = \frac{a1 \times R_{i,j}}{Y_{i,j}} \qquad (14)$$

$$S_{Y_{i,j}}^{G_{i,j}} = \frac{\Delta Y_{i,j}/Y_{i,j}}{\Delta G_{i,j}/G_{i,j}} = \frac{G_{i,j}}{Y_{i,j}} \times \frac{\Delta Y_i}{\Delta G_{i,j}} = \frac{a2 \times G_{i,j}}{Y_{i,j}} \qquad (15)$$

$$S_{Y_{i,j}}^{B_{i,j}} = \frac{\Delta Y_{i,j}/Y_{i,j}}{\Delta B_{i,j}/B_{i,j}} = \frac{B_{i,j}}{Y_{i,j}} \times \frac{\Delta Y_i}{\Delta B_{i,j}} = \frac{a3 \times B_{i,j}}{Y_{i,j}} \qquad (16)$$

After calculating the sensitivity of each primary colors for a GI image, the average sensitivity of red ($\overline{S_Y^R}$), the average sensitivity of green ($\overline{S_Y^G}$), and the average sensitivity of blue ($\overline{S_Y^B}$) are calculated by Eq.17, Eq.18 and Eq.19 for each GI images. M and N represent the width and length for a GI image, respectively. Table 3 shows the average sensitivities of red, green and blue for all tested GI images. From the calculational results, the sensitivity of blue is the slightest and hence the variation of luminance arising from the aliasing of blue is very invisible. In addition to the sensitivity of blue, the sensitivity of red is close to the one of green and thus they both have a very close influence on the variation of luminance.

$$\overline{S_Y^R} = (\frac{1}{M \times N}) \sum_{i=0}^{M-1} \sum_{j=0}^{N-1} S_{Y_{i,j}}^{R_{i,j}} \qquad (17)$$

$$\overline{S_Y^G} = (\frac{1}{M \times N}) \sum_{i=0}^{M-1} \sum_{j=0}^{N-1} S_{Y_{i,j}}^{G_{i,j}} \qquad (18)$$

$$\overline{S_Y^B} = (\frac{1}{M \times N}) \sum_{i=0}^{M-1} \sum_{j=0}^{N-1} S_{Y_{i,j}}^{B_{i,j}} \qquad (19)$$

To sum up the variance of chrominance and the sensitivity of luminance, blue is the

The Sensitivity of Primary Colors in Luminance			
Test Picture ID	S_Y^R	S_Y^G	$\overline{S_Y^B}$
1	0.49	0.43	0.08
2	0.44	0.48	0.08
3	0.55	0.39	0.06
4	0.47	0.46	0.07
5	0.45	0.47	0.08
6	0.48	0.45	0.07
7	0.52	0.42	0.06
8	0.44	0.48	0.08
9	0.51	0.43	0.06
10	0.54	0.40	0.06
11	0.55	0.39	0.06
12	0.49	0.44	0.07
Average	0.49	0.44	0.07

Table 3. The Analysis of Average Sensitivities.

most insensitive color in the GI images. Therefore, the blue component can be further downsampled without significant sharpness degradation. Moreover, comparing the red signal with the green signal, they both have a very close influence on the variation of luminance, because they have very close sensitivities. However, the chrominance of red varies more than the chrominance of green and hence the information completeness of red has higher priority than the green. Because the proposed compression coding belongs to the DCT-based image coding, the coding is processed in the spatial-frequency domain. To let the priority relationship between red and green also response in the spatial-frequency domain, the analysis of alternating current (AC) variance will be accomplished to demonstrate the inference mentioned above in the next subsection.

2.2.3 The analysis of AC variance in the 2-D DCT spatial frequency domain for gastrointestinal images

According to the analysis results from the distributions of primary colors in the RGB color space and the proportion of primary colors in the luminance for GI images, the red signal plays a decisive role in the raw image. The green signal plays a secondary role and the blue signal is very indecisive. To verify the validity of observation mentioned above, we first use the two-dimensional (2-D) 8×8 discrete cosine transform (DCT) to transfer the spatial domain into the spatial-frequency domain for each of the components, R, G1, G2 and B. The 2-D 8×8 DCT transformation can be perceived as the process of finding for each waveform in the 2-D 8×8 DCT basic functions and also can be formulated as Eq.20, Eq.21, Eq.22, Eq.23 and Eq.24 for each 8×8 block in R, G1, G2 and B subimages respectively. M and N represent the width and length for one GI image respectively. k, l=0, 1, ..., 7 and y_{kl} is the corresponding weight of DCT basic function in the kth row and the lth column. P represents the total number of pictures and B represents the total number of 8×8 blocks in the GI images.

$$R_{pb}(kl) = \frac{c(k)}{2} \sum_{i=0}^{7} [\frac{c(l)}{2} \sum_{j=0}^{7} r_{ij} cos(\frac{(2j+1)l\pi}{16})] cos(\frac{(2i+1)k\pi}{16}) \tag{20}$$

$$G_{pb}(kl) = \frac{c(k)}{2} \sum_{i=0}^{7} [\frac{c(l)}{2} \sum_{j=0}^{7} g_{ij} cos(\frac{(2j+1)l\pi}{16})] cos(\frac{(2i+1)k\pi}{16}) \tag{21}$$

$$B_{pb}(kl) = \frac{c(k)}{2} \sum_{i=0}^{7} [\frac{c(l)}{2} \sum_{j=0}^{7} b_{ij} cos(\frac{(2j+1)l\pi}{16})] cos(\frac{(2i+1)k\pi}{16}) \tag{22}$$

$$c(k) = \begin{cases} \frac{1}{\sqrt{2}}, & if\ k=0 \\ 1, & otherwise. \end{cases} \tag{23}$$

$$c(l) = \begin{cases} \frac{1}{\sqrt{2}}, & if\ l=0 \\ 1, & otherwise. \end{cases} \tag{24}$$

Next, we calculate the average energy amplitude of all alternating current (AC) coefficients of all tested GI images, in order to observe the variation of energy for each of the components R, G1, G2 and B, and the calculations are formulated as Eq.25, Eq.26, Eq.27.

$$A_R(kl) = \frac{1}{P} \sum_{p=1}^{P} [\sum_{b=0}^{B-1} |R_{pb}(kl)|] \tag{25}$$

$$A_G(kl) = \frac{1}{P} \sum_{p=1}^{P} [\sum_{b=0}^{B-1} |G_{pb}(kl)|] \tag{26}$$

$$A_B(kl) = \frac{1}{P} \sum_{p=1}^{P} [\sum_{b=0}^{B-1} |B_{pb}(kl)|] \tag{27}$$

After calculating the average energy amplitude, we convert the 2-D DCT domain into one-dimensional (1-D) signal distribution in order to conveniently observe the variation of frequency. Consequently, a tool for transforming two-dimensional signals into one dimension is needed. There are many schemes to convert 2-D into 1-D, including row-major scan, column-major scan, peano-scan, and zig-zag scan. Majority of the DCT coding schemes adopt zig-zag scan to accomplish the goal of conversion, and we use it here. The benefit of zig-zag is its property of compacting energy to low frequency regions after discrete cosine transformation. The arrangement sorts the coefficients from low to high frequency, and

Fig.4(a) shows the zig-zag scanning order for 8×8 block. Fig.4(b) shows the 1-D signal distribution after zig-zag scanning order and Fig.4(c) shows the symmetric type of frequency for the 1-D signal distribution.

Through the converting method of Fig.4, the 1-D signal distributions of each R, G1, G2, B components are shown in Fig.5. The variances of frequency are 1193, 1192, 1209 and 1244 for G1, G2, R and B respectively, and the variance of R is very close to the ones of G1 and G2 from the result. However, the datum of G are twice the datum of R based on the Bayer pattern and hence, the datum of G can be reduced to half at the most.   Based

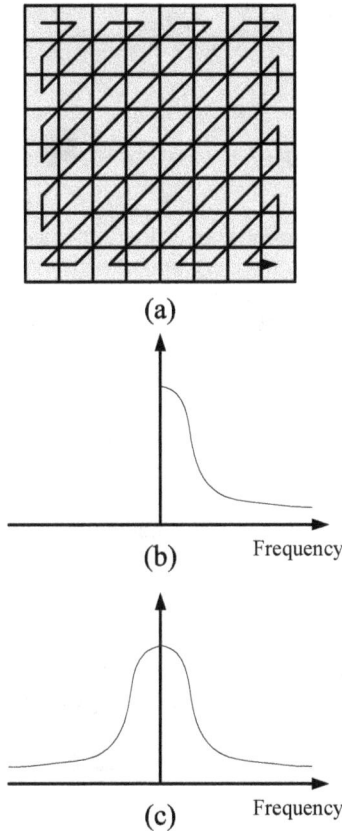

(a)

(b)   Frequency

(c)   Frequency

Fig. 4. (a) zigazg scanning for a 8×8 block (b) 1-D signal distribution after zigzag scanning order (c) The symmetric type of frequency for the 1-D signal distribution.

on the analysis result mentioned above, the R component is very decisive for GI images and it needs to be compressed completely. However, the G1, G2 and B components do not need to be compressed completely because they are of less than the R component. Therefore, in order to efficiently reduce the memory access to expend the battery life of capsule endoscopy, the datum of G1, G2 and B components should be appropriately decreased according to the proportion of their importance prior to the compression process.   In this

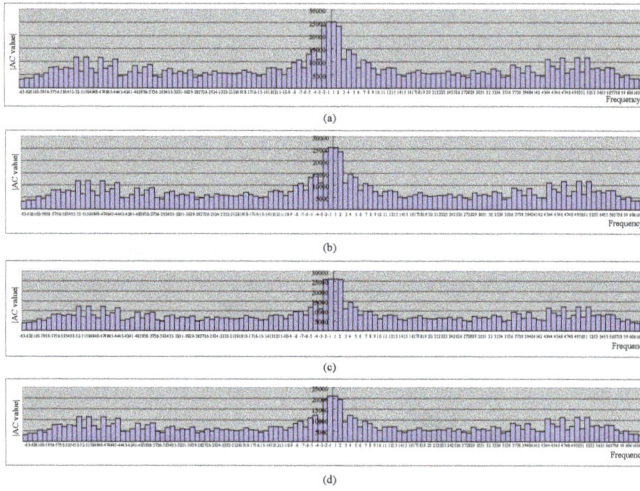

(a)

(b)

(c)

(d)

Fig. 5. (a) Spatial-frequency distribution converting into one-dimension for G1 component (b) Spatial-frequency distribution converting into one-dimension for G2 component (c) Spatial-frequency distribution converting into one-dimension for R component (d) Spatial-frequency distribution converting into one-dimension for B component.

paper, we successfully propose a subsample-based GICam image compression algorithm and the proposed algorithm firstly uses the subsample technique to reduce the incoming datum of G1, G2 and B components before the compression process. The next section will describe the proposed algorithm in detail.

2.3 The subsample-based GICam image compression algorithm

Fig.6 illustrates the GICam-II compression algorithm. For a 512×512 raw image, the raw image firstly divides into four parts, namely, R, G1, G2, and B components and each of the components has 256×256 pixels. For the R component, the incoming image size to the 2-D DCT is 256×256×8 bits, Where, the incoming image datum are completely compressed because of the importance itself in GI images. Except for the R component, the GICam-II

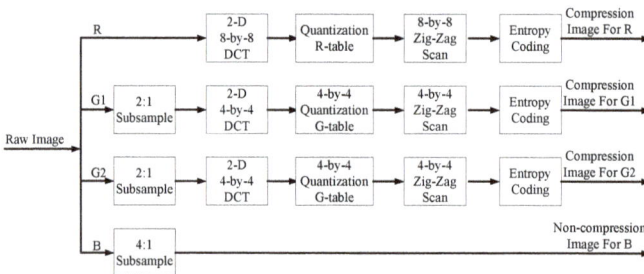

Fig. 6. The GICam-II image compression algorithm.

algorithm can use an appropriate subsample ratio to pick out the necessary image pixels into the compression process for G1, G2 and B components, and Eq.28 and Eq.29 are formulas for the subsample technique. $SM_{16:2m}$ is the subsample mask for the subsample ratio 16-to-2m as shown in Eq.28, and the subsample mask $SM_{16:2m}$ is generated from basic mask as shown in Eq.29. The type of subample direction is block-based, when certain positions in the subsample mask are one, their pixels in the same position will be compressed, or otherwise they are not processed. For the G1 and G2 components, the low subsample ratio must be assigned, considering their secondary importance in GI images. Thus, the 2:1 subsample ratio is candidate one, and the subsample pattern is shown in Fig.7 (a). Finally, for the B component, the 4:1 subsample ratio is assigned and the subsample pattern is shown in Fig.7 (b). In the GICam-II image compression algorithm, the 8×8 2-D DCT is still used to transfer the R component. However, the 4×4 2-D DCT is used for G1 and G2 components because the incoming datum are reduced by subsample technique. Moreover, the G quantization table is also modified and shown in the Fig.8. Finally, the B component is directly transmitted; not be compressed, after extremely decreasing the incoming datum. Because of the non-compression for the B component, the 8×8 and 4×4 zig-zag scanning techniques are added into the GICam-II to further increase the compression rate for R, G1 and G2 components before entering the entropy encoding. In the GICam-II, the Lempel-Ziv (LZ) coding (18) is also employed for the entropy coding because of non-look-up tables and low complex computation.

$$SM_{16:2m}\,(i,j) = BM_{16:2m}\,(i \bmod 4, j \bmod 4)$$
$$m = 1,2,3,4,5,6,7,8. \tag{28}$$

$$BM_{16:2m}\,(k,l) =$$
$$\begin{bmatrix} u\,(m-1) & u\,(m-5) & u\,(m-2) & u\,(m-6) \\ u\,(m-7) & u\,(m-3) & u\,(m-8) & u\,(m-4) \\ u\,(m-2) & u\,(m-5) & u\,(m-1) & u\,(m-6) \\ u\,(m-7) & u\,(m-3) & u\,(m-8) & u\,(m-4) \end{bmatrix} \tag{29}$$
$$\text{where } u(n) \text{ is a step function}, u\,(n) = \begin{cases} 1, & for\ n \geq 0 \\ 0, & for\ n < 0. \end{cases}$$

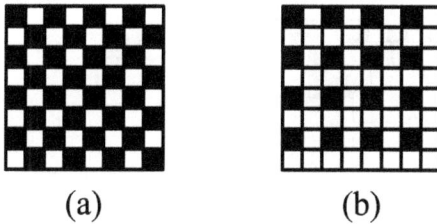

(a) (b)

Fig. 7. (a) 2:1 subsample pattern (b) 4:1 subsample pattern.

2.4 The architecture of subsample-based GICam image compressor

Fig.9 shows the architecture of the GICam-II image compressor and it faithfully executes the proposed GICam-II image compression algorithm shown in Fig.6. The window size, w,

32	32	32	32	32	32	64	64
32	16	16	32	32	64	64	128
32	16	16	32	32	64	128	128
32	32	32	32	64	64	128	256
32	32	32	64	64	128	128	256
64	64	64	128	128	128	256	256
64	128	128	128	256	256	256	256
128	128	128	256	256	256	256	512

16	16	32	32
16	16	32	64
32	32	64	64
64	64	128	128

(a)   (b)

Fig. 8. (a)The modified R quantization table (b) The modified G quantization table.

and the maximum matching length,l parameters for LZ77 encoder can be loaded into the parameter register file via a serial interface after the initial setting of the hardware reset. Similarly, coefficients of 2-D DCT and parameters of initial setting for all controllers shown in Fig.9 can be also loaded into the parameter register file. The GICam-II image compressor processes the image in the block order of G1, R, G2 and B. Because the data stream from the image sensor is block-based, the GICam-II image compressor adopts the structure of ping-pong memory to hold each block of data. The advantage of using this structure is the high parallelism between the data loading and data precessing.

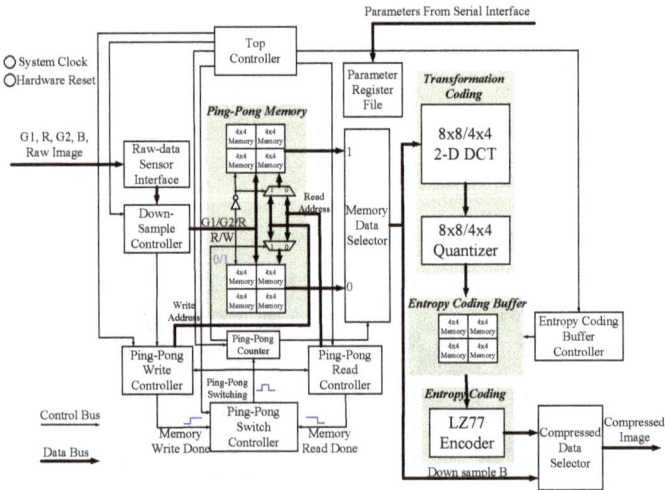

Fig. 9. The GICam-II image compressor.

When the GICam-II image compressor begins, the proposed architecture first loads the incoming image in the block order of G1, R, G2 and B from the image sensor and passes them with the valid signal control via the Raw-Data Sensor Interface. The Raw-Data Sensor Interface is a simple register structure with one clock cycle delay. This design absolutely makes sure that no any glue-logic circuits that can affects the timing of logic synthesis exists between the raw image sensor and the the GICam-II image compressor. The Down-Sample

Controller receives the valid data and then selects the candidate subsample ratio to sample the candidate image data in the block order of G1, R, G2 and B. The Ping-Pong Write Controller can accurately receive the data loading command from the Down-Sample Controller and then pushes the downsample image data into the candidate one of the ping-pong memory. At the same time, the Ping-Pong Read Controller pushes the stored image data from another memory into the Transformation Coding. The Ping-Pong Write Controller and the Ping-Pong Read Controller will issue an announcement to the Ping-Pong Switch Controller, respectively while each data-access is finished. When all announcement arrives in turn, the Ping-Pong Switch Controller will generate a pulse-type Ping-Pong Switching signal, one clock cycle, to release each announcement signal from the high-level to zero for the Ping-Pong Write Controller and the Ping-Pong Read Controller. The Ping-Pong Switch Counter also uses the Ping-Pong Switching signal to switch the read/write polarity for each memory in the structure of the Ping-Pong Memory.

The Transformation Coding consists of the 2-D DCT and the quantizer. The goal of the transformation coding is to transform processing data from the spatial domain into the spatial frequency domain and further to shorten the range in the spatial frequency domain before entropy coding in order to increasing the compression ratio. The 2-D DCT alternatively calculates row or column 1-D DCTs. The 1-D DCT is a multiplier-less implementation using the algebraic integer encoding (11). The algebraic integer encoding can minimize the number of addition operations. As regards the RG quantizer, the GICam-II image compressor utilizes the barrel shifter for power-of-two products. The power-of-two quantization table shown in Fig.8 can reduce the cost of multiplication while quality degradation is quite little. In addition, the 8-by-8 memory array between the quantizer and the LZ77 encoder is used to synchronize the operations of quantization and LZ77 encoding. Since the frame rate of GICam-II image compressor is 2 frames/second, the 2-D DCT can be folded to trade the hardware cost with the computing speed, and the other two data processing units, quantization and LZ77 encoder, can operate at low data rate. Due to non-compression for the B component, the B component is directly transmitted from the ping-pong memory, not be compressed. Finally, the LZ77 encoder is implemented by block-matching approach and the detail of each processing element and overall architecture have been also shown in paper (11).

2.5 Experimental results

We have particularly introduced the method of efficiently decreasing the incoming datum with the subsmaple technique in the GICam-II compression algorithm. The performance of the compression rate, the quality degradation and the ability of power saving will then be experimentally analyzed using the GICamm-II compressor.

2.5.1 The analysis of compression rate for GI images

In this paper, twelve GI images are tested and shown in the Fig.3. First of all, the target compression performance of the GICam-II image compression is to reduce image size by at least 75% . To meet the specification, we have to exploit the cost-optimal LZ coding parameters. There are two parameters in the LZ coding to be determined: the window size, w, and the maximum matching length,l. The larger the parameters, the higher the compression ratio will be; however,the implementation cost will be higher. In addition, there are two kinds of LZ codings in the GICam-II compressor, one is R(w, l) for R component and the other is G(w, l) for G1 and G2 components. We set the values of parameters by using a compression ratio of 4:1 as the threshold. Our goal is to determine the minimum R(w, l) and G(w, l) sets under the constraint of 4:1 compression ratio.

The compression ratio (CR) is defined as the ratio of the raw image size to the compressed image size and formulated as Eq.30. The measure of the compression ratio is the compression rate. The formula of the compression rate is calculated by Eq.31. The results in Fig.10 are shown by simulating the behavior model of GICam-II compressor; it is generated by MATLAB. As seen in Fig.10, simulating with twelve endoscopic pictures, (32, 32) and (16, 8) are the minimum R(w, l) and G(w, l) sets to meet the compression ratio requirement. The subsample technique of the GICam-II compressor initially reduces the input image size by 43.75% ((1-1/4-(1/4*1/2*2)-(1/4*1/4))*100%) before executing the entropy coding, LZ77 coding. Therefore, the overall compression ratio of GICam-II compressor minus 43.75% is the compression effect of LZ77 coding that combines with the quantization, and the simulation results are shown in Fig.11.

This research paper focuses to propose a subsample-based low-power image compressor for capsule gastrointestinal endoscopy. This obvious reddish characteristic is due to the slightly luminous intensity of LEDs and the formation of image in the capsule gastrointestinal endoscopy. The GICam-II compression algorithm is motivated on the basis of this reddish pattern. Therefore, we do not consider compressing other endoscopic images except for gastrointestinal images to avoid the confusion of topic for this research. However, general endoscopic images generated via a wired endoscopic takes on the yellow characteristic due to the vividly luminous intensity of LEDs. The yellow pattern mainly consists of red and green and it also complies with the color sensitivity result in this research work. Therefore, I believe that the proposed GICam-II still supports good compression ratio for general endoscopic images.

$$Compression\ Ratio\ (CR) = \frac{bits\ before\ compression}{bits\ after\ compression} \tag{30}$$

$$Compression\ Rate = (1 - CR^{-1}) \times 100\% \tag{31}$$

2.5.2 The analysis of compression quality for GI images

Using (32, 32) and (16, 8) as the parameter sets, in Table 4, we can see the performance in terms of the quality degradation and compression ratio. The measure of compression quality is the peak signal-to-noise ratio of luminance (PSNRY). The calculation of PSNRY is formulated as Eq.32. Where MSE is the mean square error of decompressed image and is formulated as Eq.33. In Eq.33, α_{ij} is the luminance value of original GI image and β_{ij} is the luminance value of decompressed GI image. The result shows that the degradation of decompressed images is quite low while the average PSNRY is 40.73 dB. Fig.12 illustrates the compression quality of decoded test pictures. The difference between the original image and the decompressed

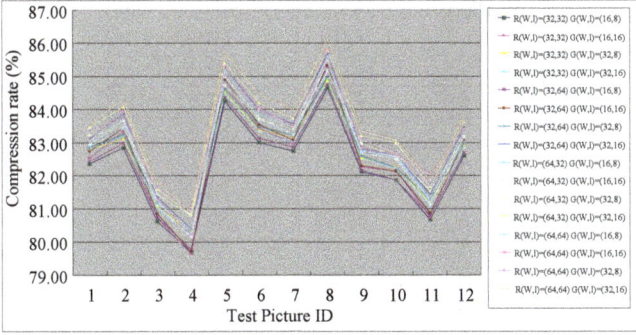

Fig. 10. The compression performance of the GICam-II image compressor.

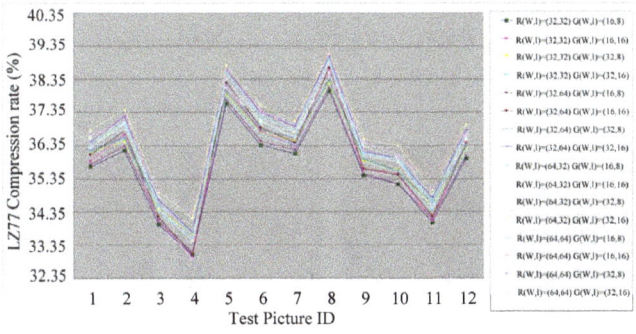

Fig. 11. The compression performance of LZ77 coding that combines with the quantization in the GICam-II image compressor .

image is invisible.

$$PSNRY = 10log_{10}(\frac{255^2}{MSE}) \qquad (32)$$

$$MSE = (\frac{1}{M \times N}) \sum_{i=0}^{M-1} \sum_{j=0}^{N-1} (\alpha_{ij} - \beta_{ij})^2 \qquad (33)$$

To demonstrate the validity of decompressed images, five professional gastroenterology doctors from the Division of Gastroenterology, Taipei Medical University Hospital are invited to verify whether or not the decoded image quality is suitable for practical diagnosis. The criterion of evaluation is shown in Table 5. The score between 0 and 2 means that the diagnosis is affected, the score between 3 and 5 means that the diagnosis is slightly affected and the score between 6 and 9 means that the diagnosis is not affected. According to the evaluation results of Fig.13, all decoded GI images are suitable for practical diagnosis because of high evaluation score and the diagnoses are absolutely not affected, except for the 5th and 8th decoded images. The degrees of diagnoses are between no affection and extremely slight affection for the 5th and the 8th decoded images because only two doctors subjectively feel their diagnoses are slightly affected. However, these two decoded images are not mistaken in

Test Picture ID	PSNRY (dB)	Compression rate (%)
1	40.76	82.36
2	41.38	82.84
3	39.39	80.62
4	38.16	79.70
5	42.56	84.25
6	41.60	83.00
7	41.03	82.74
8	43.05	84.63
9	40.21	82.11
10	40.36	81.84
11	39.39	80.66
12	40.85	82.60
Average	40.73	82.28

Table 4. The simulation results of twelve tested pictures.

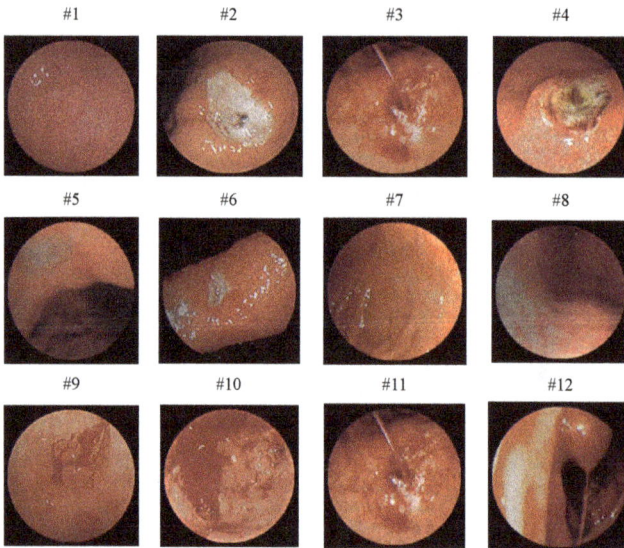

Fig. 12. Demosaicked GI images.

diagnosis for these professional gastroenterology doctors. Therefore, the PSNRY being higher than 38 dB is acceptable according to the objective criterion of gastroenterology doctors.

Score	Description
$0 \sim 2$	diagnosis is affected
$3 \sim 5$	diagnosis is slightly affected
$6 \sim 9$	diagnosis is not affected

Table 5. The criterion of evaluation.

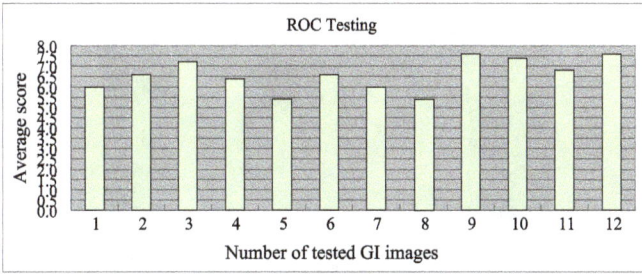

Fig. 13. The evaluation results of professional gastroenterology doctors.

2.5.3 The analysis of power saving

To validate the GICam-II image processor, we used the FPGA board of Altera APEX 2100 K to verify the function of the GICam-II image processor and the prototype is shown in Fig.14. After FPGA verification, we used the TSMC 0.18 μm 1P6M process to implement the GICam-II image compressor. When operating at 1.8 V, the power consumption of logic part is 3.88 mW, estimated by using PrimePower^{TM}. The memory blocks are generated by Artisan memory compiler and consume 5.29 mW. The total power consumption is 9.17 mW for the proposed design. When comparing the proposed GICam-II image compressor with our previous GICam one in Table 6, the power dissipation can further save 38.5% under the approximate condition of quality degradation and compression ratio because of the reduction of memory requirement for G1, G2 and B components.

The GICam-II compressor has poorer image reconstruction than JPEG and our previous GICam one because the GICam-II compressor uses the subsample scheme to down sample green and blue components according to the 2:1 and the 4:1 subsample ratios. The raw data before compression has lost some raw data information. Hence, the decoded raw data should be reconstructed (the first interpolation) before reconstructing the color images (the second interpolation). Using two level interpolations to reconstruct the color images has poorer image quality than one level interpolation. Fortunately, the decoded image quality using GICam-II compressor can be accepted and suitable for practical diagnosis and the evaluation results of professional gastroenterology doctors can be shown in the last subsection.

Finally, we compare the GICam-II image processor with other works and the comparison results are shown in the Table 7. According to the comparison results, our proposed GICam-II image compressor has lower area and lower operation frequency. It can fit into the existing designs.

3. Rank-Order Filtering (ROF) with a maskable memory

Rank-order filtering (ROF), or order-statistical filtering, has been widely applied for various speech and image processing applications [1]-[6]. Given a sequence of input samples $\{x_{i-k}, x_{i-k+1}, ..., x_i, ..., x_{i+l}\}$, the basic operation of rank order filtering is to choose the r-th largest sample as the output y_i, where r is the rank-order of the filter. This type of ROF is normally classified as *the non-recursive ROF* . Another type of ROF is called *the recursive ROF*. The difference between the recursive ROF and the non-recursive ROF is that

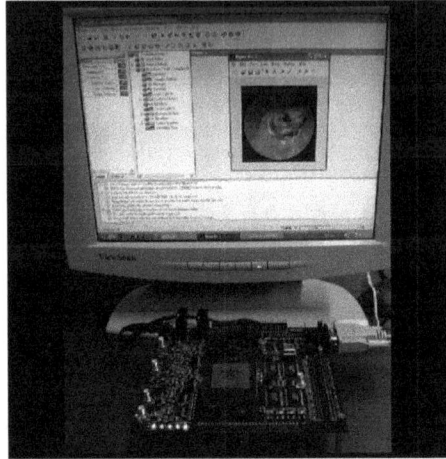

Fig. 14. The FPGA prototype of the GICam-II image compressor.

	JPEG designed by (19)	GICam image compressor (11)	Proposed GICam-II image compressor
Average PSNRY	46.37 dB	41.99 dB	40.73 dB
Average compression rate	82.20%	79.65%	82.28%
Average power dissipation	876mW	14.92 mW	9.17 mW

Table 6. The comparison Result with Previous GICam Works

	Area	Frequency (MHz)	Power (mW)	Supply Voltage (V)
GICam image compressor (11)	390k	12.58	14.92 (evaluated)	1.8
X.Xie et al.(12)*	12600k	40.0	6.2 (measured)	1.8
K.Wahid et al. (13)	325k	150	10 (evaluated)	1.8
X.Chen et al.(14)*	11200k	20	1.3 (measured)	0.95
Proposed GICam-II image compressor	318k	7.96	9.17 (evaluated)	1.8
* includes analog and transmission circuit and SRAM				

Table 7. The Comparison Results with Existing Works

the input sequence of the recursive ROF is $\{y_{i-k}, y_{i-k+1}, ..., y_{i-1}, x_i, ..., x_{i+l}\}$. Unlike linear filtering, ROF can remove sharp discontinuities of small duration without blurring the original signal; therefore, ROF becomes a key component for signal smoothing and impulsive noise elimination. To provide this key component for various signal processing applications, we intend to design a configurable rank-order filter that features low cost and high speed.

Many approaches for hardware implementation of rank-order filtering have been presented in the past decades [8, 10-24]. Many of them are based on sorting algorithm [11, 22, 23, 25-28]. They considered the operation of rank-order filtering as two steps: sorting and choosing. Papers [10, 19] have proposed systolic architectures for rank-order filtering based on sorting algorithms, such as bubble sort and bitonic sort. These architectures are fully pipelined for high throughput rate at the expense of latency, but require a large number of compare-swap units and registers. To reduce the hardware complexity, papers [8, 12, 14, 15, 29-32] present linear-array structures to maintain samples in sorted order. For a sliding window of size N, the linear-array architectures consist of N processing elements and require three steps for each iteration: finding the proper location for new coming sample, discarding the eldest one, and moving samples between the newest and eldest one position. The three-step procedure is called delete-and-insert (DI). Although the hardware complexity is reduced to $O(N)$, they require a large latency for DI steps. Paper [31] further presents a micro-programmable processor for the implementations of the median-type filters. Paper [20] presents a parallel architecture using two-phase design to improve the operating speed. In this paper, they first modified the traditional content-addressable memory (CAM) to a shiftable CAM (SCAM) processor with shiftable memory cells and comparators. Their architecture can take advantages of CAM for parallelizing the DI procedure. Then, they use two-phase design to combine delete and insert operations. Thereafter, the SCAM processor can quickly finish DI operations in parallel. Although the SCAM processor has significantly increased the speed of the linear-array architecture, it can only process a new sample at a time and cannot efficiently process 2-D data. For a window of size n-by-n, the SCAM processor needs n DI procedures for each filtering computation. To have an efficient 2-D rank-order filter, papers [12, 27] present solutions for 2-D rank-order filtering at the expense of area.

In addition to the sorting algorithm, the paper [35] applies the threshold decomposition technique for rank-order filtering. To simplify the VLSI complexity, the proposed approach uses three steps : decomposition, binary filtering, and recombination. The proposed approach significantly reduce the area complexity from exponential to linear. Papers [10, 13, 16-18, 21, 33, 34] employ the bit-sliced majority algorithm for median filtering, the most popular type of rank-order filtering. The bit-sliced algorithm [36, 37] bitwisely selects the ranked candidates and generates the ranked result one bit at a time. Basically, the bit-sliced algorithm for median filtering recursively executes two steps: majority calculation and polarization. The majority calculation, in general, dominates the execution time of median filtering. Papers [10], [17] and [37] present logic networks for implementation of majority calculation. However, the circuits are time-consuming and complex so that they cannot take full advantages of bit-sliced algorithm. Some papers claim that this type of algorithm is impractical for logic circuit implementation because of its exponential complexity [31]. Paper [21] uses an inverter as a voter for majority calculation. It significantly improves both cost and processing speed, but the noise margin will become narrow as the number of inputs increases. The narrow noise margin makes the implementation impractical and limits the configurability of rank-order filtering.

Instead of using logic circuits, this paper presents a novel memory architecture for rank-order filtering based on a generic rank-order filtering algorithm. The maskable memory structure, called dual-cell random-access memory (DCRAM), is an extended SRAM structure with maskable registers and dual cells. The maskable registers allow the architecture to selectively read or write bit-slices, and hence speed up "parallel read" and "parallel polarization" tasks. The control of maskable registers is driven by a long-instruction-word (LIW) instruction set. The LIW makes the proposed architecture programmable for various rank-order filtering algorithms, such as recursive and non-recursive ROFs. The proposed architecture has been implemented using TSMC 0.18um 1P6M technology and successfully applied for 1-D/2-D ROF applications. For 9-point 1-D and 3-by-3 2-D ROF applications, the core size is $356.1 \times 427.7um^2$. As shown in the post-layout simulation, the DCRAM-based processor can operate at 290 MHz for 3.3V supply and 256 MHz for 1.8V supply. For image processing, the performance of the proposed processor can process video clips of SVGA format in real-time.

3.1 The generic bit-sliced rank-order filtering algorithm

Let $W_i=\{x_{i-k}, x_{i-k+1}, ..., x_i, ..., x_{i+l}\}$ be a window of input samples. The binary code of each input x_j is denoted as $u_j^{B-1} \cdots u_j^1 u_j^0$. The output y_i of the r-th order filter is the r-th largest sample in the input window W_i, denoted as $v_i^{B-1} \cdots v_i^1 v_i^0$. The algorithm sequentially determines the r-th order value bit-by-bit starting from the most significant bit (MSB) to the least significant bit (LSB). To start with, we first count 1's from the MSB bit-slice of input samples and use Z_{B-1} to denote the result. The b-th bit-slice of input samples is defined as $u_{i-k}^b u_{i-k+1}^b \cdots u_i^b \cdots u_{i+l}^b$. If Z_{B-1} is greater than or equal to r, then v_i^{B-1} is 1; otherwise, v_i^{B-1} is 0. Any input sample whose MSB has the same value as v_i^{B-1} is considered as one of candidates of the r-th order sample. On the other hand, if the MSB of an input sample is not equal to v_i^{B-1}, the input sample will be considered as a non-candidate. Non-candidates will be then polarized to either the largest or smallest value. If the MSB of an input sample x_j is 1 and v_i^{B-1} is 0, the rest bits (or lower bits) of x_j are set to 1's. Contrarily, if the MSB of an input sample x_j is 0 and v_i^{B-1} is 1, the rest bits (or lower bits) of x_j are set to 0's. After the polarization, the algorithm counts 1's from the consecutive bit-slice and then repeats the polarization procedure. Consequently, the r-th order value can be obtained by recursively iterating the steps bit-by-bit. The following pseudo code illustrates the generic bit-sliced rank-order filtering algorithm:

Given the input samples, the window size $N=l+k+1$, the bitwidth B and the rank r, do:

Step 1: Set $b=B-1$.

Step 2: (Bit counting)

Calculate Z_b from $\{u_{i-k}^b, u_{i-k+1}^b, \cdots, u_i^b, \cdots, u_{i+l}^b\}$.

Step 3: (Threshold decomposition)

If $Z_b \geq r, v_i^b = 1$; otherwise $v_i^b = 0$.

Step 4: (Polarization)

If $u_j^b \neq v_i^b$, $u_j^m = u_j^b$ for $0 \leq m \leq b-1$ and $i-k \leq j \leq i+l$

Step 5: $b=b-1$.

Step 6: If $b \geq 0$ go to Step 2.

Step 7: Output y_i.

Fig.15 illustrates a bit-sliced ROF example for $N=7$, $B=4$, and $r=1$. Given that the input samples are $7(0111_2)$, $5(0101_2)$, $11(1011_2)$, $14(1110_2)$, $2(0010_2)$, $8(1000_2)$, and $3(0011_2)$, the generic algorithm will produce $14(1110_2)$ as the output result. At the beginning, the "Bit counting" step will calculate the number of 1's at MSBs, which is 3. Since the number of 1's is greater than r, the "Threshold decomposition" step sets the MSB of y_i to '1'. Then, the "Polarization" step will consider the inputs with $u_j^3 = 1$ as candidates of the ROF output and polarize the lower bits of the others to all 0's. After repeating the above steps with decreasing b, the output y_i will be $14(1110_2)$.

3.2 The dual-cell RAM architecture for rank-order filtering

As mentioned above, the generic rank-order filtering algorithm generates the rank-order value bit-by-bit without using complex sorting computations. The main advantage of this algorithm is that the calculation of rank-order filtering has low computational complexity and can be mapped to a highly parallel architecture. In the algorithm, there are three main tasks: bit counting, threshold decomposition, and polarization. To have these tasks efficiently implemented, this paper presents an ROF processor based on a novel maskable memory architecture, as shown in Fig.16. The memory structure is highly scalable with the window size increasing, by simply adding memory cells. Furthermore, with the instruction decoder and maskable memory, the proposed architecture is programmable and flexible for different kinds of ROFs.

The dual-cell random-access memory (DCRAM) plays a key role in the proposed ROF architecture. In the DCRAM, there are two fields for reusing the input data and pipelining the filtering process. For the one-dimensional (1-D) ROF, the proposed architecture receives one sample at a time. For the n-by-n two-dimensional (2-D) ROF, the architecture reads n samples into the input window within a filtering iteration. To speed up the process of rank-order filtering and pipeline the data loading and filtering calculation, the data field loads the input data while the computing field is performing bit-sliced operations. Hence, the execution of the architecture has two pipeline stages: data fetching and rank-order calculation. In each iteration, the data fetching first loads the input sample(s) into the data field and then makes copies from the data field to the computing field. After having the input window in the computing field, the rank-order calculation bitwisely accesses the computing field and executes the ROF tasks.

The computing field is in the maskable part of DCRAM. The maskable part of DCRAM performs parallel reads for bit counting and parallel writes for polarization. The read-mask register (RMR) is configured to mask unwanted bits of the computing field during read operation. The value of RMR is one-hot-encoded so that the bit-sliced values can be read from

	1: Threshold decomposition				2: Polarization				3: Threshold decomposition				4: Polarization			
7	0	1	1	1	0	0	0	0	0	0	0	0	0	0	0	0
5	0	1	0	1	0	0	0	0	0	0	0	0	0	0	0	0
11	1	0	1	1	1	0	1	1	1	0	1	1	1	0	0	0
14	1	1	1	0	1	1	1	0	1	1	1	0	1	1	1	0
2	0	0	1	0	0	0	0	0	0	0	0	0	0	0	0	0
8	1	0	0	0	1	0	0	0	1	0	0	0	1	0	0	0
3	0	0	1	1	0	0	0	0	0	0	0	0	0	0	0	0
y_i	1	X	X	X					1	1	X	X				

5: Threshold decomposition				6: Polarization				7: Threshold decomposition			
0	0	0	0	0	0	0	0	0	0	0	0
0	0	0	0	0	0	0	0	0	0	0	0
1	0	0	0	1	0	0	0	1	0	0	0
1	1	1	0	1	1	1	0	1	1	1	0
0	0	0	0	0	0	0	0	0	0	0	0
1	0	0	0	1	0	0	0	1	0	0	0
0	0	0	0	0	0	0	0	0	0	0	0
y_i 1 1 1 X								y_i 1 1 1 0			

Fig. 15. An example of the generic bit-sliced ROF algorithm for $N=7$, $B=4$, and $r=1$.

the memory in parallel. The bit-sliced values will then go to the Level-Quantizer for threshold decomposition. When the ROF performs polarization, the write-mask register (WMR) is configured to mask untouched bits and allow the polarization selector (PS) to polar lower bits of noncandidate samples. Since the structure of memory circuits is regular and the maskable scheme provides fast logic operations, the maskable memory structure features low cost and high speed. It obviously outperforms logic networks on implementation of bit counting and polarization.

To start with the algorithm, the RMR is one-hot-masked according to the value b in the generic algorithm and then the DCRAM outputs a bit-sliced value $\{u_{i-k}^b, u_{i-k+1}^b, \ldots, u_i^b, \ldots, \ldots u_{i+l}^b\}$ on "c_d". The bit-sliced value will go to both the Level-Quantizer and PS. The Level-Quantizer performs the Step 2 and Step 3 by summing up bits of the bit-sliced value to Z_b and comparing Z_b with the rank value r. The rank value r is stored in the rank register (RR). The bitwidth w of Z_b is $\lceil log_2^N \rceil$. Fig.17 illustrates the block diagram of the Level-Quantizer, where FA denotes the full adder and HA denotes the half adder. The signals "S" and "C" of each FA or HA represent sum and carry, respectively. The circuit in the dash-lined box is a comparator. The

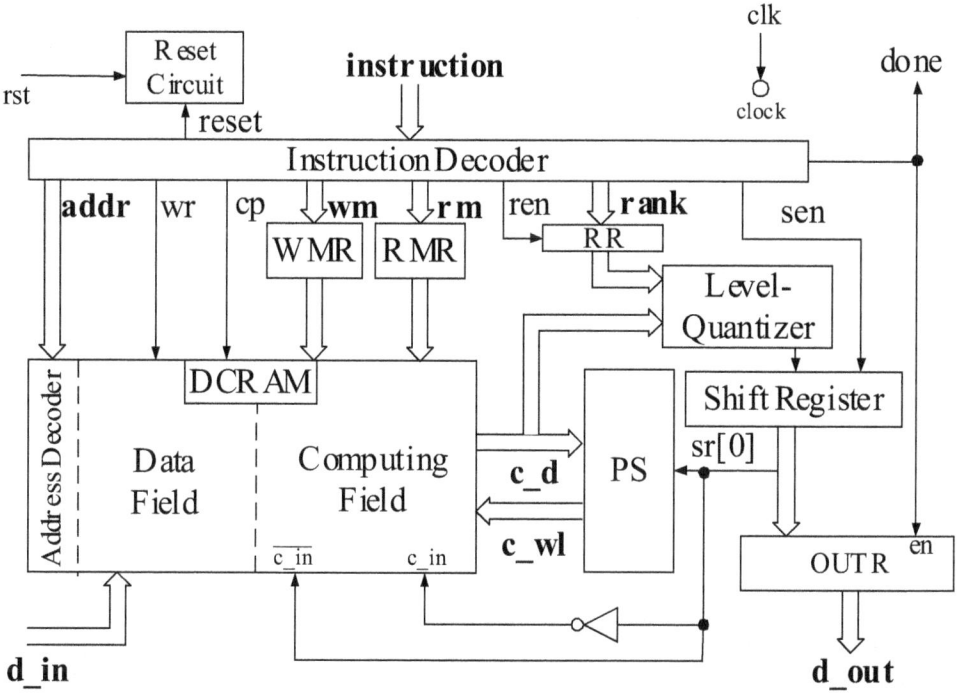

Fig. 16. The proposed rank-order filtering architecture.

comparator is implemented by a carry generator because the comparison result of Z_b and r can be obtained from the carry output of Z_b plus the two's complement of r. The carry output is the quantized value of the Level-Quantizer.

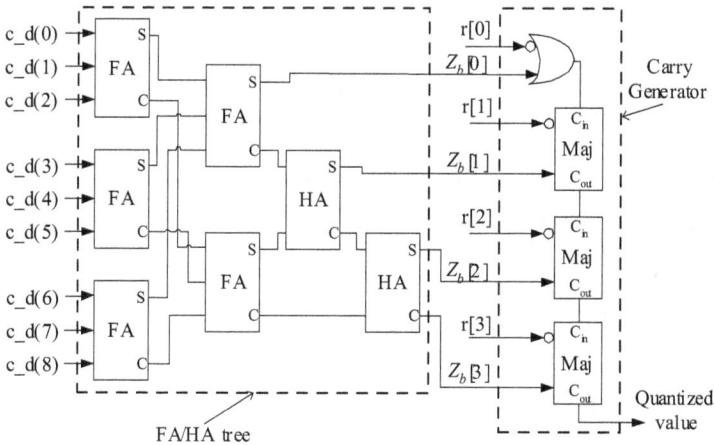

Fig. 17. The block diagram of the Level-Quantizer.

Normally, the comparison can be made by subtracting r from Z_b. Since Z_b and r are unsigned numbers, to perform the subtraction, both numbers have to reformat to two's complement numbers by adding a sign bit. In this paper, the reformated numbers of Z_b and r are expressed as $Z_{b,S}$ and r_S, respectively. Since both numbers are positive, their sign bits are equal to '0'. If $Z_{b,S}$ is less than r_S, the result of subtraction, Δ, will be negative; that is, the sign bit (or MSB) of Δ is '1'. Eq.34 shows the inequation of the comparison, where $\overline{r_S}$ denotes the one's complement of r_S and $\mathbf{1}$ denotes $(00\ldots01)_2$. Because the MSB of $Z_{b,S}$ is '0' and the MSB of $\overline{r_S}$ is '1', to satisfy Eq.34, the carry of $Z_{b,S}^{w-1} + \overline{r_S^{w-1}}$ must be equal to '0' so that the sign bit of Δ becomes '1'. To simplify the comparison circuit, instead of implementing an adder, we use the carry generator to produce the carry of $Z_{b,S}^{w-1} + \overline{r_S^{w-1}}$. Each cell of the carry generator is a majority (Maj) circuit that performs the boolean function shown in Eq.35. Furthermore, we use an OR gate at the LSB stage because of Eq.36. Thus, the dash-lined box is an optimized solution for comparison of Z_b and r without implementing the *bit-summation* and *signed-extension* parts.

$$\Delta = Z_{b,S} + \overline{r_S} + \mathbf{1} < 0. \tag{34}$$

$$Maj(A,B,C) = AB + BC + AC. \tag{35}$$

$$Z_b^0 \cdot \overline{r^0} + Z_b^0 \cdot 1 + 1 \cdot \overline{r^0} = Z_b^0 + \overline{r^0}. \tag{36}$$

After the Level-Quantizer finishes the threshold decomposition, the quantized value goes to the LSB of the shift register, "sr[0]". Then, the polarization selector (PS) uses exclusive ORs (XORs) to determine which words should be polarized, as shown in Fig.18. Obviously, the XORs can examine the condition of $u_j^b \neq v_i^b$ and select the word-under-polarization's (WUPs) accordingly. When **"c_wl"** is '1', the lower bits of selected words will be polarized; the lower bits are selected by WMR. According to the Step 4, the polarized value is u_j^b which is the inversion of v_i^b. Since v_i^b is the value of sr[0], we inverse the value of "sr[0]" to "c_in", as shown in Fig.16.

As seen in the generic algorithm, the basic ROF repeatedly executes *Bit-counting*, *Threshold decomposition*, and *Polarization* until the LSB of the ROF result being generated. Upon executing B times of three main tasks, the ROF will have the result in the Shift Register. A cycle after, the result will then go to the output register (OUTR). Doing so, the proposed architecture is able to pipeline the iterations for high-performance applications.

3.3 Implementation of dual-cell random-access memory

Fig.19 illustrates a basic element of DCRAM. Each element has two cells for data field and computing field, respectively. The data cell is basically an SRAM cell with a pair of bitlines. The SRAM cell is composed of INV1 and INV2 and stores a bit of input sample addressed by the wordline "d_wl[i]". The computing cell performs three tasks: *copy, write,* and *read*. When the copy-line "cp" is high, through INV5 and INV6, the pair of INV3 and INV4 will have the copy of the 1-bit datum in the data cell. The *copy* operation is unidirectional, and the pair of INV5 and INV6 can guarantee this directivity. When the one-bit value stored in the computing cell needs to be polarized, the "wm[j]" and "c_wl[i]" will be asserted, and the computing cell will perform the *write* operation according to the pair of bitlines "c_bl[j]" and "c_bl[j]". When the ROF reads the bit-sliced value, the computing cell uses an NMOS,

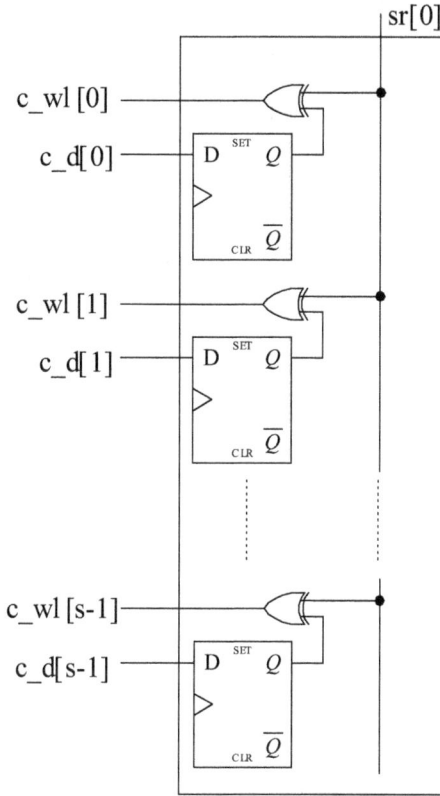

Fig. 18. The polarization selector (PS).

gated by "rm[j]", to output the complement value of the stored bit to the dataline "$\overline{c\_d[i]}$". The datalines of computing cells of each word will be then merged as a single net. Since the RMR is one-hot configured, each word has only a single bit being activated during the *read* operation.

As shown in Fig.20, the dataline "$\overline{c\_d[i]}$" finally goes to an inverter to pull up the weak '1', which is generated by the "rm[j]"-gated NMOS, and hence the signal "c_d[i]" has the value of the i-th bit of each bit-slice. Because the ROF algorithm polarizes the non-candidate words with either all zeros or all ones, the bitline pairs of computing cells are merged as a single pair of "c_in" and "$\overline{c\_in}$".

Fig.21 illustrates the implementation of DCRAM with the floorplan. Each $D_i - C_i$ pair is a maskable memory cell where D_i denotes D_cell(i) and C_i denotes C_cell(i). Each word is split into higher and lower parts for reducing the memory access time and power dissipation (64). The control block is an interface between control signals and address decoder. It controls wordlines and bitlines of DCRAM. When the write signal "wr" is not asserted, the control block will disassert all wordlines by the address decoder.

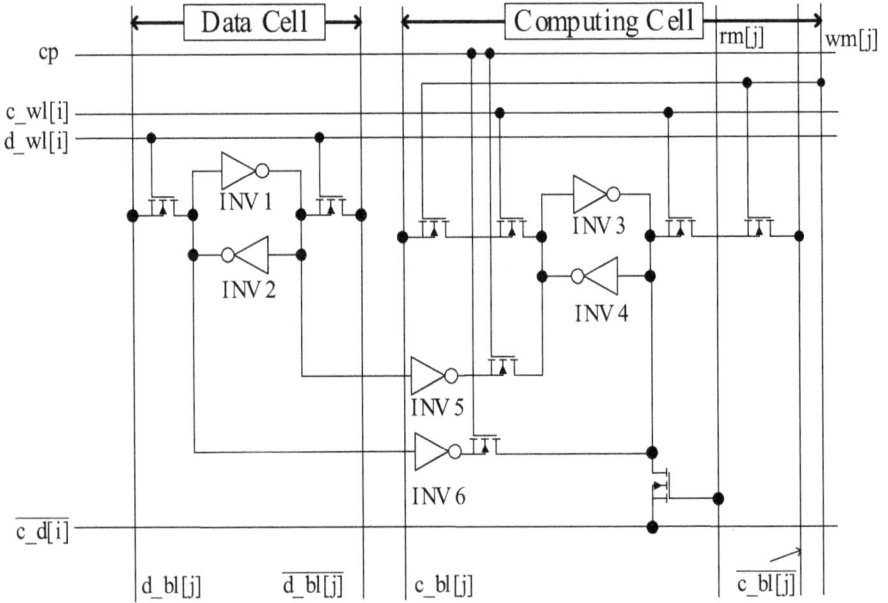

Fig. 19. A basic element of DCRAM.

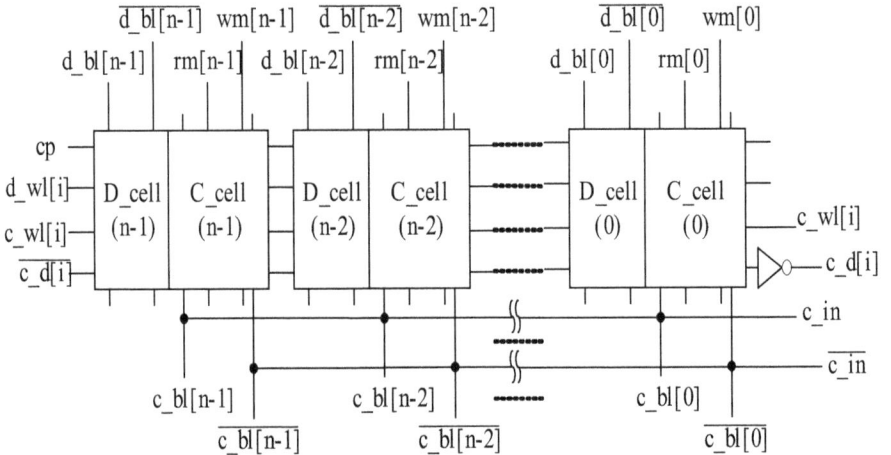

Fig. 20. A DCRAM word mixing data field and computing field. D_cell(i) denotes the data field of i-th bit and C_cell(i) denotes the computing field of i-th bit.

3.4 Instruction set of proposed ROF processor

The proposed ROF processor is a core for the impulsive noise removal and enabled by an instruction sequencer. Fig.22 illustrates the conceptual diagram of the ROF processor. The instruction sequencer is used for the generation of instruction codes and the control

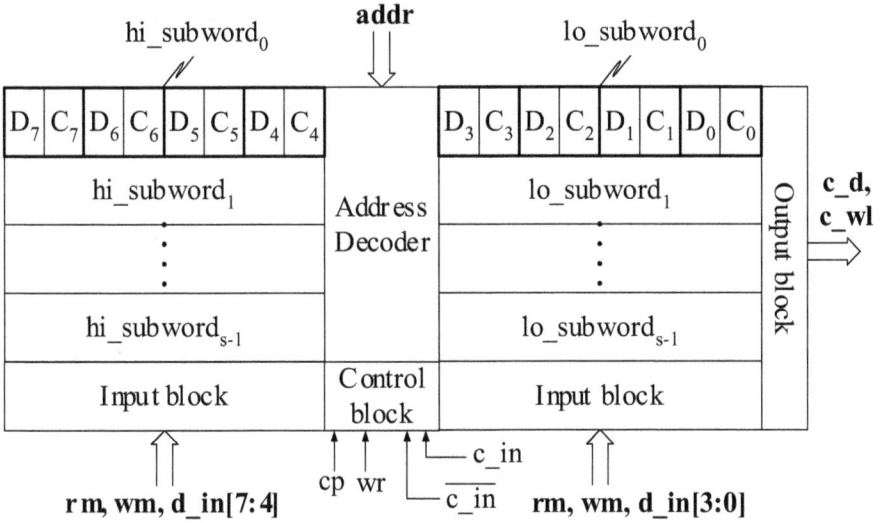

Fig. 21. The floorplan of DCRAM.

of input/output streams. The instruction sequencer can be a microprocessor or dedicated hardware.

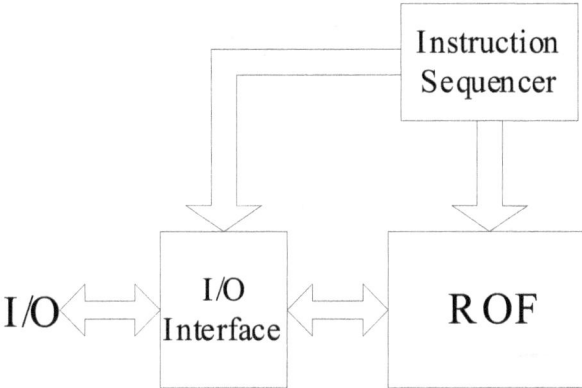

Fig. 22. The conceptual diagram of the ROF processor.

Fig.23 lists the format of the instruction set. An instruction word contains two subwords: the data field instruction and the computing field instruction. Each instruction cycle can concurrently issue two field instructions for parallelizing the data preparation and ROF execution; hence, the proposed processor can pipeline ROF iterations. When one of the field instructions performs "no operation", DF_NULL or CF_NULL will be issued. All registers in the architecture are updated a cycle after instruction issued.

The instruction SET resets all registers and set the rank register RR for a given rank-order r. The instruction LOAD loads data from "d_in" by asserting "wr" and setting "addr". The instruction COPY / DONE can perform the "COPY" operation or "DONE" operation. When the bit value of c is '1', the DCRAM will copy a window of input samples from the data field to the computing field. When the bit value of d is '1', the DCRAM wraps up an iteration by asserting "en" and puts the result into OUTR.

The instruction P_READ is issued when the ROF algorithm executes bit-sliced operations. The field $<mask>$ of P_READ is one-hot coded. It allows the DCRAM to send a bit-slice to the Level-Quantizer and PS for the *Threshold decomposition* task. The instruction P_WRITE is issued when the ROF algorithm performs the *Polarization* task. The field $<mask>$ of P_WRITE is used to set a consecutive sequence of 1's. The sequence can mask out the higher bits for polarization.

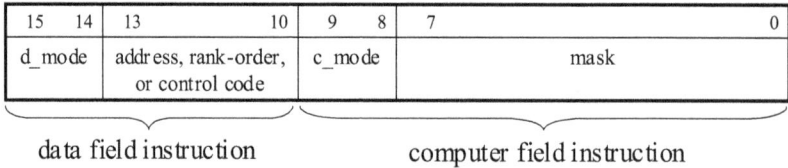

15 14	13 10	9 8	7 0
d_mode	address, rank-order, or control code	c_mode	mask

data field instruction   computer field instruction

S ET <rank>

15 14	13 10
0 0	rank-order value

LOAD <address>

15 14	13 10
0 1	address

C OP Y/D ONE

15 14	13 10
1 0	1 1 c d

DF_NULL

15 14	13 10
1 1	1 1 1 1

P_R EAD <mask>

9 8	7 0
0 0	mask

P_WRITE <mask>

9 8	7 0
0 1	mask

C F_NULL

9 8	7 0
1 1	1 1 1 1 1 1 1 1

c=1, copy; d=1, done

Fig. 23. The format of the instruction set.

To generate instructions to the ROF processor, the complete 1-D non-recursive ROF circuit includes an instruction sequencer, as shown in Fig.24.

Since the instruction set is in the format of long-instruction-word (LIW), the data fetching and ROF computing can be executed in parallel. So, the generated instruction stream can pipeline the ROF iterations, and the data fetching is hidden in each ROF latency. Fig.25 shows the reservation table of the 1-D ROF example. As seen in the reservation table, the first iteration and the second iteration are overlapped at the seventeenth, eighteenth and nineteenth clock steps. At the seventeenth clock step, the second iteration starts with loading a new sample while the first iteration processes the LSB bit-slice. At the eighteenth clock step, the second iteration copies samples from the data field to the computing field, and reads the MSB bit-slice.

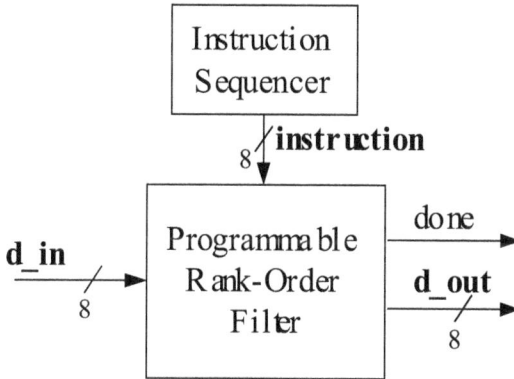

Fig. 24. Block diagram of the 1-D non-recursive ROF.

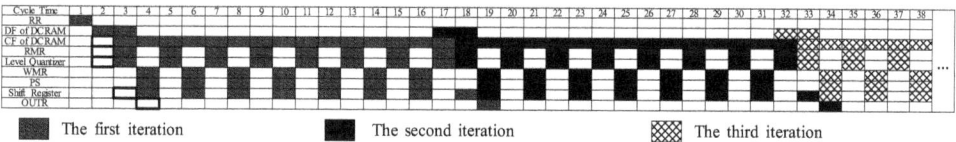

Fig. 25. Reservation table of the 1-D non-recursive ROF.

At the same time, the first iteration prepares the first ROF result for OUTR. At the nineteenth clock step, the first iteration sends the result out while the second iteration performs the first polarization. Thus, the iteration period for each iteration is 15 cycles.

3.5 Application of the proposed ROF processor

In Section 3.4, we use 1-D non-recursive ROF as an example to show the programming of the proposed ROF processor. Due to the programmable design, the proposed ROF processor can implement a variety of ROF applications. The following subsections will illustrate the optimized programs for three examples: 1-D RMF, 2-D non-recursive ROF, and 2-D RMF.

3.6 1-D recursive median filter

The recursive median filtering (RMF) has been proposed for signal smoothing and impulsive noise elimination. It can effectively remove sharp discontinuities of small duration without blurring the original signal. The RMF recursively searches for the median results from the most recent median values and input samples. So, the input window of RMF can be denoted as $\{y_{i-k}, y_{i-k+1}, ..., y_{i-1}, x_i, ..., x_{i+l}\}$, where $y_{i-k}, y_{i-k+1}, ..., y_{i-1}$ are the most recent median values and $x_i, ..., x_{i+l}$ are the input samples, and the result y_i is the $\lceil (l + k + 1)/2 \rceil$-th value of the input window.

Fig.26 demonstrates the implementation of the 1-D RMF. To recursively perform RMF with previous median values, the i-th iteration of 1-D RMF loads two inputs to the DCRAM; one is x_{i+l} and the other is y_{i-1}. As shown in Fig.26, the 2-to-1 multiplexer is used to switch the input stream to the data field, controlled by the instruction sequencer; the input stream is from either "d_in" or "d_out". When the proposed ROF processor receives the input stream, the program will arrange the data storage as shown in Fig.26. The date storage shows the data reusability of the proposed ROF processor.

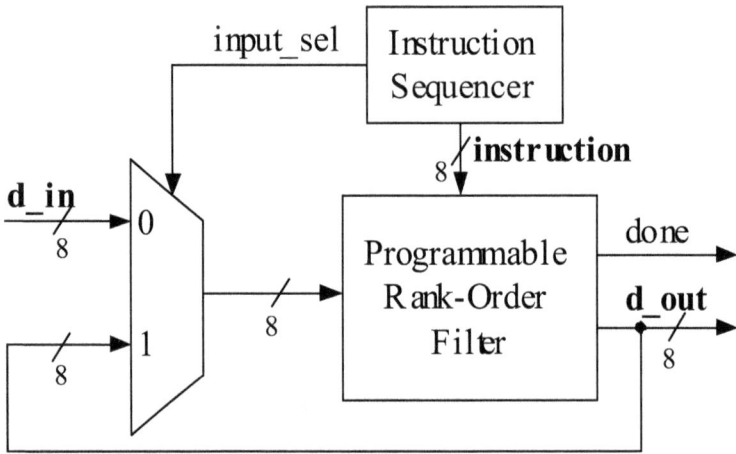

Fig. 26. Block diagram of the 1-D RMF.

As mentioned above, the input stream to the DCRAM comes from either "**d_in**"" or "**d_out**"". The statements of "set input_sel, 0;" and "set input_sel, 1;" can assert the signal "input_sel" to switch the input source accordingly. The statements of "LOAD i, CF_NULL;" and "LOAD i, CF_NULL;" is employed for the data stream, as per Fig.27. As seen in Fig.28, the throughput rate is limited by the recursive execution of the 1-D RMF; that is, the second iteration cannot load the newest median value until the first iteration generates the result to the output. However, we still optimized the throughput rate as much as possible. At the twentieth clock step, the program overlaps the first iteration and the second

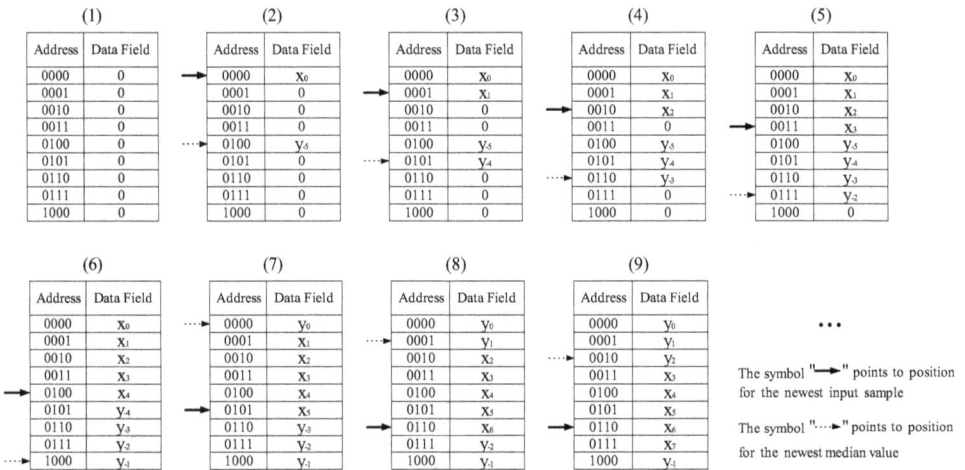

(1)

Address	Data Field
0000	0
0001	0
0010	0
0011	0
0100	0
0101	0
0110	0
0111	0
1000	0

(2)

Address	Data Field
0000	X_0
0001	0
0010	0
0011	0
0100	y_5
0101	0
0110	0
0111	0
1000	0

(3)

Address	Data Field
0000	X_0
0001	X_1
0010	0
0011	0
0100	y_5
0101	y_4
0110	0
0111	0
1000	0

(4)

Address	Data Field
0000	X_0
0001	X_1
0010	X_2
0011	0
0100	y_5
0101	y_4
0110	y_3
0111	0
1000	0

(5)

Address	Data Field
0000	X_0
0001	X_1
0010	X_2
0011	X_3
0100	y_5
0101	y_4
0110	y_3
0111	y_2
1000	0

(6)

Address	Data Field
0000	X_0
0001	X_1
0010	X_2
0011	X_3
0100	X_4
0101	y_4
0110	y_3
0111	y_2
1000	y_1

(7)

Address	Data Field
0000	y_0
0001	X_1
0010	X_2
0011	X_3
0100	X_4
0101	X_5
0110	y_3
0111	y_2
1000	y_1

(8)

Address	Data Field
0000	y_0
0001	y_1
0010	X_2
0011	X_3
0100	X_4
0101	X_5
0110	X_6
0111	y_2
1000	y_1

(9)

Address	Data Field
0000	y_0
0001	y_1
0010	y_2
0011	X_3
0100	X_4
0101	X_5
0110	X_6
0111	X_7
1000	y_1

• • •

The symbol "⟶" points to position for the newest input sample

The symbol "····►" points to position for the newest median value

Fig. 27. The flow for data storage of the 1-D RMF.

iteration so that the data fetching and result preparing can be run at the same time. As the result, the sample period is 18 cycles.

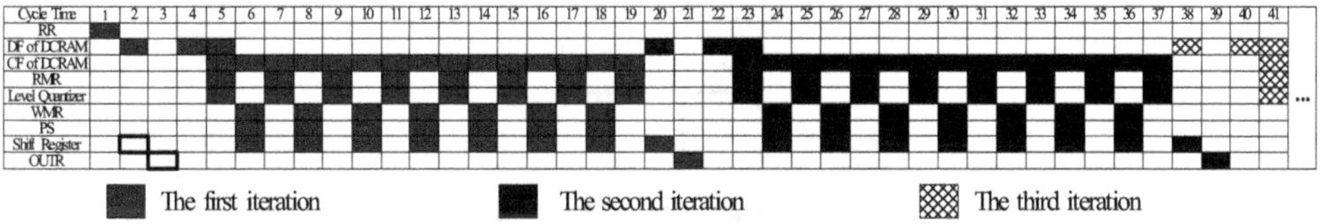

| | The first iteration | | The second iteration | | The third iteration |

Fig. 28. Reservation table of the 1-D RMF.

3.6.1 2-D non-recursive rank-order filter

Fig.29 illustrates the block diagram for the 2-D non-recursive ROF. From Fig.30, each iteration needs to update three input samples (the pixels in the shadow region) for the 3 × 3 ROF; that is, only n input samples need to be updated in each iteration for the $n \times n$ ROF. To reuse the windowing data, the data storage is arranged as shown in Fig.31. So, for the 2-D ROF, the data reusability of our process is high; each iteration updates only n input samples for an $n \times n$ window. Given a 2-D $n \times n$ ROF application with $n=3$ and $r=5$, the optimized reservation table can be scheduled as Fig.32.

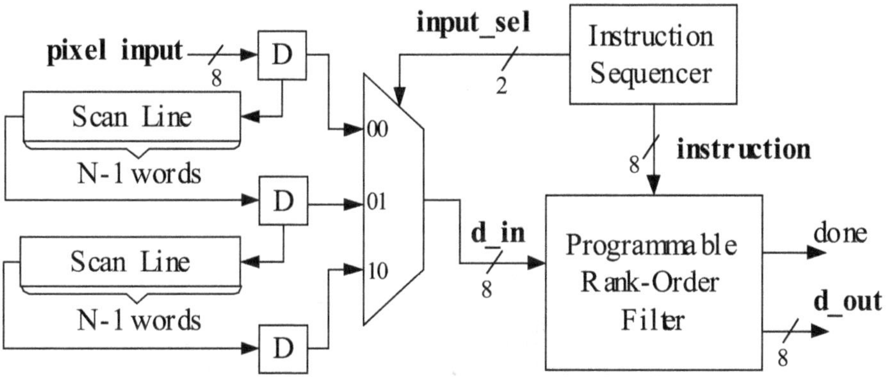

Fig. 29. Block diagram of the 2-D non-recursive ROF with 3-by-3 window.

3.6.2 2-D recursive median filter

Similar to the 1-D RMF, the two-dimensional(2-D) n-by-n RMF finds the median value from the window formed by some previous-calculated median values and input values. Fig.33(a) shows the content of the 3 × 3 window centered at (i, j). At the end of each iteration, the 2-D 3 × 3 RMF substitutes the central point of the current window with the median value. The renewed point will then be used in the next iteration. The windowing for 2-D RMF iterations is shown in Fig.33(b), where the triangles represent the previous-calculated median values and the pixels in the shadow region are updated at the beginning of each iteration. According to the windowing, Fig.34 illustrates the data storage for high degree of data reusability. Finally, we can implement the 2-D RMF as the block diagram illustrated in Fig.35. Given a 2-D 3 × 3 RMF application, the optimized reservation table can be scheduled as Fig.36.

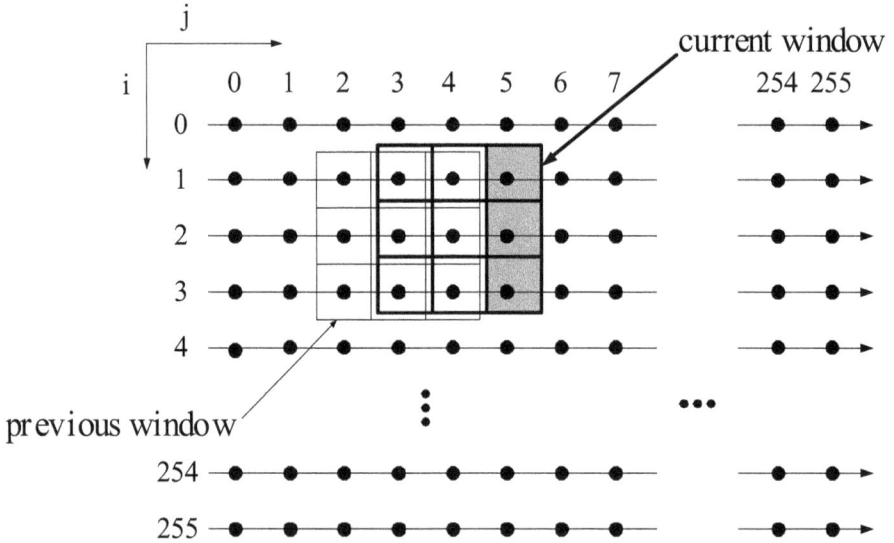

Fig. 30. The windowing of the 3 × 3 non-recursive ROF.

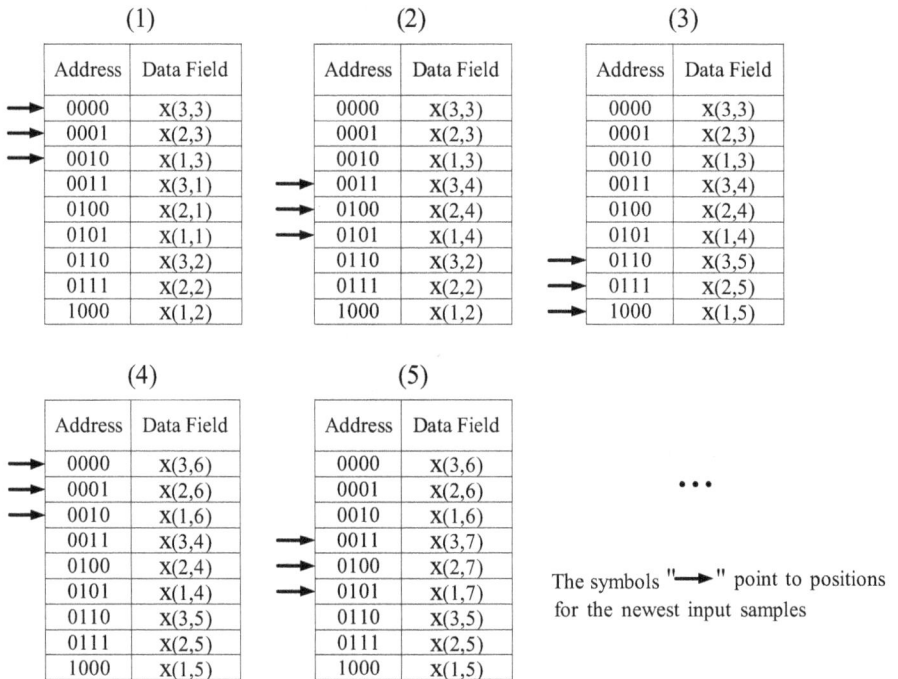

| | (1) |
Address	Data Field
→ 0000	x(3,3)
→ 0001	x(2,3)
→ 0010	x(1,3)
0011	x(3,1)
0100	x(2,1)
0101	x(1,1)
0110	x(3,2)
0111	x(2,2)
1000	x(1,2)

| | (2) |
Address	Data Field
0000	x(3,3)
0001	x(2,3)
0010	x(1,3)
→ 0011	x(3,4)
→ 0100	x(2,4)
→ 0101	x(1,4)
0110	x(3,2)
0111	x(2,2)
1000	x(1,2)

| | (3) |
Address	Data Field
0000	x(3,3)
0001	x(2,3)
0010	x(1,3)
0011	x(3,4)
0100	x(2,4)
0101	x(1,4)
→ 0110	x(3,5)
→ 0111	x(2,5)
→ 1000	x(1,5)

| | (4) |
Address	Data Field
→ 0000	x(3,6)
→ 0001	x(2,6)
→ 0010	x(1,6)
0011	x(3,4)
0100	x(2,4)
0101	x(1,4)
0110	x(3,5)
0111	x(2,5)
1000	x(1,5)

| | (5) |
Address	Data Field
0000	x(3,6)
0001	x(2,6)
0010	x(1,6)
→ 0011	x(3,7)
→ 0100	x(2,7)
→ 0101	x(1,7)
0110	x(3,5)
0111	x(2,5)
1000	x(1,5)

• • •

The symbols "→" point to positions
for the newest input samples

Fig. 31. The data storage of the 2-D non-recursive ROF.

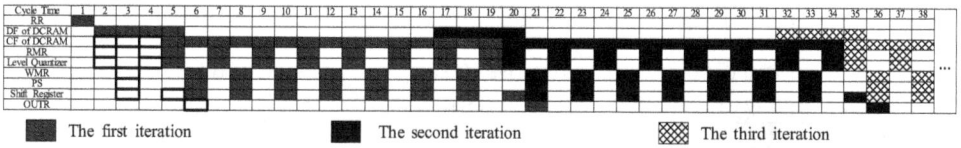

Fig. 32. Reservation table of the 2-D ROF.

y(i-1,j-1)	y(i-1,j)	y(i-1,j+1)
y(i,j-1)	x(i,j)	x(i,j+1)
x(i+1,j-1)	x(i+1,j)	x(i+1,j+1)

(a)

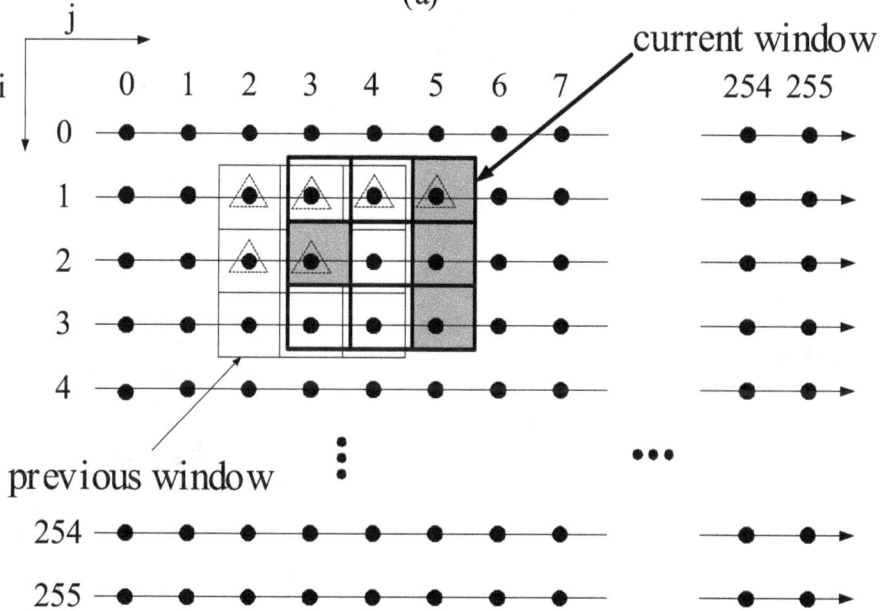

(b)

Fig. 33. (a) The content of the 3×3 window centered at (i, j). (b) The windowing of the 2-D RMF.

3.7 The fully-pipelined DCRAM-based ROF architecture

As seen in Section 3.5, the reservation tables are not tightly scheduled because the dependency of bit-slicing read, threshold decomposition, and polarization forms a cycle. The dependency cycle limits the schedulability of ROF tasks. To increase the schedulability, we further extended the ROF architecture to a fully-pipelined version at the expense of area. The

(1)

Address	Data Field
0000	x(3,3)
0001	x(2,3)
0010	y(1,3)
0011	x(3,1)
0100	y(2,1)
0101	y(1,1)
0110	x(3,2)
0111	x(2,2)
1000	y(1,2)

(2)

Address	Data Field
0000	x(3,3)
0001	x(2,3)
0010	y(1,3)
0011	x(3,4)
0100	x(2,4)
0101	y(1,4)
0110	x(3,2)
0111	y(2,2)
1000	y(1,2)

(3)

Address	Data Field
0000	x(3,3)
0001	y(2,3)
0010	y(1,3)
0011	x(3,4)
0100	x(2,4)
0101	y(1,4)
0110	x(3,5)
0111	x(2,5)
1000	y(1,5)

(4)

Address	Data Field
0000	x(3,6)
0001	x(2,6)
0010	y(1,6)
0011	x(3,4)
0100	y(2,4)
0101	y(1,4)
0110	x(3,5)
0111	x(2,5)
1000	y(1,5)

(5)

Address	Data Field
0000	x(3,6)
0001	x(2,6)
0010	y(1,6)
0011	x(3,7)
0100	x(2,7)
0101	y(1,7)
0110	x(3,5)
0111	y(2,5)
1000	y(1,5)

• • •

The symbols "──►" point to positions for the newest input samples

The symbols "····►" point to positions for the windowing median values

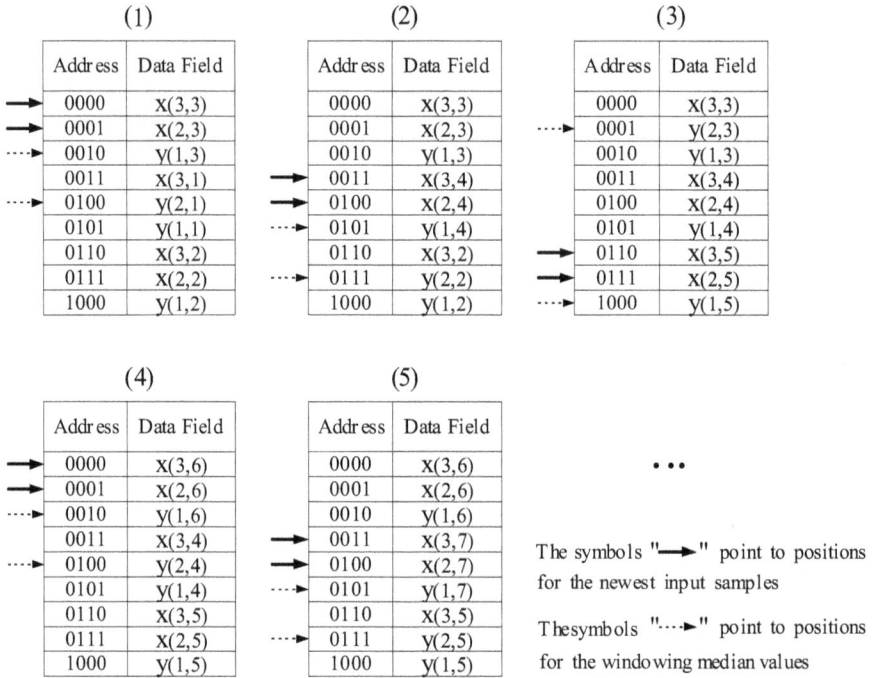

Fig. 34. The data storage of 2-D RMF.

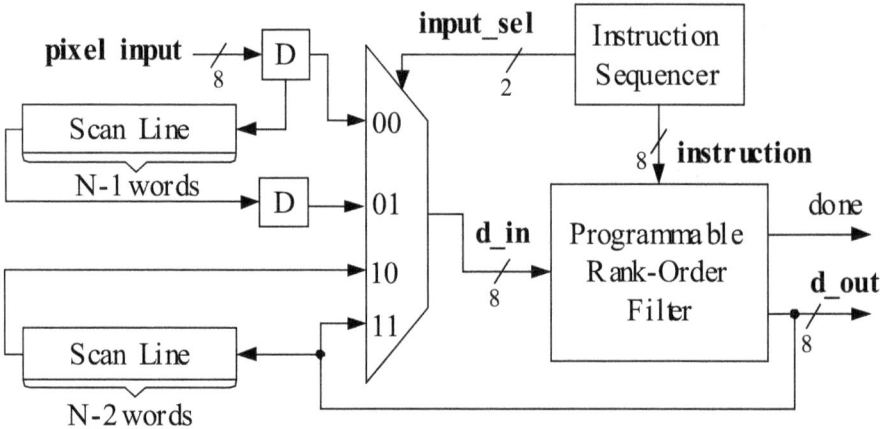

Fig. 35. Block diagram of the 2-D RMF with 3-by-3 window.

fully-pipelined ROF architecture interleaves three ROF iterations with triple computing fields. As shown in Fig. 37, there are three computing fields which process three tasks alternatively. To have the tightest schedule, we pipelined the Level-Quantizer into two stages, LQ1 and LQ2, so the loop (computing field, Level-Quantizer, Shift Register) has three pipeline stages

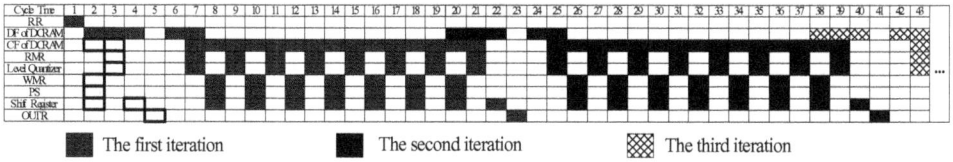

Fig. 36. Reservation table of optimal pipeline for 2-D recursive median filter.

for the highest degree of parallelism. The LQ1 is the FA/HA tree and the LQ2 is the carry generator.

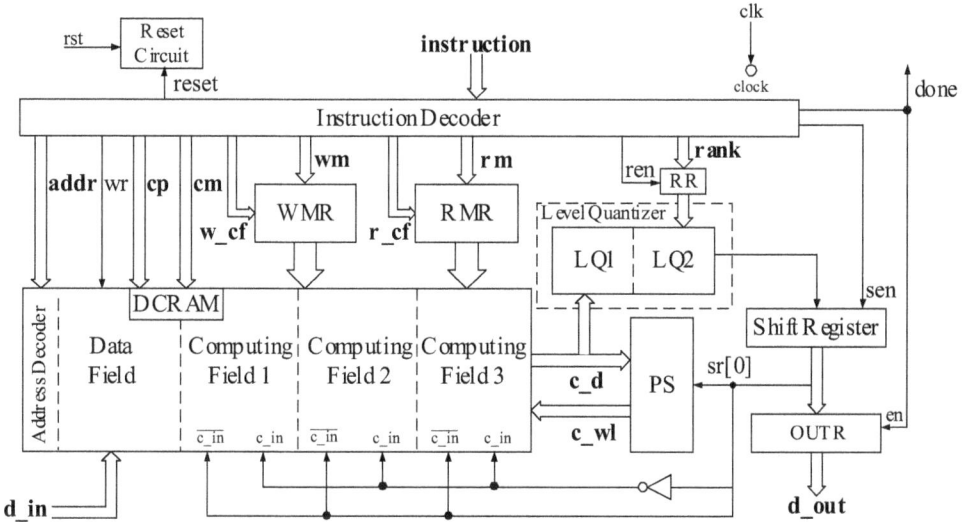

Fig. 37. The fully-pipelined ROF architecture.

Since there exists three iterations being processed simultaneously, a larger memory is required for two more iterations. Hence, we extended the DCRAM to an $(N + 2\delta)$-word memory, where N is the window size of ROF and δ is the number of updating samples for each iteration. The value of δ is 1 for 1-D ROF, and n for 2-D n-by-n ROF. To correctly access the right samples for each iteration, the signal "**cm**" is added to mask the unwanted samples during the copy operation. In each computing field, the unwanted samples are stored as all zeros. Doing so, the unwanted samples will not affect the rank-order results. Fig.38 illustrates the modified computing cell for fully-pipelined ROF. The INV5 and INV6 are replaced with GATE1 and GATE2. When "cm[i]" is '0' the computing cell will store '0'; otherwise, the computing cell will have the copy of the selected sample from the data cell. Finally, we use "**cp**", "**w_cf**", and "**r_cf**" to selectively perform *read, write,* or *copy* on computing fields. To efficiently program the fully-pipelined architecture, the instruction set is defined as shown in Fig.39. The fields $<c\_cf>$ of COPY, $<w\_cf>$ of P_WRITE, and $<r\_cf>$ of P_READ are used to control "**cp**", "**w_cf**", and "**r_cf**". Given a 1-D non-recursive rank order filter application with N=9 and r=3, the reservation table can be optimized as shown in Fig.40.

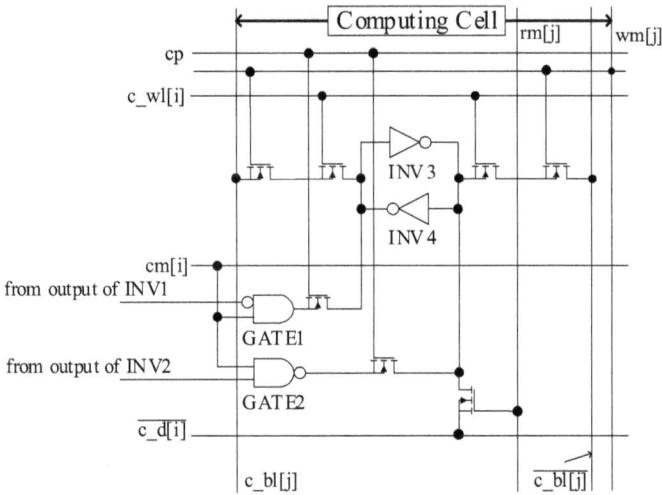

Fig. 38. A modified circuit of computing cell for fully-pipelined ROF.

41 26	25 24	23 12	11 0
Sub instruction 1	Sub instruction 2	Sub instruction 3	Sub instruction 4

Data field instruction

Sub instruction 1

SET <rank value>

41	39	38	30	29	26
0	0 0	1 1 1 1 1 1 1 1 1		rank value	

LOAD <address>

41	39	38	30	29	26
0	0 1	1 1 1 1 1 1 1 1		address	

COPY <c_cf> <cp_mask>

41	39	38	37	36	26
0	1 0	c_cf		cp_mask	

SI1_NULL

41	39	38	26
1	1 1	1 1 1 1 1 1 1 1 1 1 1 1 1	

Sub instruction 2

DONE

25	24
0	d

SI2_NULL

25	24
1	1

Computing field instruction

Sub instruction 3

P_WRITE <w_cf> <mask>

23	22	21	20	19	12
0	0	w_cf		mask	

SI3_NULL

23	22	21	12
1	1	1 1 1 1 1 1 1 1 1 1 1	

Sub instruction 4

P_READ <r_cf> <mask>

11	10	9	8	7	0
0	0	r_cf		mask	

SI4_NULL

11	10	9	0
1	1	1 1 1 1 1 1 1 1 1 1 1	

Fig. 39. The format of the extended instruction set for the fully-pipelined ROF architecture.

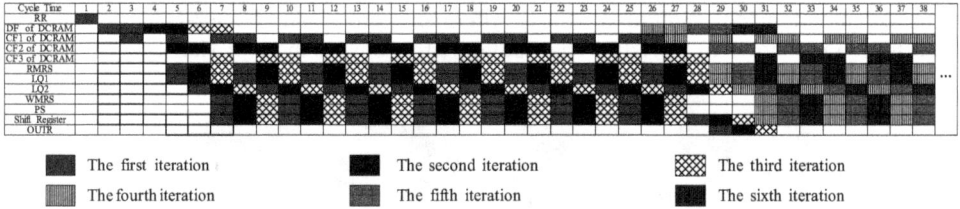

Cycle Time	1	2	3	4	5	6	7	8	9	10	11	12	13	14	15	16	17	18	19	20	21	22	23	24	25	26	27	28	29	30	31	32	33	34	35	36	37	38
RR																																						
DF of DCRAM																																						
CF1 of DCRAM																																						
CF2 of DCRAM																																						
CF3 of DCRAM																																						
RMRS																																						
LQ1																																						
LQ2																																						
WMRS																																						
PS																																						
Shift Register																																						
OUTR																																						

- ■ The first iteration
- ■ The second iteration
- ▨ The third iteration
- ▦ The fourth iteration
- ▨ The fifth iteration
- ■ The sixth iteration

Fig. 40. Reservation table of the 1-D non-recursive ROF for fully-pipelined ROF architecture.

3.8 Chip design and simulation results

To exercise the proposed architecture, we have implemented the ROF architecture, shown in Fig.16, using TSMC 0.18 μm 1P6M technology. First, we verified the hardware in VHDL at the behavior level. The behavior VHDL model is cycle-accurate. As the result of simulation, the implementations of the above examples are valid. Fig.41 and Fig.42 demonstrate the results of VHDL simulations for the 2-D ROF and RMF, respectively. Fig.41(a) is a noisy "Lena" image corrupted by 8% of impulsive noise. After being processed by 2-D ROFs with r=4, 5, and 6, the denoise results are shown in Fig.41(b), (c), and (d), respectively. Fig.42(a) is a noisy "Lena"

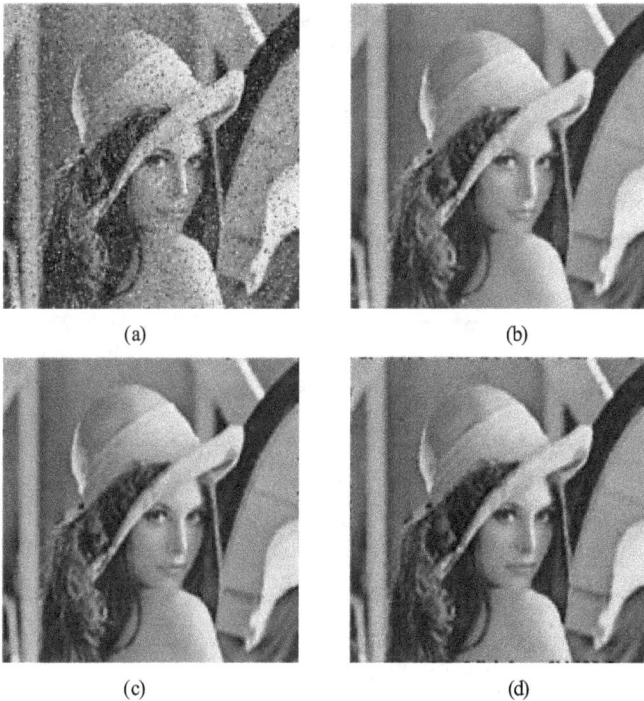

(a)

(b)

(c)

(d)

Fig. 41. Simulation results of a 2-D ROF application. (a) The noisy "Lena" image corrupted by 8% of impulsive noise. (b) The "Lena" image processed by the 3 × 3 4th-order filtering. (c) The "Lena" image processed by the 3 × 3 5th-order filtering. (d) The "Lena" image processed by the 3 × 3 6th-order filtering.

image corrupted by 9% of impulsive noise. After being processed by the 2-D 3 × 3 RMF, the denoise result is shown in Fig.42(b). The results are the same as those of Matlab simulation.

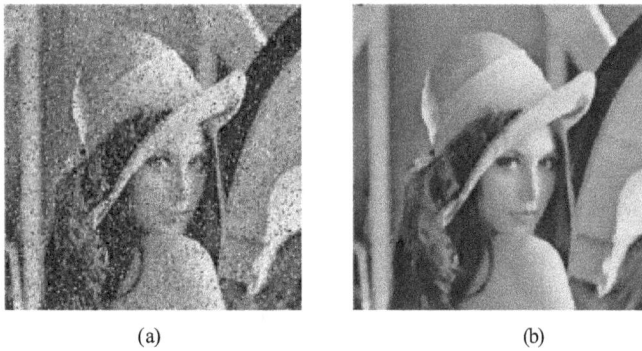

(a)   (b)

Fig. 42. Simulation results of a 2-D RMF application. (a) The noisy "Lena" image corrupted by 9% of impulsive noise. (b) The "Lena" image processed by the 3 × 3 RMF.

Upon verifying the proposed ROF processor using the cycle-accurate behavior model, we then implemented the processor in the fully-custom design methodology. Because of high regularity of memory, the proposed memory-based architecture saves the routing area while comparing with the logic-based solutions. Fig.43 (a) shows the overall chip layout and the dash-lined region is the core. The die size is $1063.57 \times 1069.21 \mu m^2$ and the pinout count is 40. Fig.43 (b) illustrates the detail layout of the ROF core. The core size is $356.1 \times 427.7 \mu m^2$ and the total transistor count is 3942. Fig.43 (c) illustrates the floorplan and placement. The physical implementation has been verified by the post-layout simulation. Table 8 shows the result of timing analysis, obtained from NanoSim. As seen in the table, the critical path is the path 3 and the maximum clock rate can be 290 MHz at 3.3V and 256 MHz at 1.8V. As the result of post-layout simulation, the power dissipation of the proposed ROF is quite low. For the 1-D/2-D ROFs, the average power consumption of the core is 29mW at 290MHz or 7mW at 256MHz. The performance sufficiently satisfies the real-time requirement of video applications in the formats of QCIF, CIF, VGA, and SVGA. The chip is submitting to Chip Implementation Center (CIC), Taiwan for the fabrication.

Path	Description	1.8V supply	3.3V supply
1	From the output of RR to the input of the shift register.	1.2 ns	0.78 ns
2	From the output of RMR, thru DCRAM to the input of the PS.	1.8 ns	1.1 ns
3	From the output of RMR, thru DCRAM and the Level-Quantizer, to the input of the shift register.	3.9 ns	3.44 ns
4	From the shift register, thru the inverter connected to "c_in", to the SRAM cell of the computing field.	3.02 ns	1.96 ns
5	From "d_in" to the SRAM cell of the data field.	3.05 ns	1.85 ns
6	From the SRAM cell of the data field to the SRAM cell of the computing field.	1.24 ns	1.09 ns

Table 8. Timing analysis of the proposed ROF processor.

(a)   (b)

(c)

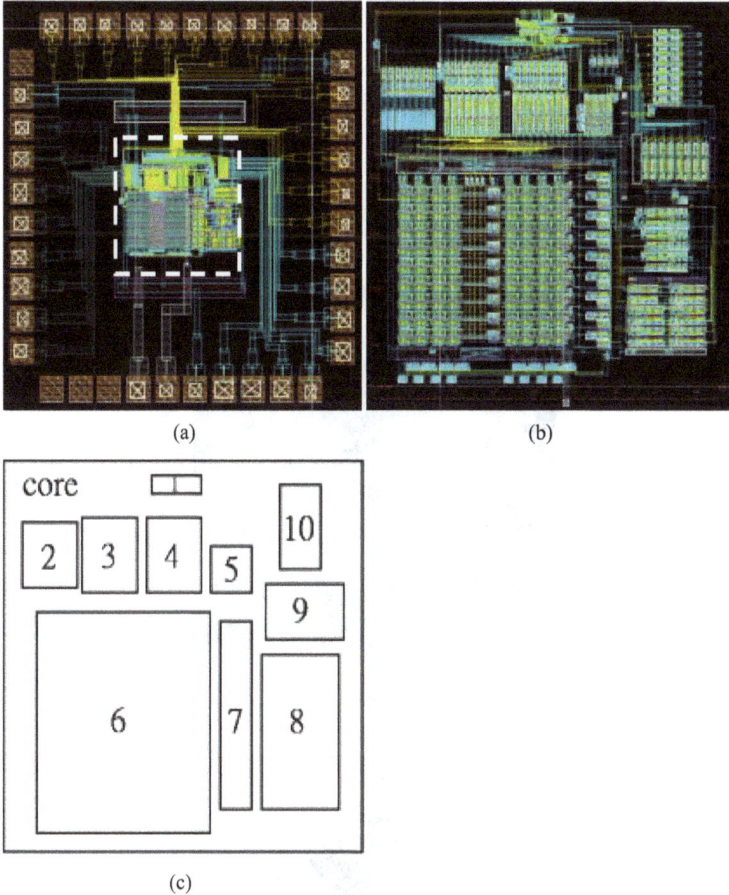

Fig. 43. The result of chip design using TSMC 0.18um 1P6M technology. (a) The chip layout of proposed rank-order filter. (b) The core of the proposed ROF processor. (c) The floorplan and placement of (b). (1: Instruction decoder; 2: Reset circuit, 3: WMR, 4: RMR, 5: RR, 6: DCRAM, 7: PS; 8: Level Quantizer; 9: Shift Register; 10: OUTR.)

Furthermore, We have successfully built a prototype which is composed of a FPGA board and DCRAM chips to validate the proposed architecture before fabricating the custom designed chip. The FPGA board is made by Altera and the FPGA type is APEX EP20K. The FPGA board can operate at 60 MHz at the maximum. The DCRAM chip was designed by full-custom CMOS technology. Fig.44(a) shows the micrograph of the DCRAM chip. The chip implements a subword part of DCRAM and the Fig.44(b) illustrates the chip layout. The fabricated DCRAM chip was layouted by full-custom design flow using TSMC 0.35 2P4M technology. As shown in Fig.45, with the supply voltage of 3.3V, the DCRAM chip can operate at 25 MHz. Finally, we successfully integrated the FPGA board and the DCRAM chips into a prototype as shown in Fig.46 The prototype was validated with ROF algorithms mentioned above.

(a)

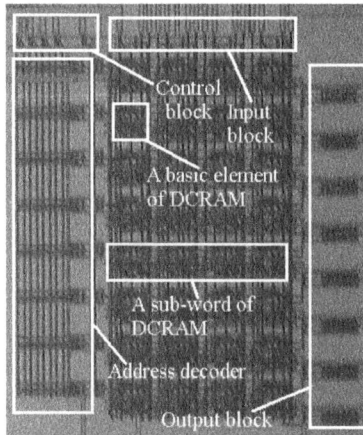

Control
block   Input
        block

A basic element
of DCRAM

A sub-word of
DCRAM

Address decoder

Output block

(b)

Fig. 44. (a) The micrograph of DCRAM chip. (b) The layout of the DCRAM chip.

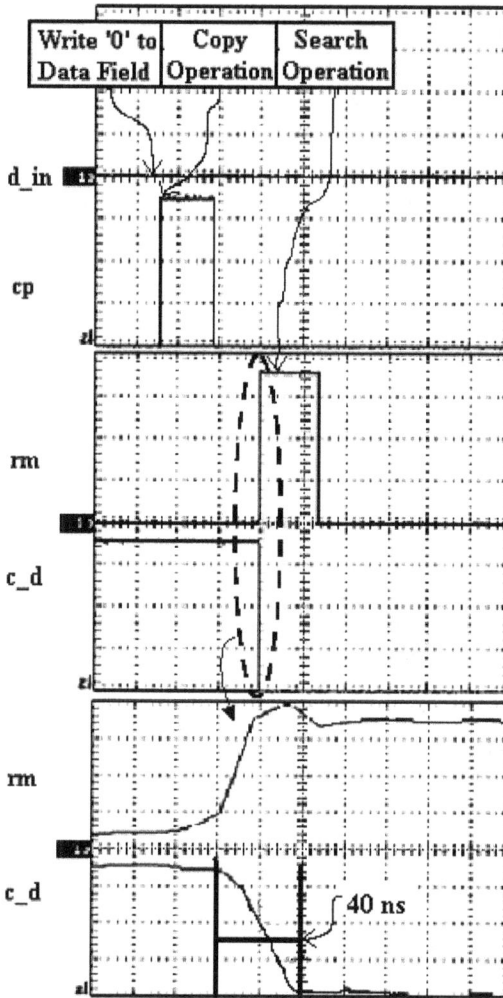

Fig. 45. Measured waveform of the DCRAM chip.

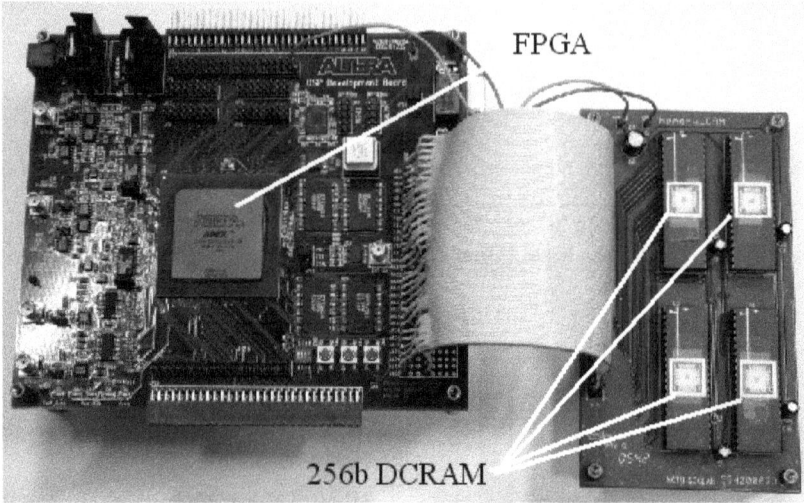

Fig. 46. The system prototype of rank-order filtering processor.

4. Conclusion

In order to further extend the battery life of capsule endoscope, this paper mainly focus on a series of mathematical statistics to systematically analyze the color sensitivity in GI images from the RGB color space domain to the 2-D DCT spatial frequency domain. According to the analysis results, an improved ultra-low-power subsample-based GICam image compression processor are proposed for capsule endoscope or swallowable imaging capsules. we make use of the subsample technique to reduce the memory requirements of G1, G2 and B components according to the analysis results of DC/AC coefficients in 2-D DCT domain. As shown in the simulation result, the proposed image compressor can efficiently save 38.5% more power consumption than previous GICam one (11), and can efficiently reduce image size by 75% at least for each sampled gastrointestinal image. Therefore, video sequences totally reduce size by 75% at least. Furthermore, the proposed image compressor has lower area and lower operation frequency according to the comparison results. It can fit into the existing designs.

Forthemore, we have proposed an architecture based on a maskable memory for rank-order filtering. This paper is the first literature using maskable memory to realize ROF. Driving by the generic rank-order filtering algorithm, the memory-based architecture features high degree of flexibility and regularity while the cost is low and the performance is high. With the LIW instruction set, this architecture can be applied for arbitrary ranks and a variety of ROF applications, including recursive and non-recursive algorithms. As shown in the implementation results, the core of the processor has high performance and low cost. The post-layout simulation shows that the power consumption can be as low as 7 mW at 256 MHz. The processing speed can meet the real-time requirement of image applications in the QCIF, CIF, VGA, or SVGA formats.

5. References

[1] G. Iddan, G. Meron, A. Glukhovsky, and P. Swain, "Wireless capsule endoscopy," Nature, vol. 405, pp. 417-418, May 25, 2000.

[2] Shinya Itoha, Shoji Kawahitob, Tomoyuki Akahoric, and Susumu Terakawad, "Design and implementation of a one-chip wireless camera device for a capsule endoscope," SPIE, vol. 5677, pp. 108-118, 2005.

[3] F. Gong. P. Swain, and T. Mills, "Wireless endoscopy," Gastrointestinal Endoscopy, vol.51, no. 6, pp. 725-729, June 2000.

[4] H. J.Park, H.W. Nam, B.S. Song, J.L. Choi, H.C. Choi, J.C. Park, M.N. Kim, J.T. Lee, and J.H. Cho, "Design of bi-directional and multi-channel miniaturized telemetry module for wireless endoscopy," in Proc. of the 2nd Annual Intl IEEE-EMBS Special Topic Conference on Microtechnologies in Medicine and Biology, May 2-4, 2002, Madison, USA, pp. 273-276.

[5] http://www.givenimaging.com/Cultures/en-US/given/english

[6] http://www.rfsystemlab.com/

[7] M. Sendoh, k. Ishiyama, and K.-I. Arai, "Fabrication of Magnetic Actuator for Use in a Capsule Endoscope," IEEE Trans. on Magnetics, vol. 39, no. 5, pp. 3232-3234, September 2003.

[8] Louis Phee, Dino Accoto, Arianna Menciassi*, Cesare Stefanini, Maria Chiara Carrozza, and Paolo Dario, "Analysis and Development of Locomotion Devices for the Gastrointestinal Tract," IEEE Trans. on Biomedical Engineering, vol. 49, no. 6, JUNE 2002.

[9] Shaou-Gang Miaou, Shih-Tse Chen, and Fu-Sheng Ke, "Capsule Endoscopy Image Coding Using Wavelet-Based Adaptive Vector Quantization without Codebook Training," International Conference on Information Technology and Applications (ICITA), vol. 2, pp. 634-637, July 2005.

[10] Shaou-Gang Miaou, Shih-Tse Chen, and Chih-Hong Hsiao, "A wavelet-based compression method with fast quality controlling capability for long sequence of capsule endoscopy images," IEEE-EURASIP Workshop on Nonlinear Signal and Image Processing (NSIP), pp. 34-34, 2005.

[11] M. Lin, L. Dung, and P. Weng, "An Ultra Low Power Image Compressor for capsule Endoscope," BioMedical Engineering Online 2006, vol. 5:14.

[12] X. Xie, G. Li, X. Chen, X. Li, and Z. Wang, "A Low Power Digital IC Design Inside the Wireless Endoscopic Capsule", IEEE Journal of Solid-State Circuits, vol. 41, no. 11, pp. 2390-2400, November 2006.

[13] K. Wahid, S-B. Ko, and D. Teng, "Efficient Hardware Implementation of an Image Compressor for Wireless Capsule Endoscopy Applications", Proceedings of the IEEE International Joint Conference on Neural Networks pp. 2762-2766, 2008;

[14] Xinkai Chen; Xiaoyu Zhang; Linwei Zhang; Xiaowen Li; Nan Qi; Hanjun Jiang; Zhihua Wang, "A Wireless Capsule Endoscope System With Low-Power Controlling and Processing ASIC", IEEE Trans. on Biomedical Circuits and Systems, vol. 3, no. 1, pp.11-22, Feb. 2009.

[15] H.A. Peterson, H. Peng, J. H. Morgan, and W. B. Pennebaker, "Quantization of color image components in the DCT domain" SPIE , Human Vision, Visual Processing, and Digital Display II, vol.1453, 1991.

[16] Dr. R.W.G. Hunt, "Measuring Colour," Fountain Press, 1998.

[17] Henry R. Kang "Color Technology For Electronic Imaging Devices," SPIE Optical Engineering Press, 1997.

[18] J.Ziv and A. Lempel, "A universal algorithm for sequential data compression," IEEE Trans. on Inform. Theory, vol. 23, pp. 337-343, May 1977.

[19] Vasudev Bhaskaran, and Konstantinos Kon stantinides "Images and Video Compression Standards : Alogorithms and Architectures, Second edition," Kliwer Academic Publishers.

[20] Meng-Chun Lin, Lan-Rong Dung, and Ping-Kuo Weng "An Improved Ultra-Low-Power Subsample based Image Compressor for Capsule Endoscope," Medical Informatics Symposium in Taiwan (MIST), 2006.

[21] Gi-Shih Lien, Chih-Wen Liu, Ming-Tsung Teng, and Yan-Min Huang, " Integration of Two Optical Image Modules and Locomotion Functions in Capsule Endoscope Applications," The 13th IEEE International Symposium on Consumer Electronics, pp.828-829, 2009.

[22] Mao Li, Chao Hu, Shuang Song, Houde Dai, and Max Q.-H. Meng, "Detection of Weak Magnetic Signal for Magnetic Localization and Orientation in Capsule Endoscope," Proceedings of the IEEE International Conference on Automation and Logistics Shenyang, China, pp.900-905, August 2009.

[23] Chao Hu, Max Q.-H. Meng, Li Liu, Yinzi Pan, and Zhiyong Liu "Image Representation and Compression for Capsule Endoscope Robot," Proceedings of the 2009 IEEE International Conference on Information and Automation, pp.506-511, June 2009.

[24] Jing Wu, and Ye Li, "Low-complexity Video Compression for Capsule Endoscope Based on Compressed Sensing Theory," 31st Annual International Conference of the IEEE EMBS Minneapolis, pp.3727-3730, Sep. 2009.

[25] Jinlong Hou, Yongxin Zhu, Le Zhang, Yuzhuo Fu, Feng Zhao, Li Yang, and Guoguang Rong, "Design and Implementation of a High Resolution Localization System for In-vivo Capsule Endoscopy," 2009 Eighth IEEE International Conference on Dependable, Autonomic and Secure Computing, pp.209-214, 2009.

[26] Chang Cheng, Zhiyong Liu and Chao Hu, and Max Q.-H. Meng "A Novel Wireless Capsule Endoscope With JPEG Compression Engine," Proceedings of the 2010 IEEE International Conference on Automation and Logistics, pp.553-558, Aug. 2010.

[27] D.H. Kang, J.H. Choi, Y.H. Lee, and C. Lee, "Applications of a DPCM system with median predictors for image coding," IEEE Trans. Consumer Electronics, vol.38, no.3, pp.429-435, Aug. 1992.

[28] H. Rantanen, M. Karlsson, P. Pohjala, and S. Kalli, "Color video signal processing with median filters," IEEE Trans. Consumer Electron., vol.38, no.3, pp.157-161, Apr. 1992.

[29] T. Viero, K. Oistamo, and Y. Neuvo, "Three-dimensional median-related filters for color image sequence filtering," IEEE Trans. Circuits Syst. Video Technol., vol.4, no.2, pp.129-142, Apr. 1994.

[30] X. Song, L. Yin, and Y. Neuvo, "Image sequence coding using adaptive weighted median prediction," Signal Processing VI, EUSIPCO-92, Brussels, pp.1307-1310, Aug. 1992.

[31] K. Oistamo and Y. Neuvo, "A motion insensitive method for scan rate conversion and cross error cancellation," IEEE Trans. Consumer Electron., vol.37, pp.296-302, Aug. 1991.

[32] P. Zamperoni, "Variation on the rank-order filtering theme for grey-tone and binary image enhancement," IEEE Int. Conf. Acoust., Speech, Signal Processing, pp.1401-1404, 1989.

[33] C.T. Chen and L.G. Chen, "A self-adjusting weighted median filter for removing impulse noise in images," Int. Conf. Image Processing, pp.16-19, Sept. 1996.

[34] D. Yang and C. Chen, "Data dependence analysis and bit-level systolic arrays of the median filter," IEEE Trans. Circuits and Systems for Video Technology, vol.8, no.8, pp.1015-1024, Dec. 1998.

[35] T. Ikenaga and T. Ogura, "CAM2: A highly-parallel two-dimensional cellular automation architecture," IEEE Trans. Computers, vol.47, no.7, pp.788-801, July 1998.

[36] L. Breveglieri and V. Piuri, "Digital median filter," Journal of VLSI Signal Processing, vol.31, pp.191-206, 2002.

[37] C. Chakrabarti, "Sorting network based architectures for median filters," IEEE Trans. Circuits ans Systems II: Analog and Digital Signal Processing, vol.40, pp.723-727, Nov. 1993.

[38] C. Chakrabarti, "High sample rate architectures for median filters," IEEE Trans. Signal Processing, vol.42, no.3, pp.707-712, March 1994.

[39] L. Chang and J. Lin, "Bit-level systolic array for median filter," IEEE Trans. Signal Processing, vol.40, no.8, pp.2079-2083, Aug. 1992.

[40] C. Chen, L. Chen, and J. Hsiao, "VLSI implementation of a selective median filter," IEEE Trans. Consumer Electronics, vol.42, no.1, pp.33-42, Feb. 1996.

[41] M.R. Hakami, P.J. Warter, and C.C. Boncelet, Jr., "A new VLSI architecture suitable for multidimensional order statistic filtering," IEEE Trans. Signal Processing, vol.42, pp.991-993, April 1994.

[42] Hatirnaz, F.K. Gurkaynak, and Y. Leblebici, "A compact modular architecture for the realization of high-speed binary sorting engines based on rank ordering," IEEE Inter. Symp. Circuits and Syst., Geneva, Switzerland, pp.685-688, May 2000.

[43] A.A. Hiasat, M.M. Al-lbrahim, and K.M. Gharailbeh, "Design and implementation of a new efficient median filtering algorithm," IEE Proc. Image Signal Processing, vol.146, no.5, pp.273-278, Oct. 1999.

[44] R.T. Hoctor and S.A. Kassam, "An algorithm and a pipelined architecture for order-statistic determination and L-filtering," IEEE Trans. Circuits and Systems, vol.36, no.3, pp.344-352, March 1989.

[45] M. Karaman, L. Onural, and A. Atalar, "Design and implementation of a general-purpose median filter unit in CMOS VLSI," IEEE Journal of Solid State Circuits, vol.25, no.2, pp.505-513, April 1990.

[46] C. Lee, P. Hsieh, and J, Tsai, "High-speed median filter designs using shiftable content-addressable memory," IEEE Trans. Circuits and Systems for Video Technology, vol.4, pp.544-549, Dec. 1994.

[47] C.L. Lee and C. Jen, "Bit-sliced median filter design based on majority gate," IEE Proc.-G Circuits, Devices and Systems, vol.139, no.1, pp.63-71, Feb. 1992.

[48] L.E. Lucke and K.K. Parchi, "Parallel processing architecture for rank order and stack filter," IEEE Trans. Signal Processing, vol.42, no.5, pp.1178-1189, May 1994.

[49] K. Oazer, "Design and implementation of a single-chip 1-D median filter," IEEE Trans. Acoust., Speech, Signal Processing, vol.ASSP-31, no.4, pp.1164-1168, Oct. 1983.

[50] D.S. Richards, "VLSI median filters," IEEE Trans. Acoust., Speech, and Signal Processing, vol.38, pp.145-153, January 1990.

[51] G.G. Boncelet, Jr., "Recursive algorithm and VLSI implementation for median filtering," IEEE Int. Sym. on Circuits and Systems, pp.1745-1747, June 1988.

[52] C. Henning and T.G. Noll, "Architecture and implementation of a bitserial sorter for weighted median filtering," IEEE Custom Integrated Circuits Conference, pp.189-192, May 1998.

[53] C.C Lin and C.J. Kuo, "Fast response 2-D rank order filter by using max-min sorting network," Int. Conf. Image Processing, pp.403-406, Sept. 1996.

[54] M. Karaman, L. Onural, and A. Atalar, "Design and implementaion of a general purpose VLSI median filter unit and its application," IEEE Int. Conf. Acoustics, Speech, and Signal Processing, pp.2548-2551, May 1989.

[55] J. Hwang and J. Jong, "Systolic architecture for 2-D rank order filtering," Int. Conf. Application-Specific Array Processors, pp.90-99, Sept. 1990.

[56] I. Pitas, "Fast algorithms for running ordering and max/min calculation," IEEE Trans. Circuits and Systems, vol.36, no.6, pp.795-804, June 1989.

[57] O. Vainio, Y. Neuvo, and S.E. Butner, "A signal processor for median-based algorithm," IEEE Trans. Acoustics, Speech, and Signal Processing, vol.37, no.9, pp.1406-1414, Sept. 1989.

[58] H. Yu, J. Lee, and J, Cho, "A fast VLSI implementation of sorting algorithm for standard median filters," IEEE Int. ASIC/SOC Conference, pp.387-390, Sptember 1999.

[59] J.P. Fitch, "Software and VLSI algorithm for generalized renked order filtering," IEEE Trans. Circuits and Systems, vol.CAS-34, no.5, pp.553-559, May 1987.

[60] M.Karaman and L. Onural, "New radix-2-based algorithm for fast median filtering," Electron. Lett., vol.25, pp.723-724, May 1989.

[61] J.P. Fitch, "Software and VLSI Algorithms for Generalized Ranked Order Filtering," IEEE Trans. Circuits and Syst., vol.CAS-34, no.5, pp.553-559, May 1987.

[62] B.K. Kar and D.K. Pradhan, "A new algorithm for order statistic and sorting," IEEE Trans. Signal Processing, vol.41, no.8, pp.2688-2694, Aug. 1993.

[63] V.A. Pedroni, "Compact hamming-comparator-based rank order filter for digital VLSI and FPGA implementations," IEEE Int. Sym. on Circuits and Systems, vol.2, pp.585-588, May 2004.

[64] Shobha Singh, Shamsi Azmi, Nutan Agrawal, Penaka Phani and Ansuman Rout, "Architecture and design of a high performance SRAM for SOC design," IEEE Int. Sym. on VLSI Design, pp.447-451, Jan 2002.

12

Low Cost Prototype of an Outdoor Dual Patch Antenna Array for the Openly TV Frequency Ranges in Mexico

M. Tecpoyotl-Torres[1], J. A. Damián Morales[1], J. G. Vera Dimas[1],
R. Cabello Ruiz[1], J. Escobedo-Alatorre[1],
C. A. Castillo-Milián[2] and R. Vargas-Bernal[3]
[1]*Centro de Investigación en Ingeniería y Ciencias Aplicadas (CIICAp-UAEM)*
[2]*Facultad de Ciencias Químicas e Ingeniería (FCQeI-UAEM)*
[3]*Instituto Tecnológico Superior de Irapuato (ITESI)*
Mexico

1. Introduction

In this research, we developed a dual antenna array for the household reception of openly analogical television (TV) frequencies. This array was designed on Flame Retardant-4 (FR-4) as substrate, in order to obtain a low cost prototype.

The interest in this area is because of the fact that the openly TV is one of the most important communication media in our country. From information supplied in 2009 in the National Survey over availability and use of Technologies, it reveals that the 72.8% of the population uses the services of the openly TV (Instituto Nacional de Estadística, Geografía e Informática [INEGI], 2009). A TV can be found in almost all homes of the country, but only 13.6% correspond to digital technology, while only a half of homes with a digital TV requires signal payment. The availability of TVs in Mexican homes in 2010 remained without severe changes (INEGI, 2010).

Since the operation frequencies ranges are not so high, and then the antenna sizes, obtained directly from the design equations are very large. Therefore, the scaling is a necessary step in order to reduce the antenna sizes to achieve its easy manipulation.

As it is well-known, a very simple common example of antennas used for household reception of TV is the Yagi-Uda (or Yagi) array, where the length of the dipoles established the phase of the individually received signals. An example of a commercial Yagi-Uda antenna designed for channels 2-13 can be found in (Balanis, 2005). The TV transmission has the polarization vector in the horizontal plane so that the array must also be horizontal (Melissinos, 1990).

The evolution of the antennas designed in order to improve the openly TV reception has notably changed in the last years, in such a way, for outdoor use it is possible to find the large aerial antennas, fixed or with an integrated rotor, under different geometries, such as single dipoles and combinations of Yagi arrays. A decrease in sizes is noted in some cases. An example of design development of antenna for TV transmission for outdoor broadcasts can be found in (Rathod, 2010), at 750 MHz as the center frequency, where the antenna was fabricated on FR-4, with substrate sizes of 40x40 cm². This antenna was designed for the study of rural areas in India.

For indoor use, there are also several options of relative small sizes, and under different geometries. Recently, new commercial options based on patch or microstrip antennas have been proposed, which can be located on the rear part of the TV display that means, hidden to the user. But there is not available technical information about its design. An UHF planar O-shaped antenna has been proposed and studied (Barir & Hamid, 2010), which was fabricated on FR-4, with a dimension area of 20x20 cm², with an enough bandwidth to cover Indonesian broadcasters.

Other special case is formed by the antennas for TV reception in cars. In (Neelakanta & Chatterjee, 2003), a V-structure dipole, which is part of the dipole families, has been conceived for the purpose of TV reception (VHF/UHF bands), which gives a directional pattern with horizontal polarization. An active loop antenna suitable as automobile television receiving antenna, for channels 13-62 (from 470-770 MHz in Japan) can be found in (Taguchi et al., 1996).

In (Wang & Lecours, 1999), an antenna array with orthogonal polarization finds applications in Direct Broadcasting Systems (DBS), Personal Communication Services (PCS) and Indoor Communication Systems (ICS). As the current DBS technology uses both horizontal and vertical polarizations, and then the microstrip arrays with orthogonal polarizations are needed. While in PCS and ICS, waves are scattered by the environment and the signal takes several paths from a transmitter to a receiver, with resulting fluctuations in amplitude because of multipath fading effect. To overcome this effect, it is necessary to implement a polarization diversity technique, for which antenna arrays with orthogonal polarizations and very low cross couplings are needed.

On the other hand, it is recognized as a common problem in TV to the multipath reception, where signals from the same station can reach the reception antenna by two or more distinct paths which differ significantly in length (web site: http://www. electusdistribution.com.au/images_uploaded/tvrecepe.pdf, May 2011).

In (Brown et al., 2007) it was shown that dipoles and other linear antennas can sometimes, although not always, have a large degree of polarization diversity if they have different polarization orientations. A typical configuration of polarization diversity system consists of one transmit and one dual-polarized receive antenna (i.e., maximal diversity order of two) (Kapinas et al., 2007).

Dual linear polarization is characterized by two orthogonal linear polarizations on the same antenna. Dual polarization antennas have the benefit of allowing two signals, with different orientations, to be broadcast or received on the same antenna (Smith, 2008).

1.1 Mexican TV system

The TV channels in Mexico, in the VHF band are divided in two sub-bands: From 2 to 6, they are in the range from 54 MHz to 88 MHz, and from 7 to 13 are transmitted from 174

MHz to 216 MHz. Some channels are divided between the two most important television companies as follows: the broadcast channels of Televisa are 2, 4, 5 and 9 (a repetition of channel 2); and the corresponding of TV Azteca are 7 and 13 (channels in operation in D.F. in 2006 (Jalife & Sosa, 2007)). Some channels can be transmitted in different frequencies depending of the corresponding Mexican states.

In addition, in each Mexican state, there are additional channels by concession, for example in Morelos, channel 6 corresponds to the Instituto Politécnico Nacional, 3 to the Government of the Morelos State, 11 to Radio Televisora de Mexico Norte S. A. de C. V., 28 to TV Azteca, and 22 to the Presidencia Municipal de Zacatepec (Comisión Federal de Telecomunicaciones [COFETEL], 2008). From channels 14 to 83, they correspond to UHF band.

In our country, for some analogical active channels, temporary it is assigned an additional channel (mirror) to transmit the same information, but with a digital format, until the transition to the digital terrestrial television in Mexico concludes. The last period of transition was planned from 2019 up to 2021, but recently it has been established until December 31, 2015. The temporary digital channels assigned are shown in Table 1. The analogical channels will be returned when the transition will be finished.

TV analogical channel	TV digital channel
2	48
4	49
5	50
7	24
9	44
13	25

Table 1. Digital assigned channels as mirrors.

In this works, our interest is focused in a patch antenna array prototype for openly TV frequency ranges in Mexico, with polarization diversity, for outdoor use. In Section 2, the design of the antenna array will be described and the corresponding simulations will be provided in Section 3. The first tests results are discussed in Section 4, and finally, in Section 5, some concluding remarks are given.

2. Antenna array design

Considering only the two current VHF frequency ranges of the openly TV in Mexico, we chose as operation frequency to 71 MHz for the design of the rectangular patch antenna, which will be used as the base of the prototype patch antenna array. This frequency corresponds to the central frequency of the first sub-range of frequency of VHF. The rectangular antenna was designed using the well-known equations (Balanis, 2005 and Garg et al., 2001):

For the patch width:

$$W = \frac{c}{2f_0\sqrt{\frac{\varepsilon_r + 1}{2}}} \tag{1}$$

where c is the constant speed of light in vacuum, ε_r, the dielectric constant substrate and f_0, the operating frequency equal to 71 MHz.

The effective dielectric constant:

$$\varepsilon_{reff} = \frac{\varepsilon_r + 1}{2} + \frac{\varepsilon_r - 1}{2}\left(1 + 12\frac{h}{W}\right)^{-1/2} \quad \text{if } \frac{W}{h} > 1 \tag{2}$$

The effective length is calculated using:

$$L_{eff} = \frac{c}{2 f_0 \sqrt{\varepsilon_{reff}}} \tag{3}$$

The two increments in the length, which are generated by the fringing fields, make electrical length slightly larger than the physical length of the patch:

$$\Delta L = 0.412 h \frac{\left(\varepsilon_{reff} + 0.3\right)\left(\frac{W}{h} + 0.264\right)}{\left(\varepsilon_{reff} - 0.258\right)\left(\frac{W}{h} + 0.8\right)} \tag{4}$$

The patch length is given by:

$$L = L_{eff} - 2\Delta L \tag{5}$$

The length and width of ground plane (and the substrate), are given by:

$$L_g = 6h + L \quad \text{and} \quad W_g = 6h + W \tag{6}$$

The rectangular patch designed at 71 MHz is shown in Figure 1, considering a reduction factor of 8, required to decrease the patch and substrate sizes. The sizes of the patch are given in Table 2 and the feed point location in Table 3. FR-4 was used as substrate; it has a height of 0.0016 m and a dielectric permittivity of 4.2.

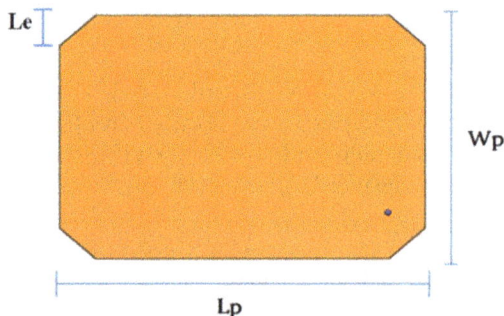

Fig. 1. Single rectangular patch antenna.

The wave group of length λ_g determines to the Length edge (Le, in Figure 1) of the cuts. Le has a value of $\lambda_g/8$ (see Figure 1), which in this case is of 0.0161184 m. .

With the reduced sizes, the antenna array has been designed using a superposition of two rectangular patches, as it can be seen in Figure 2. Some adjustments in length were required in order to locate both central frequencies of the VHF TV channels transmission, with a minimal error (a shift of 0.14% in 71 MHz and of 0.1% in 195 MHz). As it can be noted, the orthogonal lengths have the same length ($W=L+L1$), as it is desirable for dual polarized antennas (Smith, 2008). The sizes of the rectangular antenna array are given in Table 4.

Patch antenna array sizes (m)	
Wp	0.1272564
Lp	0.1621136

Table 2. Sizes of the individual patch antenna.

Feed point location (m)	
X	0.065
Y	-0.04

Table 3. Feed point location of the individual patch antenna(coordinates are considered as they are established by FEKO program) .

Sizes	W	L	W_1	L_1	W_2
Patch	0.167	0.132	0.017	0.035	0.127
Substrate	0.176	0.142	0.027	0.045	0.1366

Table 4. The rectangular antenna array sizes (in meters).

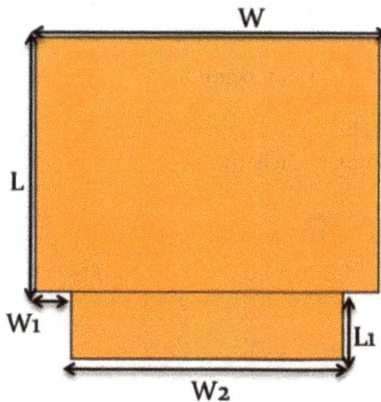

Fig. 2. Geometry of the rectangular patch antenna array.

Cuts on the array were also implemented in order to increase the corresponding gain. The sizes of the array with the cuts implemented on its corner are given in Table 5. The length edge of the cuts is also given by $\lambda_g/8$, as in the case of the single antenna.

Patch antenna array sizes (m)			
Wp	0.16366496	Lp	0.1272564
Wp1	0.01742811	Lp1	0.0165981
Wp2	0.1272564		

Table 5. Sizes of the patch antenna array with cuts.

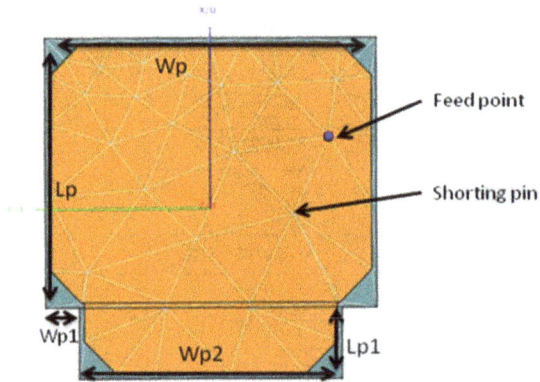

Fig. 3. Geometry of the antenna array (T shape), with cuts on its corners.

The feed point and shorting pin location are shown in Figure 3 and its coordinates in Table 6.

Location, (m).			
Feed point		Shorting pin	
X	0.035	X	-0.0025
Y	-0.0595	Y	-0.04175
Z	0	Z	0

Table 6. The feed point and shorting pin location.

Before to obtain the geometry shown in Figure 3, other two geometries were realized (Figure 4), but there were some problems in each one.

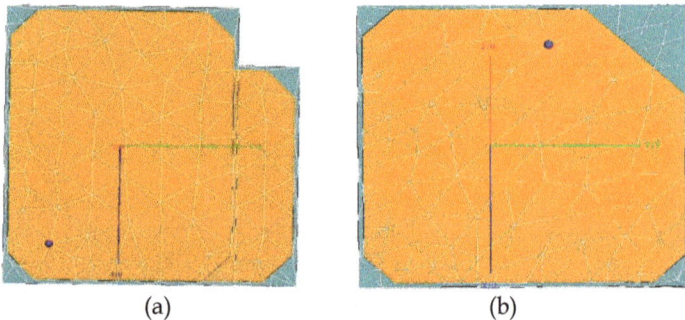

(a)   (b)

Fig. 4. First geometries implemented (L shape) and (b) irregular cuts.

For Figure 4(a), the high return loss were obtained (bigger than -5dB), and for (b) the operation frequencies were so far from one to each other. These were the reasons to choose the T geometry to realize the prototype.

3. Simulation results

The 3D far electrical field magnitude patterns as a function of frequency are shown in Figure 5. The electrical far field components, in polar coordinates are shown in Figure 6.

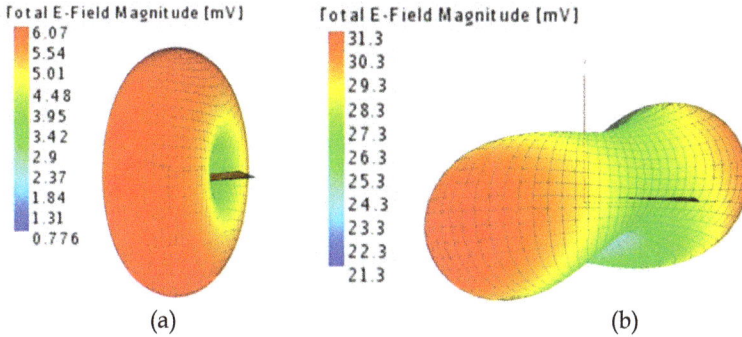

(a)   (b)

Fig. 5. Radiation pattern of the far electrical field magnitude at (a) 70.98 MHz and (b) 194.8 MHz.

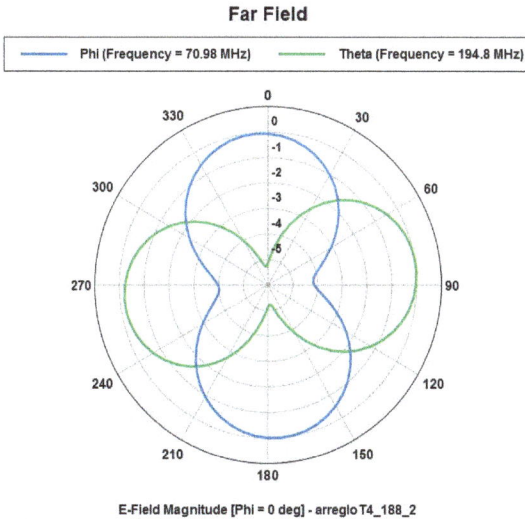

E-Field Magnitude [Phi = 0 deg] - arreglo T4_188_2

Fig. 6. Components of the electrical Far Field.

As it can be observed from Figure 5a, the radiation pattern at 70.98 MHz, corresponds to an omnidireccional antenna, with horizontal polarization, while in Figure 5b, the radiation pattern at 194.8 MHz corresponds to a directive antenna directed on the X-axis, with maxima on both directions.

The Reflection Coefficient magnitude of the antenna is shown in Figure 7, where the peaks of response are located at 70.98 MHz and 194.8 MHz, very near to the selected design operation frequencies (71 MHz and 195 MHz, which correspond to the central frequencies of the two sub-ranges). A zoom at both frequencies is presented in Figures 8 and 9.

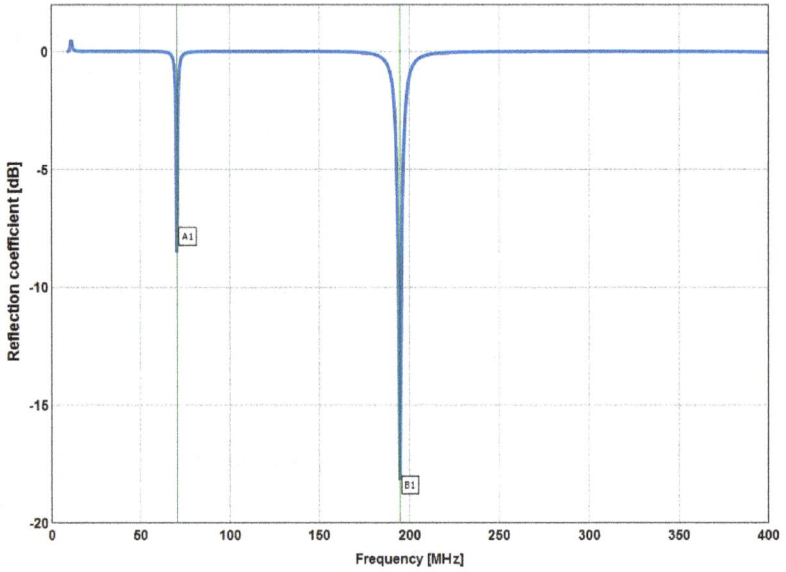

Fig. 7. Reflection coefficient magnitude.

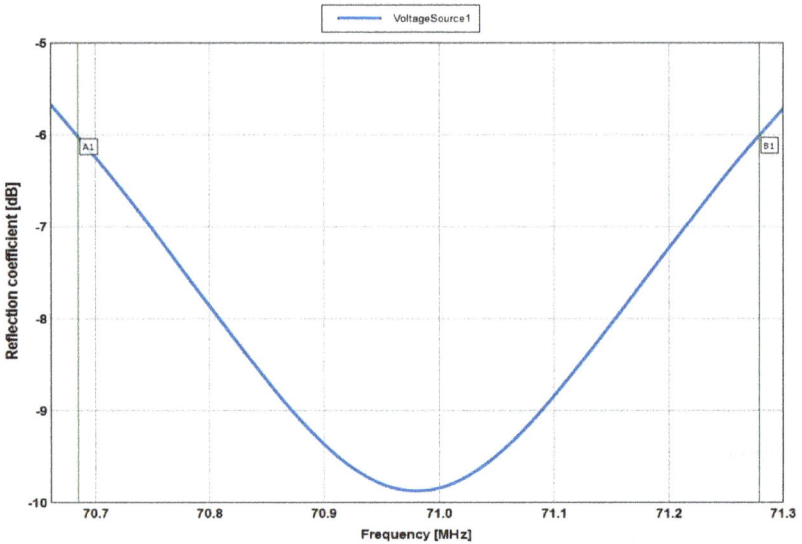

Fig. 8. Reflection coefficient magnitude centered at 70.98 MHz, with a minimum return loss of -9.86 dB.

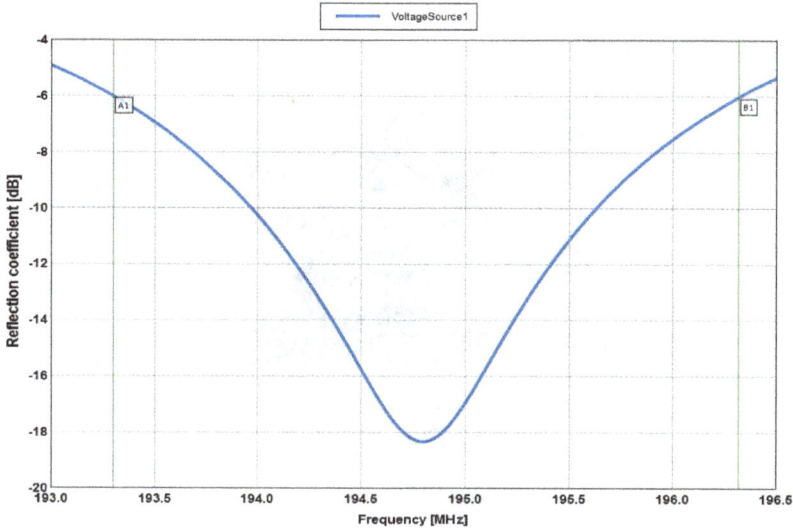

Fig. 9. Reflection coefficient magnitude centered at 194.8 MHz, with a minimum return loss of -18.4 dB.

4. Experimental and practical results

On the base of simulation results, the prototype was fabricated on FR-4 and coupled with coaxial cable of 75 ohms (see Figure 10). In Figure 11, the spectrum analyzer displays the two ranges frequencies received with the prototype: (58.75 MHz, 109.37 MHz) and (155.5MHz, 238.75 MHz), the maximum peak response has a value of -38.5 dBm. Even the primary results shown here, more experimental analysis must be still realized, but it must be recognized that our laboratory equipment is limited.

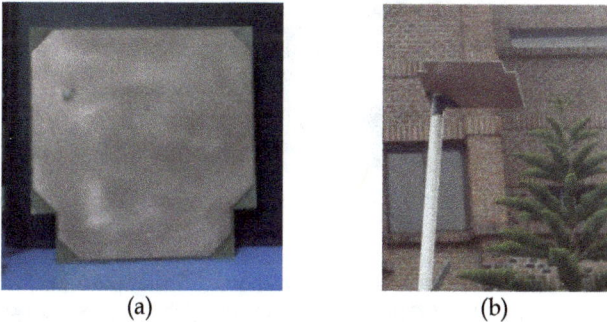

(a)   (b)

Fig. 10. (a) Patch antenna array prototype. (b) Prototype mounted on a PVC base, outdoors of the CIICAp building.

The received range frequencies for the case of a rabbit-ear antenna are shown in Figure 12, where it can be also noted two frequency ranges: (88 Mhz, 108.25 MHz) and (171.25MHz, 182.5 MHz), with a maximum peak of -50 dBm. From these photographs, the bigger receptions of the patch antenna array are clearly noted, as well as the received power. Even the primary results shown here, more experimental analysis must be still realized.

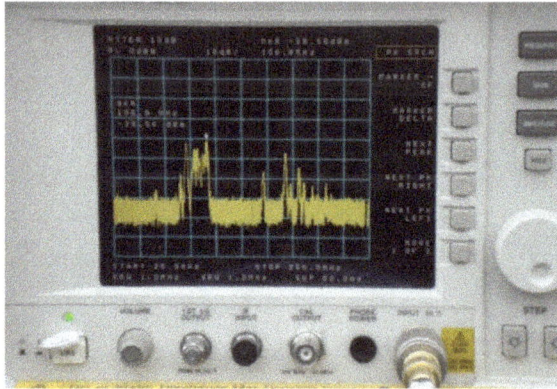

Fig. 11. Two received range frequencies with the spectrum analyzer with the prototype of the antenna array.

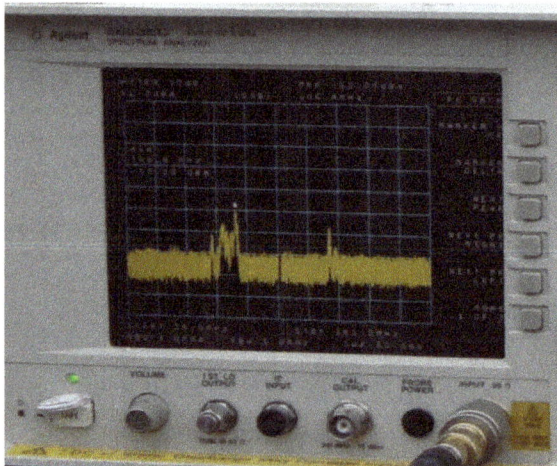

Fig. 12. Two received range frequencies with the spectrum analyzer with a rabbit-ear antenna.

On the other hand, practical tests were also realized on different places of Morelos State. The first ones were realized outside of CIICAp building (18°58'56" N, 99°14'1.9" WO), with the antenna located approximately at only 2 meters of the ground. The photographs are shown in Table 7.

Using an analogical TV, a comparison of the reception of the most common channels, considering two commercial antennas (a mini-combined antenna (Figure 13) and a rabbit-ear antenna) and the prototype was also realized.

Reception with the prototype	

Channel 5	Channel 9
	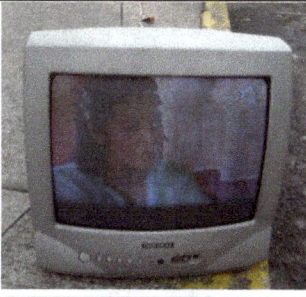
Channel 7	Channel 13
Reception with the commercial combined indoor antenna	

Channel 5	Channel 9

Channel 7	Channel 13

Reception with the Rabbit-ear antenna

Channel 5	Channel 9

Channel 7	Channel 13

Table 7. TV channels reception of openly TV using a scanning TV.

Fig. 13. Mini-combined antenna.

As it can be observed from Table 7, the best reception was achieved with the patch antenna array, followed by the reception with the combined antenna, where some problems were appreciated in channels 5 and 13. The antenna with more reception problems was the rabbit-ear antenna.

On the other side, the antenna prototype was also used on a house roof, in three different places using digital TVs. At first, the tests were realized for a household reception in Temixco, Morelos (located at 18. 51° N, 99°13'48'' WO, with a height over sea level: 1,280 m). In Table 8, photographs of four analogycal representative channels are shown, with a considerable sharpness in all them. In this place, two High Definition (HD) channels were also received (11 and 13; see table 8).

Reception with the prototype	
Channel 2	Channel 5
Channel 7	Channel 13
Channel 11 HD	Channel 13 HD

Table 8. TV channels reception using a LCD TV.

As second case, other tests were realized considering our prototype and a rabbit-ear antenna for a household reception in Monte Casino (located at 19°00'53.7'' N, 99°14'50.6'' WO, height over sea level: 2250 m). With the prototype the reception of the analogical channels was possible without problems, with better sharpness that in the case of the rabbit-ear antenna.

Special attention was focused on signal level of the HD reception of the single channel received there, due to this attribute of the TV. In Tables 9 and 10 the signal intensities are shown for both cases.

TV channel	Current level (%)	Maximum level (%)
13.1	44	48

Table 9. HD reception with a rabbit-ear antenna.

TV channel	Current level (%)	Maximum level (%)
13.1	50	55

Table 10. HD reception with the patch antenna prototype.

When the prototype was used as outdoor antenna, the maximum intensity reception was obtained (see Table 11 and Figure 14).

TV channel	Current level (%)	Maximum level (%)
13.1	59	59

Table 11. HD reception with the patch antenna prototype located on the house's roof.

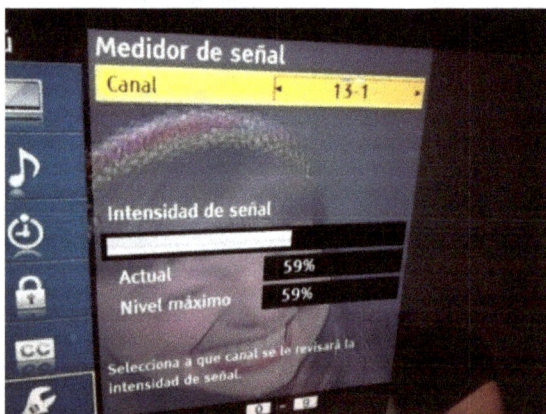

Fig. 14. Visualized images of HDTV reception antenna using the patch prototype, which is located on the house's roof.

Finally, the third place where the tests were realized was Morelia, Michoacan (19°43' N, 101°12' WO), another Mexican State. The tests were realized using a LCD TV with our prototype and a commercial aerial antenna (Figure 15). The results are shown in Table 12.

From Table 12, it can be observed that the TV reception is better in the case of the prototype not only for the analogical channels, receiving an additional one, but also for the single HD channel received (1.1). It must be noted that the location of both antennas is not the best (Figure 15), but several homes in our country have similar conditions. The sizes of the antenna array prototype are considerable smaller than the available aerial antenna.

Aerial antenna

Patch antenna prototype

Fig. 15. Location of the antennas used on a house roof in Morelia, Mich.

Reception with the prototype	

Channel 4	Channel 8

Channel 10	Channel 21 (Repetition of channel 10)
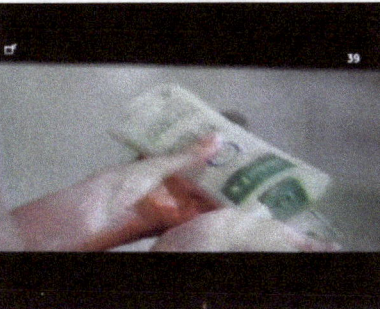	
Channel 39	Channel 45

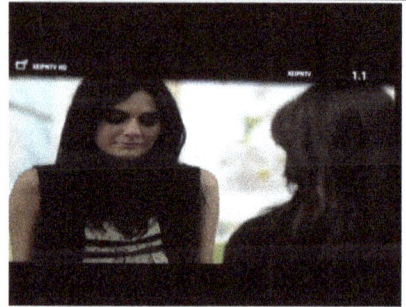

| Channel 13 | (HD) Channel 1.1 |

Reception with the commercial aerial antenna

| Channel 4 | Channel 8 |

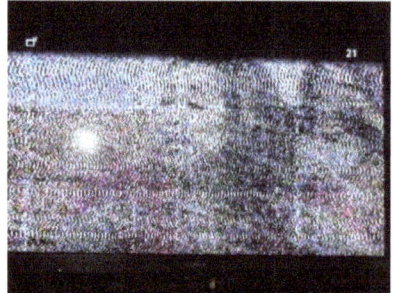

| Channel 10 | Channel 21 (Repetition of channel 10) |

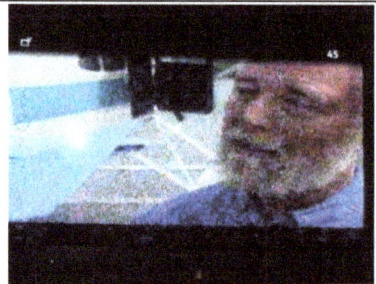

| Channel 39 | Channel 45 |

Not available	
Channel 13	(HD) Channel 1.1

Table 12. Visualized images of TV received with antennas located on the house's roof.

5. Conclusions

The simulations of the patch antenna array show its dual frequency performance. In spite of the inherent narrow broadband of the microstrip antennas, the practical reception, realized in different geographical sites of Morelos and in Morelia, Michoacan, has also confirmed the feasibility of its use for household reception of openly TV frequency ranges.

The difference in the TV reception can be attributed to the proximity of the repeaters antennas, and to the elevation conditions. The vegetation is also relevant.

The experimental and practical tests show an acceptable reception of channels in both VHF sub-ranges of frequencies.

In TVs that accounts with graphical signal meter, it was possible to observe that with the antenna on the house's roof, the current signal meter obtained its maximal level, for the case of HD channels, which is certainly a very good practical result, and it constitutes a base to suggest its use as outdoor antenna.

In the three tests realized using digital TVs on different places, it must be mentioned that unfortunately the available TVs have different attributes, but in all cases the better reception was obtained with our prototype. The comparison with aerial antennas was only possible where they were available.

The prototype sizes make it a competitive option compared with some commercial aerial antennas available in the market. Additionally, for the case of digital TVs, it does not require of an amplifier or a rotor, once it was properly directed. For the case of analogical TVs the reception improvement is considerable compared with the options shown here.

As future work, it is planned to design an appropriate radome for protection of the prototype to the weather.

The implementation of geometry modifications is also been considered in order to increment the broadband of the antenna array. Its sizes can be also reduced using another material, but its costs will be incremented.

6. Acknowledgment

The authors wish to thank to EM Software & Systems (USA) Inc., for FEKO license. J. G. Vera-Dimas and J. A. Damián Morales acknowledge financial support from CONACYT scholarships under grants 270210/219230 and 336781/235572, respectively. R. Vargas-Bernal acknowledges partial financial support from PROMEP.

7. References

Balanis, Constantine. (2005). *Antenna Theory*. Third edition. Wiley-Interscience, Hoboken, ISBN: 047166782X, New Jersey.

Barir, Syfaul & Hamid, Sofian (2010), A Novel Microstrip Patch antenna for 470-890 MHz Indoor Digital Television Receiver. *Digital Proc. Of IMMAC 2010*. 2010 Indonesia-Malaysia Microwave Antenna Conference, Depok, Indonesia, June 11-12.

Brown, T. W. C.; Saunders, S. R.; Stavrou, S. & Fiacco, M. (2007), Characterization of Polarization Diversity at the Mobile. *IEEE Transactions on Vehicular Technology*, Vol. 56, No. 5, pp. 2440-2447.

COFETEL, (2008), Infraestructura de Estaciones de Televisión. August 2008. Available from: <http://www.cofetel.gob.mx/es/Cofetel_2008/Cofe_estaciones_de_television_in>

Garg, Ramesh; Bhartia Prakash; Bahl, Inder & Ittipiboon, Apisak. (2001). *Microstrip Antenna Design Handbook*, Artech House Inc., ISBN:0890065136, Boston. London.

INEGI, (2009), Encuesta Nacional sobre disponibilidad y uso de tecnologías de información y comunicaciones en los hogares. August 2011. Available from <http://www.inegi.org.mx/prod_serv/contenidos/espanol/bvinegi/productos/e ncuestas/especiales/endutih/ENDUTIH_2009.pdf>.

INEGI, (2010), Population and Housing Census 2010: Private homes with television. September 2011. Available from: <http://www.inegi.org.mx/sistemas/mexicocifras/MexicoCifras.aspx?e=0&m=0 &sec=M&ind=1003000021&ent=0&enn=Mexico&ani=2010&i=i&src=0>.

Kapinas, Vasilios M.; Ilić, Maja; Karagiannidis, George K., & Pejanović-Đurišić, Milica. (2007), Aspects on Space and Polarization Diversity in Wireless Communication Systems. *15th Telecommunications forum TELFOR 2007*, pp. 183-186, Serbia, Belgrade, November 20-22, 2007.

Melissinos, Adrian C. (1990). *Principles of Modern Technology*. Cambridge University Press, ISBN 9780521352499, Cambridge.

Neelakanta, Perambur S. & Chatterjee, Rajeswari. (2003). *Antennas for Information Super Skyways: An Exposition on Outdoor and Indoor Wireless Antennas*. Research Studies Press LTD, ISBN 0-86380-267-2, Baldock, Hertfordshire, England.

Rathod, Jagdish M. (2010), Design Development of Antenna for TV Transmission for Connecting Outdoor Broadcasts Van to the Studio for Rural Areas. *International Journal of Computer and Electrical Engineering*, Vol. 2, No. 2, pp. 251-256.

Jalife, Villalón Salma & Sosa, Plata Gabriel (2007), Radiodifusión y Telecomunicaciones: Aspectos técnicos fundamentales. August 2010. Available from: http://www.senado.gob.mx/telecom_radiodifusion/content/sesiones_trabajo/ docs/Salma_Gabriel.pdf>

Smith, Christopher B. (2008), Wideband Dual-Linear Polarized Microstrip Patch Antenna, *Thesis of Master of Science*, Texas A&M University, December 2008, United States of America.

Taguchi, Mitsuo Nakamura; Takuya, Fujimoto Takafumi & Tanaka, Kazumasa (1996), CPW Fed Active loop Antenna for Television Receivers. *Proceedings of ISAP'96*, pp. 521-524, Chiba, Japan.

Wang, Qingyuan & Lecours Michel (1999), Dual-frequency microstrip antenna with orthogonal polarization. Laboratoire de Radiocommunications et de Traitement du Signal. *Rapport annuel d'activités 1998-1999*, pp. 85-90.

Permissions

The contributors of this book come from diverse backgrounds, making this book a truly international effort. This book will bring forth new frontiers with its revolutionizing research information and detailed analysis of the nascent developments around the world.

We would like to thank Prof. Esteban Tlelo-Cuautle and Prof. Sheldon X.-D. Tan, for lending their expertise to make the book truly unique. They have played a crucial role in the development of this book. Without their invaluable contribution this book wouldn't have been possible. They have made vital efforts to compile up to date information on the varied aspects of this subject to make this book a valuable addition to the collection of many professionals and students.

This book was conceptualized with the vision of imparting up-to-date information and advanced data in this field. To ensure the same, a matchless editorial board was set up. Every individual on the board went through rigorous rounds of assessment to prove their worth. After which they invested a large part of their time researching and compiling the most relevant data for our readers. Conferences and sessions were held from time to time between the editorial board and the contributing authors to present the data in the most comprehensible form. The editorial team has worked tirelessly to provide valuable and valid information to help people across the globe.

Every chapter published in this book has been scrutinized by our experts. Their significance has been extensively debated. The topics covered herein carry significant findings which will fuel the growth of the discipline. They may even be implemented as practical applications or may be referred to as a beginning point for another development. Chapters in this book were first published by InTech; hereby published with permission under the Creative Commons Attribution License or equivalent.

The editorial board has been involved in producing this book since its inception. They have spent rigorous hours researching and exploring the diverse topics which have resulted in the successful publishing of this book. They have passed on their knowledge of decades through this book. To expedite this challenging task, the publisher supported the team at every step. A small team of assistant editors was also appointed to further simplify the editing procedure and attain best results for the readers.

Our editorial team has been hand-picked from every corner of the world. Their multi-ethnicity adds dynamic inputs to the discussions which result in innovative outcomes. These outcomes are then further discussed with the researchers and contributors who give their valuable feedback and opinion regarding the same. The feedback is then collaborated with the researches and they are edited in a comprehensive manner to aid the understanding of the subject.

Apart from the editorial board, the designing team has also invested a significant amount of their time in understanding the subject and creating the most relevant covers. They scrutinized every image to scout for the most suitable representation of the subject and create an appropriate cover for the book.

The publishing team has been involved in this book since its early stages. They were actively engaged in every process, be it collecting the data, connecting with the contributors or procuring relevant information. The team has been an ardent support to the editorial, designing and production team. Their endless efforts to recruit the best for this project, has resulted in the accomplishment of this book. They are a veteran in the field of academics and their pool of knowledge is as vast as their experience in printing. Their expertise and guidance has proved useful at every step. Their uncompromising quality standards have made this book an exceptional effort. Their encouragement from time to time has been an inspiration for everyone.

The publisher and the editorial board hope that this book will prove to be a valuable piece of knowledge for researchers, students, practitioners and scholars across the globe.

List of Contributors

Louiza Sellami
Electrical and Computer Engineering Department, US Naval Academy, Annapolis, MD, USA

Robert W. Newcomb
Electrical and Computer Engineering Department, University of Maryland, College Park, MD, USA

Díaz Méndez J. Alejandro and Arroyo Huerta J. Erasmo
National Institute for Astrophysics, Optics and Electronics, Mexico

López Delgadillo Edgar
Universidad Autónoma de Aguascalientes, Mexico

Rafael Vargas-Bernal and Gabriel Herrera-Pérez
Instituto Tecnológico Superior de Irapuato (ITESI), México

Chiao-Ling Lung and Shih-Chieh Chang
Department of Computer Science, National Tsing Hua University, Taiwan

Chiao-Ling Lung, Jui-Hung Chien, Yung-Fa Chou and Ding-Ming Kwai
Information and Communications Research Laboratories, Industrial Technology Research Institute, Taiwan

Ana D. Martínez, Joel Ramírez, Jesús S. Orea and Víctor M. Jiménez-Fernández
Universidad Veracruzana/Facultad de Instrumentación Electrónica, México

Omar Alba
Instituto Tecnológico Superior de Xalapa/ Departamento de Electrónica, México

Pedro Julián, Juan A. Rodríguez, Osvaldo Agamennoni and Omar D. Lifschitz
Universidad Nacional del Sur/ Departamento de Ingeniería Eléctrica y de Computadoras, Argentina

K.A. Sumithra Devi
R.V. College of Engineering, Bangalore, India
Vishweshvaraya Technological University, Karnataka, India

Sheldon X.-D. Tan, Xue-Xin Liu and Eric Mlinar
Department of Electrical Engineering, University of California, Riverside, CA 92521, USA

Esteban Tlelo-Cuautle
Department of Electronics, INAOE, Mexico

Waqar Ahmad and Hannu Tenhunen
KTH Royal Institute of Technology KTH/ICT/ECS/ESD, Stockholm, Sweden

Noureddine Bouhmala
Vestfold University College, Norway

Mu-Shun Matt Lee and Chien-Nan Jimmy Liu
National Central University, Taiwan (ROC)

Meng-Chun Lin
Department of Computer National Chengchi University, Taiwan

M. Tecpoyotl-Torres, J. A. Damián Morales, J. G. Vera Dimas, R. Cabello Ruiz and J. Escobedo-Alatorre
Centro de Investigación en Ingeniería y Ciencias Aplicadas (CIICAp-UAEM), Mexico

C. A. Castillo-Milián
Facultad de Ciencias Químicas e Ingeniería (FCQeI-UAEM), Mexico

R. Vargas-Bernal
Instituto Tecnológico Superior de Irapuato (ITESI), Mexico

www.ingramcontent.com/pod-product-compliance
Lightning Source LLC
Chambersburg PA
CBHW070737190326
41458CB00004B/1213